Mechanical Property Research of Advanced Asphalt-Based Materials

Mechanical Property Research of Advanced Asphalt-Based Materials

Guest Editors

Bo Li

Zhuangzhuang Liu

Pan Pan

Xunhao Ding

MDPI

Basel • Beijing • Wuhan • Barcelona • Belgrade • Novi Sad • Cluj • Manchester

Guest Editors

Bo Li
School of Civil Science and Engineering
Yangzhou University
Yangzhou
China

Zhuangzhuang Liu
School of Highway
Chang'an University
Xi'an
China

Pan Pan
School of Resource and Civil Engineering
Wuhan Institute of Technology
Wuhan
China

Xunhao Ding
School of Transportation
Southeast University
Nanjing
China

Editorial Office
MDPI AG
Grosspeteranlage 5
4052 Basel, Switzerland

This is a reprint of the Special Issue, published open access by the journal *Materials* (ISSN 1996-1944), freely accessible at: www.mdpi.com/journal/materials/special_issues/W0I7U6ZJ08.

For citation purposes, cite each article independently as indicated on the article page online and using the guide below:

Lastname, A.A.; Lastname, B.B. Article Title. *Journal Name* **Year**, *Volume Number*, Page Range.

ISBN 978-3-7258-3900-1 (Hbk)
ISBN 978-3-7258-3899-8 (PDF)
https://doi.org/10.3390/books978-3-7258-3899-8

Contents

materials

MDPI

Article

Evaluation of Cracking Resistance of SMA-13 Hot Recycling Asphalt Mixtures Reinforced by Basalt Fiber

Yu Zhang [1], Yao Zhang [1,2], Bo Li [1,2,*], Aihong Kang [1] and Yu Wang [1]

[1] College of Civil Science and Engineering, Yangzhou University, Yangzhou 225127, China; zy502499180@163.com (Y.Z.); yaozhang@yzu.edu.cn (Y.Z.); aihongkang@yzu.edu.cn (A.K.); wangyu@163.com (Y.W.)

[2] Research Center for Basalt Fiber Composite Construction Materials, Yangzhou 225127, China

* Correspondence: libo@yzu.edu.cn; Tel.: +86-0514-8797-9418

Abstract: In the context of green and low-carbon development, energy saving, and emission reduction, hot recycling technology (RT) has been researched, which is divided into hot central plant RT and hot in-place RT. However, due to the aged asphalt binders, the shortcomings of hot recycled asphalt mixtures have become apparent, as in comparison to new asphalt mixtures, their resistance to cracking was inferior and the cracking resistance deteriorated more rapidly. Therefore, it was very necessary to focus on the improvement of crack resistance of hot recycled asphalt mixtures. Basalt fiber has been proved to be able to effectively improve the comprehensive road performance of new asphalt mixtures. Therefore, this paper introduced basalt fiber to hot central plant recycled and hot in-place recycled asphalt mixtures, in order to improve the crack resistance of asphalt as a new type of fiber stabilizer. Firstly, six types of SMA-13 fiber asphalt mixtures were designed and prepared, i.e., hot mixtures with basalt fiber or lignin fiber, hot central plant recycled mixtures with basalt fiber or lignin fiber, and hot in-place recycled mixtures with basalt fiber or lignin fiber. Secondly, the trabecular bending test, low-temperature creep test, semi-circular bending test, and IDEAL-CT were used to comparatively study the changing patterns of low and intermediate temperature cracking resistance of hot recycled mixtures with conventional lignin fibers or basalt fibers. Finally, Pearson's correlation coefficient was used to analyze the correlation of the different cracking resistance indicators. The results show that the low and intermediate temperature cracking resistance of hot central plant recycled mixtures increased by 45.6% (dissipative energy ratio, W_d/W_s) and 74.8% (flexibility index, FI), respectively. And the corresponding cracking resistance of hot in-place recycled mixture increased by 105.4% (W_d/W_s) and 55.7% (FI). The trabecular bending test was more suitable for testing the low-temperature cracking resistance of hot recycled asphalt mixtures, while the IDEAL-CT was more suitable for testing the intermediate-temperature cracking resistance. The results can provide useful references for the utilization of basalt fiber in the hot recycling of SMA-13 asphalt mixtures.

Keywords: asphalt mixture; hot recycling; cracking resistance; basalt fiber; low-temperature creep

Citation: Zhang, Y.; Zhang, Y.; Li, B.; Kang, A.; Wang, Y. Evaluation of Cracking Resistance of SMA-13 Hot Recycling Asphalt Mixtures Reinforced by Basalt Fiber. *Materials* **2024**, *17*, 1762. https://doi.org/10.3390/ma17081762

Academic Editor: Giovanni Polacco

Received: 18 March 2024
Revised: 2 April 2024
Accepted: 9 April 2024
Published: 11 April 2024

1. Introduction

Asphalt pavement hot recycling technologies, encompassing hot central plant recycling and hot in-place recycling, have gained increasing popularity due to their numerous advantages. These advantages include rapid execution, cost-effectiveness, and resource conservation. By utilizing these technologies, the efficiency of resource utilization could be significantly enhanced, thereby promoting green and low-carbon development [1–3]. In Europe, more than 70% of the annual recycled asphalt pavement material (RAP) is recycled [4]. According to statistics from the Federal Highway Administration (FHWA) and the National Asphalt Pavement Association (NAPA), the majority of recycled waste asphalt pavement materials in the United States are utilized in the production of hot-mix asphalt mixtures, representing a significant proportion ranging from 83 to 95 percent of the total recycled materials [5]. Comparatively, in China, the recycling rate of asphalt pavement materials stands at 30 percent, indicating a notable gap in the utilization of recycled asphalt materials between the two countries [6].

However, performance deterioration of hot recycled asphalt mixtures, especially the insufficiency of cracking resistance, has been noticed, though the essential performance meets the requirements [7–9]. Through low-temperature crack resistance testing, researchers have discovered that a hot recycled asphalt mixture exhibits increased viscosity and brittleness compared to the original asphalt mixture. This finding suggests that the recycled material may have altered the physical properties of the asphalt, potentially affecting its performance and durability under low-temperature conditions [10–12]. The interface between a new asphalt mixture and a reclaimed asphalt mixture will increase the concentrated stress, resulting in a tendency of cracking in low-temperature environments [13–15]. Indeed, various methods have been employed to assess the cracking resistance of recycled asphalt mixtures. T. Mandal et al. conducted measurements to assess the impact of binder replacement from recycled asphalt pavement, binder modification, and low-temperature binder grade on the DCT test results. However, it was discovered that the significance of these factors was insufficient to account for the full range of DCT response variables [16]. S. Xiang et al. conducted an evaluation of the impact of reclaimed asphalt pavement (RAP) on the cracking resistance of asphalt mixtures using the SCB test. Their findings indicate that, while RAP generally enhances the SCB tensile strength of the mixtures, it significantly reduces their post-failure tenacity. Furthermore, the J-integral (fracture criterion in elastoplastic fracture mechanics) of the asphalt mixture decreased with the addition of RAP, resulting in a weakened cracking resistance [17]. Z. Fu et al. conducted a comparison between IDEAL-CT results and field cracking data gathered from various sources, including the Federal Highway Administration's accelerated load facility, Texas SH15 and SH62, as well as MnROAD. The IDEAL-CT indicators demonstrated a strong correlation with field performance in respect to fatigue, reflective and thermal cracking [18]. To improve the low-temperature cracking resistance of recycled asphalt mixtures, H. Jing et al. determined the optimum design of recycled asphalt mixtures through the trabecular bending test [19]. K. Moon et al. conducted a study on the impact of incorporating both reclaimed asphalt pavement (RAP) and recycled asphalt shingles (RAS) into virgin asphalt mixtures through a simple low-temperature creep test utilizing asphalt mixture beams. The findings revealed that most mixtures formulated with combinations of RAP and RAS exhibited comparable performance to standard mixtures at low temperatures [20].

Scholars are also searching for measures to improve the cracking resistance of hot recycled mixtures. The normally used ones include improving the mixing process of hot recycled asphalt mixtures [21]; using external mixing additives [22] such as glass fibers, polyester fibers, high modulus agents, rutting resistance agents, etc.; adding recycled plastics [23]; using rejuvenators [17]; and so on.

Currently, basalt fiber is extensively employed to boost the overall performance of asphalt mixtures, resulting in a notable enhancement in crack resistance [24–27]. Z. Yao et al. examined a basalt fiber asphalt mixture through mid- to low-temperature (25 °C and −10 °C) indoor cracking tests, including the trabecular bending test, IDEAL-CT test, and SCB test. They delved into the correlation between crack resistance and morphological parameters, and further explored the impact of fiber length on the fracture characteristics of the basalt fiber asphalt mixture [24]. By incorporating basalt fibers into asphalt mixtures, H. Ying et al. discovered that it effectively enhanced the performance characteristics of the mixture, including its cracking resistance, rutting resistance, and fatigue resistance [28]. The potential for improved cracking resistance in hot recycled asphalt mixtures was indicated by the inference that their performance may be enhanced through the incorporation of basalt fibers.

The purpose of this study is to explore the impact of basalt fibers on the cracking resistance performance of both hot central plant recycled and hot in-place recycled asphalt mixtures. The intended target is the gradation of Stone Matrix Asphalt with a nominal maximum aggregate size (NMAS) of 13.2 mm, referred to as SMA-13. Trabecular bending tests and creep testes are used to evaluate the low-temperature performance, while semi-circular bending tests and IDEAL cracking tests are conducted to test the intermediate-temperature cracking properties.

2. Materials and Methods

2.1. Materials

2.1.1. Reclaimed Asphalt Pavement (RAP)

The RAP used in this paper was taken from the Wuxi section of the Shanghai–Nanjing expressway. The upper layer was with the gradation of SMA-13. The old asphalt was extracted according to the specification JTG 5142-2019 [29], including extraction, high-speed centrifugation, and rotary evaporation. Then, the extracted old asphalt was tested for penetration, softening point, ductility, and viscosity. The test results are shown in Table 1.

Table 1. Results for properties of extracted old asphalt.

Property	Old Asphalt	Requirements for SBS Modified Asphalt	Test Method JTG E20 [30]
Penetration (25 °C)/0.1 mm	39	50–80	T0604
Softening point/°C	69	>60	T0606
Ductility (5 °C)/cm	7.8	>30	T0605
Viscosity (135 °C)/Pa·s	2.33	≤3	T0613

In addition, the asphalt content and mineral gradation of the old pavement were determined according to the methods specified in JTG E20-2011 (T0722-1993) and (T0725-2000). The asphalt content of RAP was 5.48%, and the mineral gradation is shown in Table 2.

Table 2. Mineral gradation of RAP.

Sieve Size/mm	16.0	13.2	9.5	4.75	2.36	1.18	0.6	0.3	0.15	0.075
Upper limit	100	100	75	34	26	24	20	16	15	12
Lower limit	100	90	50	20	15	14	12	10	9	8
Median	100	95	62.5	27	20.5	19	16	13	12	10
RAP	100.0	96.3	64.3	30.1	24.5	19.7	17.0	14.8	12.6	10.2

2.1.2. Rejuvenating Agent

The rejuvenating agent RA-102, manufactured by Subote New Materials Co., (Yangzhou, China) was employed in this study. The specific technical indexes pertaining to RA-102 are detailed in Table 3.

Table 3. Technical index of the rejuvenating agent RA-102.

Technical Index	RA-102	Requirements in JTG/T 5521-2019 [31]	Test Method JTG E20
Viscosity at 60 °C/cP	4000	——	T0619
Flashpoint/°C	248	≥220	T0633
Saturated fraction content/%	25.6	≤30	T0618
Aromatic content/%	53	≥30	T0618
Viscosity ratio (RTFOT pre to post)	1.34	≤3	T0610
Mass change (%, RTFOT pre to post)	1.02	≤4%	T0603

2.1.3. New Aggregates and New Asphalt

For the aggregates, basaltic coarse and fine aggregates were chosen, and limestone mineral powder was selected as the filler. All the aggregates and fillers underwent testing according to the corresponding requirements of the Test Procedure for Aggregates in Highway Engineering (JTG E42) [32]. The results of these tests were found to be compliant with the specified standards.

The SBS-modified asphalt with a PG grade of 76-22 was chosen in this study. It had a needle penetration of 54 dmm, a softening point of 80 °C, and a rotational viscosity of 2.35 Pa·s at 135 °C. The corresponding requirements of the test specification were met by the test results.

2.1.4. Basalt Fiber

The basalt fiber utilized in this study was sourced from Jiangsu Tianlong Basalt Continuous Fiber Co., Ltd. (Yizheng, China), and its technical properties are presented in Table 4.

Table 4. Technical index of basalt fiber.

Index	Results	Test Method
Diameter/μm	17	GB/T 7690.5 [33]
Length range/mm	3–9 mixed length	JT/T 776.1 [34]
Fracture strength/MPa	2430	GB/T 20310 [35]
Elongation at break/%	3.0	GB/T 20310
Modulus of elasticity/GPa	84.5	GB/T 20310
Heat resistance, retained fracture strength/%	94.8	JT/T 776.1
(Fe_2O_3 + FeO) content/%	9.68	GB/T 1549 [36]
Acidity factor	6.1	GB/T 1549

2.1.5. Lignin Fiber

Hot recycled asphalt mixtures with lignin fiber were prepared and utilized as comparison groups to those containing basalt fiber. The technical properties of the lignin fiber are detailed in Table 5.

Table 5. Technical index of lignin fiber.

Index	Requirements	Results
Ash content/%	18 ± 5	19.3
PH Value	7.5 ± 1.0	7.7
Oil absorption	5 times the fiber quality	6.2

2.2. *Mixture Design*

2.2.1. Gradation

Three SMA-13 asphalt mixture types were studied: (1) hot mix (0% RAP), (2) hot central plant mix (30% RAP), and (3) hot in-place mix (80% RAP). For each type of mixture, two sets of samples were fabricated either with lignin fiber or with basalt fiber. The synthetic gradation of the new asphalt mixes, hot in-central plant RT mixes, and hot in place RT mixes are shown in Figure 1.

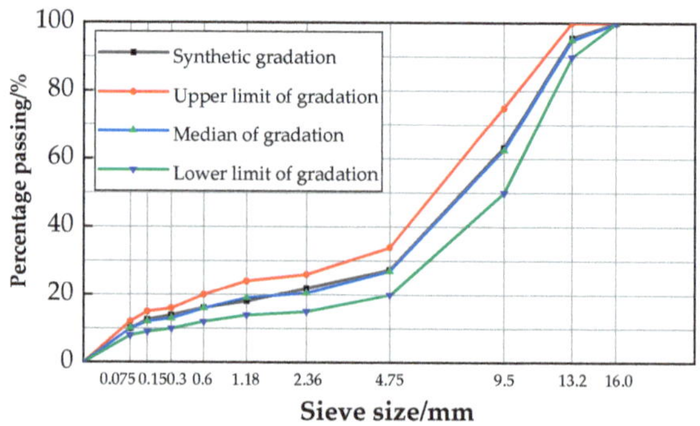

(**a**)

Figure 1. *Cont.*

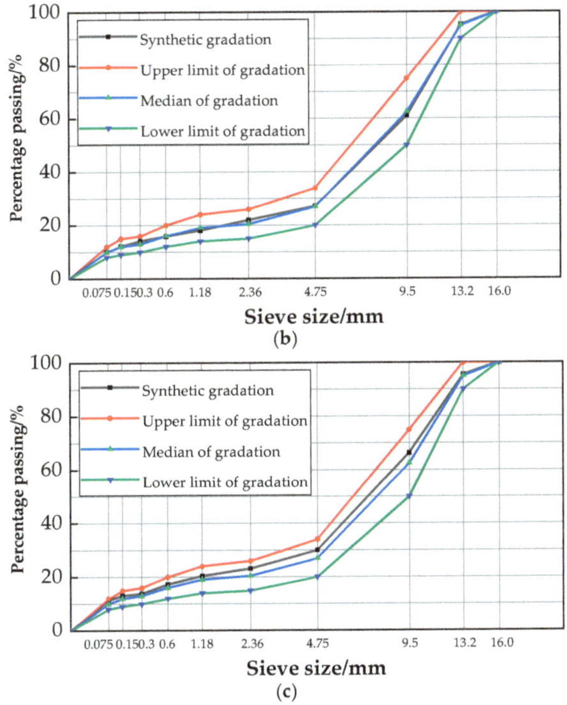

Figure 1. Synthetic gradation: (**a**) hot central plant recycled asphalt mixture; (**b**) hot in-place recycled asphalt mixture; (**c**) hot mix asphalt mixture.

2.2.2. Dosage of Rejuvenating Agent

To produce the recycled asphalt, dosages of the rejuvenating agent were set at 4%, 6%, 8%, and 10% by weight of old asphalt. Conventional performance evaluations were conducted through penetration, softening point, and ductility tests. The test results, presented in Table 6, indicated that when the rejuvenating agent dosage reached 6%, the penetration and softening point of the recycled asphalt nearly reached comparable levels to that of new asphalt, while the ductility of the recycled asphalt recovered to a certain extent. Based on these findings, a rejuvenating agent dosage of 6% was chosen for this study.

Table 6. Properties of recycled asphalt with different rejuvenating agent dosages.

Property	Rejuvenating Agent Dosage					New Asphalt	Test Method JTG E20
	0	4	6	8	10		
Penetration (0.1 mm, 25 °C)	39	60	68	74	78	71	T0604
Softening point (°C)	69	65	63	61	56	64	T0606
Ductility (cm, 5 °C)	7.8	22.4	28.6	31.4	34.6	48	T0605

2.2.3. Optimum Oil/Aggregate Ratio

Optimum oil/aggregate ratio of each type mixture was determined according to JTG E20. The final optimum asphalt content and the corresponding volumetric parameters are summarized in Table 7. For both hot in-place recycled mixes and hot central plant recycled mixes, the oil/aggregate ratio of basalt fiber mixtures was 0.2 percentage points less than that of the lignin fiber mixtures. As far as hot in-place recycled mixtures were concerned, the oil/aggregate ratio was the same for both mixtures with basalt fiber or lignin fiber.

Table 7. Design results of mix proportions.

Gradation Type	Fiber Type	Fiber Dosage/‰	Oil/Aggregate Ratio/%	Air Voids/%	Voids in Mineral Aggregate/%	Voids Filled with Asphalt/%	Marshall Stability/kN
Hot mixed SMA-13	Lignin fiber	3	6.0	3.8	17.2	74.8	8.4
	Basalt fiber (6 mm)	3	5.8	4.3	16.8	74.4	11.5
Hot central plant recycled SMA-13	Lignin fiber	3	6.1	4.4	17.5	74.9	10.3
	Basalt fiber (6 mm)	3	5.9	4.1	17.1	76.0.	12.1
Hot in-place recycled SMA-13	Lignin fiber	1	6.0	4.0	17.5	78.3	11.3
	Basalt fiber (6 mm)	3	6.0	4.1	17.5	76.9	12.5
Requirements	——	——	——	3~4.5	≥16.5	70~85	≥6

note: fiber content was calculated by the weight of recycled asphalt mixture.

According to the specifications in JTG E20-2011, T0732, and T0733, Schellenberg leakage tests and Fort Kentucky flyaway tests were conducted to verify whether the amount of asphalt in the mixture was excessive and whether the cohesion between aggregate and asphalt was strong enough. The asphalt content of each mixture type was deemed adequate based on the results of the tests conducted on the six types of mixtures, which are summarized in Table 8. These findings were in compliance with the established requirements.

Table 8. Results of leakage and flyaway tests.

Gradation Type	Fiber Type	Flyaway Losses/%	Leakage Losses/%
Hot mixed SMA-13	Lignin fiber	6.8	0.09
Hot mixed SMA-13	Basalt fiber (6 mm)	4.5	0.08
Hot central plant recycled SMA-13	Lignin fiber	7.3	0.08
Hot central plant recycled SMA-13	Basalt fiber (6 mm)	4.9	0.07
Hot in-place recycled SMA-13	Lignin fiber	7.9	0.09
Hot in-place recycled SMA-13	Basalt fiber (6 mm)	5.4	0.08
Requirements	——	≤15	≤0.1

2.2.4. Preparation of Recycled Asphalt Mixture

Both hot central plant recycled and hot in-place recycled asphalt mixtures were prepared according the corresponding construction processes, as shown in Figures 2 and 3. The new fibers, either lignin fibers or basalt fibers, were mixed with the heated new aggregates, which is called the dry mixing process. And then, the recycled mixtures could be fabricated by mixing the new asphalt mixture and preheated RAP. It can be noticed from Figures 2 and 3 that the heating temperatures were obviously different. The heating temperatures for the new asphalt mixture and RAP were set at 180 °C and 130 °C, respectively, for the hot central plant recycled mixture. For the hot in-place recycled mixture, the corresponding heating temperatures were 170 °C and 160 °C.

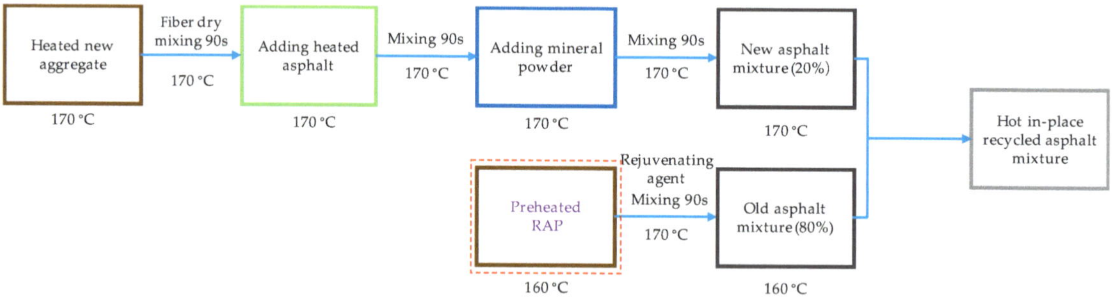

Figure 2. Indoor simulation preparation process of hot in-place recycled mixtures.

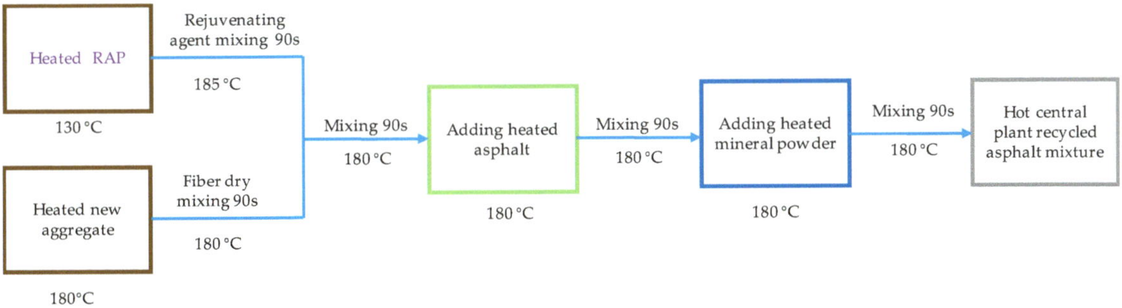

Figure 3. Indoor simulation preparation process of hot central plant recycled mixtures.

2.3. Test Method

2.3.1. Trabecular Bending Test

In accordance with the specifications in JTG E20-2011, the trabecular bending test was conducted to analyze the low-temperature cracking resistance of recycled asphalt mixtures. Prismatic beam samples measuring 250 mm × 30 mm × 35 mm were prepared for the testing. Four duplicate samples per group were tested at a temperature of −10 °C and a loading rate of 50 mm/min. The bending tensile strength (R_B/MPa), maximum failure strain ($\varepsilon_B/\mu\varepsilon$), and stiffness modulus (S_B/MPa) were then calculated using Equations (1)–(3). The test procedure is shown in Figure 4.

$$R_B = \frac{3 \times L \times P_B}{2 \times b \times h^2} \tag{1}$$

$$\varepsilon_B = \frac{6 \times h \times d}{L^2} \tag{2}$$

$$S_B = \frac{R_B}{\varepsilon_B} \tag{3}$$

where b is the width of the specimen, mm; h is the height of specimen, mm; L is the length of the specimen, mm; P_B is the maximum load, N; and d is the span deflection when the specimen failed, mm.

Figure 4. Illustration of trabecular bending test.

2.3.2. Low-Temperature Creep Test

The low-temperature bending creep test, as stipulated in the JTG E20-2011 specification, was an effective means to evaluate the low-temperature performance of asphalt mixtures, taking into account factors like the creep rate and dissipation energy. The specimens used for this test were prism-shaped, with dimensions of 250 mm ± 2 mm in length, 30 mm ± 2 mm in width, and 35 mm ± 2 mm in height.

Due to the typical viscoelastic characteristics, the Burgers model was usually used to characterize the mechanical properties of the asphalt mixtures. The model was composed of the Maxwell model and the Kelvin model. The model consisted of two elastic elements (E_1, E_2) and two viscous elements (η_1, η_2), as shown in Figure 5.

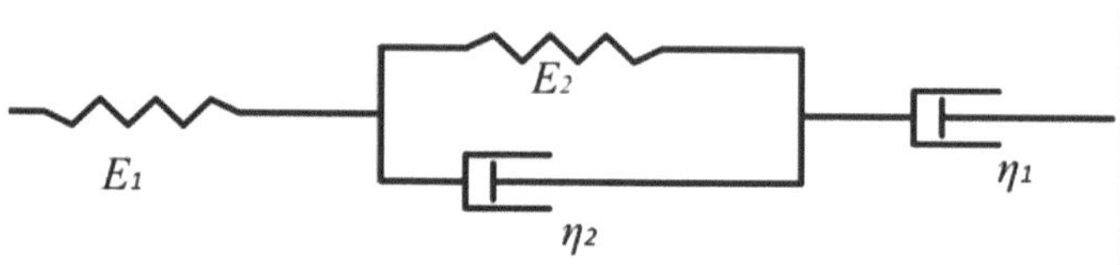

Figure 5. Illustration of the Burgers model.

The four-parameter creep equation of the Burgers model can be expressed by Equation (4).

$$\varepsilon(\mathrm{t}) = \sigma_0\left[\frac{1}{E_1} + \frac{t}{\eta_1} + \frac{1}{E_2}(1 - e^{-\frac{E_2}{\eta_2}t})\right] \quad (4)$$

where $\varepsilon(t)$ is the bending tensile strain of the beam bottom; σ_0 is the creep bending tensile stress of specimen, MPa; t is the load time, s; E_1, E_2 are the elastic moduli; and η_1, η_2 are the coefficients of viscosity.

The optimal solution of the four parameters in the Burgers model could be obtained by nonlinear fitting using the global optimization algorithm through the changeable data of mid-span deflection with time collected by the test. Once the optimal solution for the four parameters in the Burgers model was obtained, they were utilized to calculate the dissipation energy (W_d), storage energy (W_s), bending stiffness modulus $S(t)$, and creep rate $m(t)$ for each mixture type. These calculations served to characterize the low-temperature performance of each mixture. The essence of each element in the Burgers model was the spring and damper, corresponding to the storage energy and dissipation energy, respectively. According to the literature [20], the dissipative energy ratio of W_d/W_s can accurately represent the low-temperature performance of the material, with a higher value indicating better low-temperature performance. Moreover, contradictory phenomena frequently arose when relying solely on a single bending stiffness modulus and creep rate. However, by utilizing the creep rate per unit stiffness, $m(t)/S(t)$, the contradictory phenomenon could be effectively avoided, thereby accurately reflecting the strengths and weaknesses of the material's low-temperature performance [21]. Specifically, a higher value of $m(t)/S(t)$ indicates better low-temperature performance. The specific calculation equations for $m(t)/S(t)$, W_d, and W_s are presented in Equations (5), (6), and (7), respectively.

$$\frac{m(t)}{S(t)} = \left(\frac{1}{\eta_1} + \frac{1}{\eta_2}e^{-\frac{E_2}{\eta_2}t}\right) \quad (5)$$

$$W_d(\mathrm{t}) = {\sigma_0}^2\left[\frac{t}{\eta_1} + \frac{1}{2E_2}(1 - e^{-\frac{2E_2}{\eta_2}t})\right] \quad (6)$$

$$W_s(t) = {\sigma_0}^2\left[\frac{1}{E_1} + \frac{1}{2E_2}(1 - 2e^{-\frac{E_2}{\eta_2}t} + e^{-\frac{2E_2}{\eta_2}t})\right] \quad (7)$$

where W_d is the stored energy; W_s is the dissipative energy; $S(t)$ is the bending modulus of strength; and $m(t)$ is the creep rate.

2.3.3. Semi-Circular Bending Test

To assess the intermediate anti-cracking capability of asphalt mixtures, the semi-circular bending (SCB) test was conducted in accordance with the standard test method outlined in AASHTO TP 124-16 [37]. During this test, the load and displacement curves were carefully recorded. The fracture work generated throughout the cracking process was represented by the area enclosed by these curves. The fracture energy, defined by Equations (8)–(10), was calculated as the ratio of the fracture work to the fracture area. A higher value of fracture energy indicates superior anti-cracking performance. For accurate results, four duplicate samples were used for each group.

$$G_f = \frac{W_f}{Area_{lig}} \times 10^6 \tag{8}$$

$$W_f = \int P du \tag{9}$$

$$A_{lig} = (r - a)t \tag{10}$$

where G_f is the fracture energy, J/m^2; W_f is the fracture work, J; A_{lig} is the area of fracture area, mm^2; $r - a$ is the length of the fracture zone; and t is the thickness of specimen, mm.

The flexibility index (FI) was introduced as a metric to reflect and characterize the crack propagation rate, as depicted in Figure 6a. Equation (11) provides the mathematical expression for this FI index. Furthermore, Figure 6b outlines the test procedure.

$$FI = \frac{G_f}{|m|} \times A \tag{11}$$

where *FI* is the flexibility index; $|m|$ is the absolute value of the post-peak slope, kN/mm; and *A* is the unit conversion coefficient, which is 0.01.

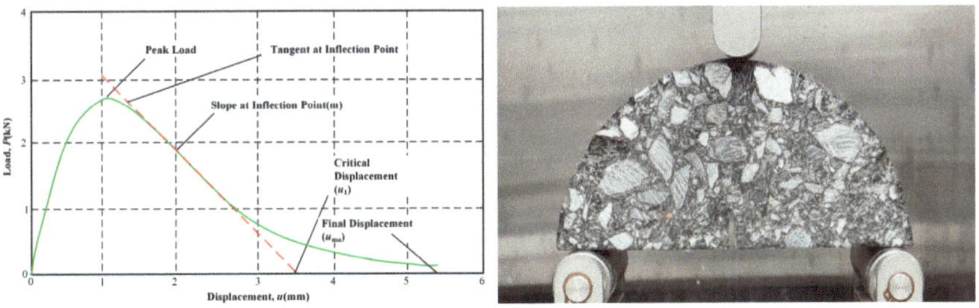

(**a**) (**b**)

Figure 6. Illustration of semi-circular bending test: (**a**) load (P) vs. displacement (u) curve; (**b**) test procedure.

2.3.4. IDEAL Cracking Test

The IDEAL-CT test was one of the methods employed to assess the crack resistance of asphalt mixtures. In this test, a cylindrical asphalt mixture specimen measuring 62 mm in height and 150 mm in diameter was used [18]. The CT_{index} proposed in this test served as a metric to evaluate the anti-cracking performance of the asphalt mixtures. This index reflected the asphalt mixtures' resistance to crack propagation, and its calculation formulas are detailed in Equations (12) and (13). The test procedure is illustrated in Figure 7.

$$|m_{75}| = |(p_{85} - p_{65})/(l_{85} - l_{65})| \tag{12}$$

$$CT_{Index} = \frac{G_f}{|m_{75}|} \times \left(\frac{l_{75}}{D}\right) \tag{13}$$

where CT_{index} is the cracking test index; t is the specimen thickness, mm; G_f is the fracture energy, J/m^2; m_{75} is the absolute value of the slope at 75% peak in the post-peak section; l_{75} is the displacement at 75% of the peak value in the post-peak section, mm; and D is the specimen diameter, mm.

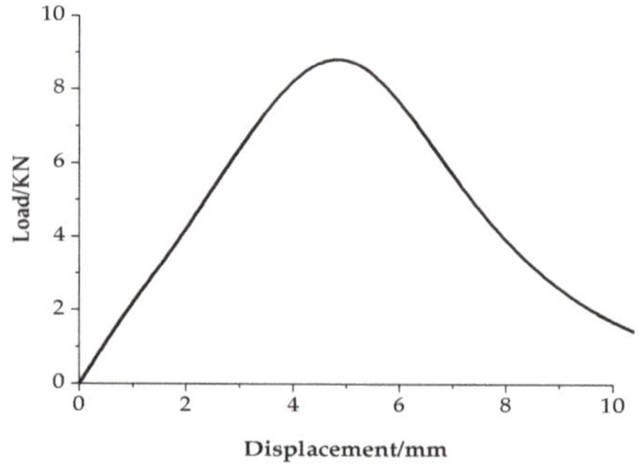

(a)

(b)

Figure 7. The test procedure of IDEAL-CT: (**a**) test conditions; (**b**) specimen and fixture.

3. Results and Discussion

3.1. Results of Trabecular Bending Test

The results of the stiffness modulus and maximum failure strain obtained from the trabecular bending tests are graphically represented in Figure 8. It is evident from Figure 8 that, for a given fiber type, as the RAP content increases, the stiffness modulus of the recycled asphalt mixture also increases, whereas the maximum failure strain exhibits a decreasing trend. Especially in the case of the hot in-place recycled mixture with 80% RAP, when the conventional lignin fiber was used, the maximum failure strain only reached 1846 $\mu\varepsilon$, which was far less than the specification requirement of not less than 2500 $\mu\varepsilon$. This means that the addition of a large content of RAP would have an adverse effect on the low-temperature performance of the recycled mixture. This is because the aromatics and saturates were oxidized into resins and asphaltenes, resulting in the more brittle and less tough characteristics of the aged asphalt in RAP [38]. Even if the incorporation of rejuvenating agent softened the aged asphalt to a certain extent, its low-temperature performance still presents a certain gap to the specification requirements.

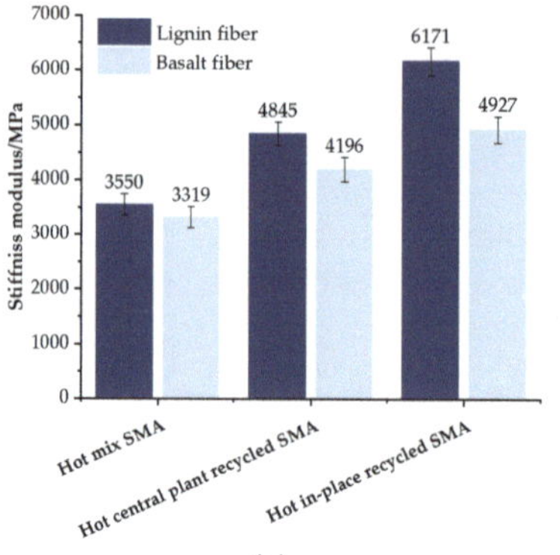

(a)

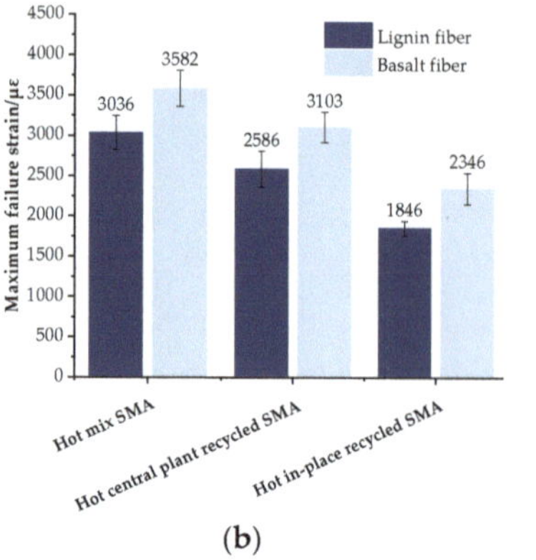

(b)

Figure 8. Results of trabecular bending test: (**a**) stiffness modulus; (**b**) failure stain.

When the mixture type was the same, the stiffness modulus of the recycled asphalt mixture reinforced by basalt fiber decreased greatly, while the maximum failure strain was significantly improved compared with that of the lignin fiber recycled one. When the RAP content was 0%, 30%, and 80%, the maximum failure strain of the corresponding mixture increased by 18%, 20%, and 27%, respectively. This observation can be primarily attributed to the inherent characteristics of basalt fiber, which possesses high strength and a high modulus [17]. At low temperatures, basalt fiber is closely combined with asphalt and has strong integrity, which enhances the elasticity of the basalt fiber recycled mixture, thus improving its anti-cracking ability at low temperatures.

3.2. Results of Low-Temperature Creep Test

According to the results from the low-temperature bending creep test, the ratio of W_d/W_s and $m(t)/S(t)$ at t = 3000 s was calculated, and the calculated results of W_d/W_s and $m(t)/S(t)$ are presented in Figure 9. As shown in Figure 9, whether basalt fiber or lignin fiber was used, the W_d/W_s and $m(t)/S(t)$ of recycled mixtures show a downward trend with the increase in RAP content, indicating that RAP weakens the stress relaxation ability of the recycled mixture and makes it difficult to dissipate the cumulative stress timely, resulting in the inferior low-temperature crack resistance. These results are also consistent with the test results of the trabecular bending test.

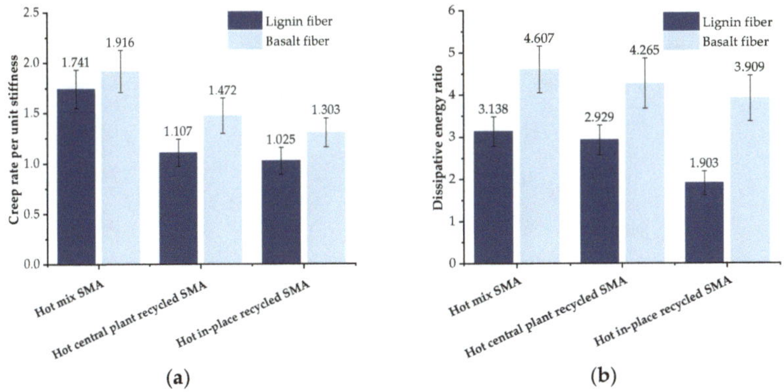

Figure 9. Results of low-temperature creep test: (**a**) creep rate per unit stiffness ($m(t)/S(t)$); (**b**) dissipative energy ratio (W_d/W_s).

When comparing the mixture types, it was observed that the basalt fiber recycled asphalt mixture exhibited significantly improved values for $m(t)/S(t)$ and W_d/W_s compared to the lignin fiber recycled asphalt mixture. Specifically, at a RAP content of 30%, the basalt fiber recycled mixture demonstrated an increase of 33% in $m(t)/S(t)$ and 45.6% in W_d/W_s. When the content of RAP was 80%, the $m(t)/S(t)$and W_d/W_s of the basalt fiber recycled mixture increased by 27.1% and 105.4%, respectively. This was consistent with the findings of L. Co et al. [39] who showed that the effect of basalt fibers was still relatively significant even in the case of different types of asphalt mixtures. According to the composite material theory, when basalt fiber is added into the recycled asphalt mixture, a three dimensional network structure will form in the mixture. A large number of new interfaces will also be generated, which can reduce the temperature sensitivity of asphalt [39]. Meanwhile, the network structure formed in the mixture could play a role in dissipating stress concentration to a certain extent [40], thereby improving the low-temperature cracking resistance of the mixture.

3.3. Results of Semi-Circular Bending Test

Figure 10 presents the results obtained from the semi-circular bending tests. A careful analysis of Figure 10 indicates that, as the RAP content increased, the fracture energy (G_f) and flexibility index (FI) of the recycled asphalt mixture decreased for a particular fiber type. This observation implies that RAP had a detrimental effect on the recycled mixtures' resistance to cracking at intermediate temperatures. This is because the old asphalt becomes hard and brittle after aging, which reduces the overall relaxation ability of the recycled asphalt mixture. Therefore, with a high RAP content, the asphalt pavement is more prone to cracking. When the mixture type was the same, compared with the lignin fiber recycled asphalt mixture, the G_f of the basalt fiber recycled asphalt mixture was greatly improved. At 0%, 30%, and 80% RAP content, the G_f of basalt fiber mixture increased by 14.5%, 31.8%, and 19.9%, respectively, indicating that the mixture could absorb more cracking energy after adding basalt fiber. The FI of the corresponding mixture increased by 50.8%, 74.8%, and 55.7%, respectively, indicating that basalt fiber could delay the propagation of cracks. This trend was consistent with the findings of W. Bang et al. [41]. This is due to the load dispersion and transmission ability of basalt fiber, which can reduce the stress concentration, thereby improving the anti-crack propagation ability of the mixture.

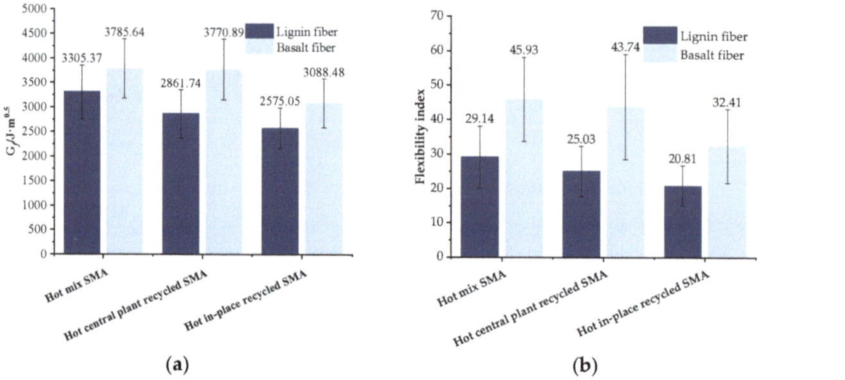

Figure 10. Results of semi-circular bending test: (**a**) fracture energy (G_f); (**b**) flexibility index (FI) of recycled asphalt mixture.

3.4. Results of IDEAL-CT Cracking Test

The results of the IDEAL-CT cracking tests are shown in Figure 11. It is evident from Figure 11 that as the RAP content increased, the fracture energy (G_{f0}) and cracking index of asphalt mixtures with the same fiber type displayed a downward trend. This trend suggests that the intermediate-temperature crack resistance of the mixture is weakened by RAP, making the asphalt mixture more prone to brittleness. When the mixture type was the same, compared with the lignin fiber recycled asphalt mixture, the fracture energy (G_{f0}) and CT_{index} of the basalt fiber recycled asphalt mixture were significantly improved. As for the hot central plant recycled mixture (30% RAP) the G_{f0} and CT_{index} increased by 18.5% and 37.2%, respectively. In terms of the hot in-place recycled one (80% RAP), the G_{f0} and CT_{index} increased by 28.6% and 19.9%, respectively. This infers that basalt fiber presents superior enhancing ability to the crack resistance of recycled asphalt mixture than lignin fiber.

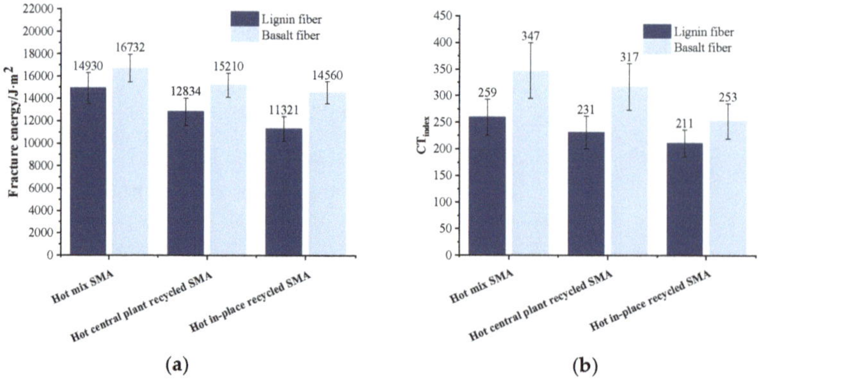

Figure 11. Results of IDEAL-CT cracking test: (**a**) fracture energy (G_{f0}); (**b**) crack resistance index (CT_{index}).

3.5. Correlation Analysis

Differences in the crack resistance of hot central plant recycled and hot in-place recycled asphalt mixtures at low and intermediate temperatures have been demonstrated through the preceding discussions. To gain a deeper understanding of how the crack resistance of hot recycled asphalt mixtures depended on various fibers, a correlation analysis of mixture detection indexes was conducted. In this study, Pearson's correlation coefficient was used to investigate the correlation between the results of each crack resistance index, with the mix type (RAP content) as the variable. Through IBM SPSS Statistics 27.0, the anti-cracking indexes (maximum failure strain, G_f, FI, G_{f0}, CT_{index}, W_d/W_s, and $m(t)/S(t)$) were imported into the SPSS software, and the quotient of the product of covariance and standard deviation between variables was calculated and defined as Pearson's correlation coefficient 'r'. The calculation formula is expressed by Equation (14). The value range of Pearson's correlation coefficient, denoted as 'r', lies between −1 and 1. The degrees and interpretation of Pearson's correlation coefficient are shown in Table 9 [42].

$$r = \frac{\sum_{i=1}^{n}\left(x_i - \overline{X}\right)\left(y_i - \overline{Y}\right)}{\sqrt{\sum_{i=1}^{n}\left(x_i - \overline{X}\right)^2}\sqrt{\sum_{i=1}^{n}\left(x_i - \overline{Y}\right)^2}} \tag{14}$$

where x_i, y_i are the corresponding cracking indicators; $\overline{X}, \overline{Y}$ are the average values of the crack resistance index; and r is Pearson's correlation coefficient.

Table 9. Pearson's correlation coefficients and their degrees and interpretation.

	Size of Correlation	Interpretation
Pearson's correlation coefficients	0.9 to 1.0 (−0.9 to −1.0)	Very high positive (negative) correlation
	0.7 to 0.9 (−0.7 to −0.9)	High positive (negative) correlation
	0.5 to 0.7 (−0.5 to −0.7)	Moderate positive (negative) correlation
	0.3 to 0.5 (−0.3 to −0.5)	Low positive (negative) correlation
	0.0 to 0.3 (−0.0 to −0.3)	Negligible correlation

The horizontal and vertical coordinates in Figure 12 show the test metrics for each cracking test. Figure 12a displays the correlation of each cracking indicator for the lignin fiber asphalt mixture, while Figure 12b presents the corresponding correlation for the basalt fiber asphalt mixture. In the visualization of the correlation matrix, a positive correlation is represented by orange. Conversely, green represents a negative correlation. The size of the square within the matrix corresponds to the degree of correlation between the respective variables x_i and y_i. The greater the correlation, the larger the area of the square, and vice versa. According to Figure 12, comparing the indexes of low-temperature cracking resistance (maximum failure strain, W_d/W_s, and $m(t)/S(t)$), the correlation between maximum failure strain and other indicators was better. From a mathematical perspective, the trabecular bending test is more effective in evaluating the low-temperature cracking resistance of hot recycled asphalt mixtures. When comparing indices of mid-temperature cracking resistance such as Gf, FI, G_{f0}, and CT_{index}, it was observed that the correlation between G_{f0} (and/or CT_{index}) and the other indicators was stronger. Therefore, the IDEAL-CT test is more suitable for assessing the mid-temperature cracking resistance of hot recycled asphalt mixtures.

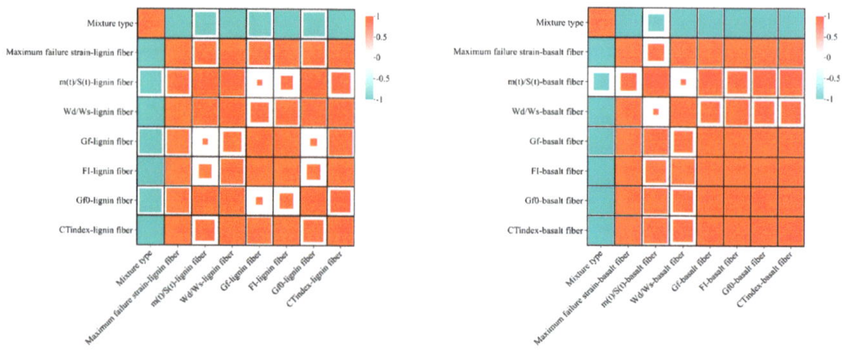

(**a**) (**b**)

Figure 12. Correlation analysis of indicators: (**a**) lignin fiber hot recycled asphalt mixture; (**b**) basalt fiber hot recycled asphalt mixture.

4. Conclusions

This study examined the impact of basalt fiber on the cracking resistance of hot recycled asphalt mixtures through various low- and intermediate-temperature (−10 °C and 25 °C) cracking tests. Based on these tests, the following conclusions can be drawn:

(1) Compared to the hot mix SMA-13 asphalt mixture, the low and intermediate-temperature cracking resistance of the hot recycled asphalt mixture deteriorated. The amplitude of the hot in-place recycled asphalt mixtures decreased more due to the higher percentage of RAP. In terms of intermediate-temperature cracking resistance, the hot in-place recycled mixture decreased by 22.1% and 28.6% (in terms of G_f and FI, respectively), while the hot central plant recycled mixture decreased by 13.4% and 14.1%. In terms of low-temperature cracking resistance, the hot in-place recycled mixture decreased by 41.1% and 39.4% (in terms of $m(t)/S(t)$ and W_d/W_s, respectively), while the hot central plant recycled mixture decreased by 36.4% and 6.7%.

(2) Basalt fiber improved the low and intermediate-temperature cracking resistance of recycled mixtures compared to lignin fiber. The enhancing amplitude in cracking resistance was greater for the hot in-place recycled mixture. In terms of intermediate-temperature cracking resistance, the hot in-place recycled mixture increased by 19.9% and 55.7% (in terms of G_f and FI, respectively), while the hot central plant recycled mixture increased by 31.8% and 74.8%. In terms of low-temperature cracking resistance, the hot in-place recycled mixture decreased by 27.1% and 105.4% (in terms of $m(t)/S(t)$ and W_d/W_s, respectively), while the hot central plant recycled mixture decreased by 33.0% and 45.6%. It was shown that basalt fibers could improve toughness and enhance energy dissipation.

(3) The results of the above four sets of cracking tests show that basalt fiber could effectively improve the cracking resistance of hot in-place recycled mixtures and hot central plant recycled mixtures, and the test results supplement the research on the application of basalt fiber in asphalt mixtures, which could provide an effective basis for the recycling of pavements.

(4) There were differences in the cracking resistance properties derived from different experimental methods, which may reflect different cracking mechanisms. Based on the correlation analysis, the results indicated that the trabecular bending test was more apt for evaluating the low-temperature cracking resistance of hot recycled asphalt mixtures, whereas the IDEAL-CT test was more suitable for assessing the intermediate-temperature cracking resistance of the mixtures.

Author Contributions: Conceptualization, Y.Z. (Yu Zhang) and B.L.; methodology, Y.Z. (Yao Zhang) and A.K.; formal analysis, Y.Z. (Yu Zhang) and B.L.; validation, B.L. and Y.W.; resources, A.K. and Y.Z. (Yao Zhang); data curation, Y.W. and Y.Z. (Yu Zhang); writing—original draft preparation, Y.Z. (Yu Zhang) and B.L.; writing—review and editing, Y.Z. (Yu Zhang) and B.L.; project administration, Y.Z. (Yao Zhang) and B.L.; funding acquisition, Y.Z. (Yao Zhang). All authors have read and agreed to the published version of the manuscript.

Funding: This research was funded by the National Natural Science Foundation of China, grant numbers 52178439 and 52108422 and the Postgraduate Research and Practice Innovation Program of Jiangsu Province (Yangzhou University) (SJCX22_1751). In addition, this research was funded by the postgraduate education and teaching reform and practice project of Yangzhou University, grant number XJGKT22_004.

Institutional Review Board Statement: Not applicable.

Informed Consent Statement: Not applicable.

Data Availability Statement: Data are contained within the article.

Conflicts of Interest: The authors declare no conflicts of interest.

References

1. Aravind, K.; Das, A. Pavement design with central plant hot-mix recycled asphalt mixes. *Constr. Build. Mater.* **2007**, *21*, 928–936. [CrossRef]
2. Bowers, B.F.; Powell, R.B. Use of a hot-mix asphalt plant to produce a cold central plant recycled mix: Production method and performance. *Transp. Res. Rec.* **2021**, *2675*, 451–459. [CrossRef]
3. Moon, B.; Bozorgzad, A.; Lee, H.; Kwon, S.; Jeong, K.D.; Cho, N.J. Development of warm in-place recycling technique as an eco-friendly asphalt rehabilitation method. *Infrastructure* **2021**, *6*, 101. [CrossRef]
4. Chiara, R.; Augusto, C.; Massimo, L.; Michael, P. Rheological modeling of asphalt binder and asphalt mortar containing recycled asphalt material. *Mater. Struct.* **2016**, *49*, 4167–4183.
5. Kent, R.; Hansen, A. Asphalt pavement industry survey on recycled materials and warm-mix asphalt usage 2016 (Information Series 138) 7th annual survey. *NAPA* **2017**, 2015. [CrossRef]
6. Xiang, M.; Jia, W.; Yin, X. Investigation on the Effects of RAP Proportions on the Pavement Performance of Recycled Asphalt Mixtures. *Front. Mater.* **2022**, *8*, 841809.
7. Kocak, S.; Haider, S.W. Performance and cost evaluations of 100% recycled hot asphalt mixtures for pothole patching applications in flexible pavements. *Constr. Build. Mater.* **2023**, *392*, 131921. [CrossRef]
8. Nosetti, A.; Pérez-Madrigal, D.; Pérez-Jiménez, F.; Martínez, A.H. Effect of the recycling process and binder type on bituminous mixtures with 100% reclaimed asphalt pavement. *Constr. Build. Mater.* **2018**, *167*, 404–448. [CrossRef]
9. Mojtaba, M.; Hesham, A.; James, A.; Musse, M.; Gregory, A.; Aidin, M. The effect of aging on the cracking resistance of recycled asphalt. *Adv. Civ. Eng.* **2017**, *2017*, 7240462.
10. Bouraima, M.B.; Zhang, X.H.; Rahman, A.; Qiu, Y.J. A comparative study on asphalt binder and mixture performance of two traffic lanes during hot in-place recycling (HIR) procedure. *Constr. Build. Mater.* **2019**, *223*, 33–43. [CrossRef]
11. Sivilevicius, H.; Braziunas, J.; Prentkovskis, O. Technologies and principles of hot recycling and investigation of preheated reclaimed asphalt pavement batching process in an asphalt mixing plant. *Appl. Sci.* **2017**, *7*, 1104. [CrossRef]
12. Seyed, A.Z.; Fereidoon, M.N.; Mohammad, D.; Mansour, F. Evaluation of rheological and mechanical properties of hot and warm mix asphalt mixtures containing Electric Arc Furnace Slag using gyratory compactor. *Constr. Build. Mater.* **2023**, *378*, 131042.
13. Tran, N.H.; Julian, G.; Taylor, A.J.; Willis, R.; Hunt, D. Effect of geosynthetic material in reclaimed asphalt pavement on performance properties of asphalt mixtures. *Transp. Res. Rec.* **2012**, *2294*, 26–33. [CrossRef]
14. Noto, S.; Mangiafico, S.; Sauzéat, C.; Di Benedetto, H.; Romeo, E.; Tebaldi, G. Diffusion phenomenon between two different bitumens from mechanical analysis. *J. Mater. Civil. Eng.* **2022**, *34*, 04021476. [CrossRef]
15. Nahar, S.N.; Mohajeri, M.; Schmets, A.J.M.; Scarpas, A.; Ven, M.F.C.; Schitter, G. First observation of blending-zone morphology at interface of reclaimed asphalt binder and virgin bitumen. *Transport Res. Rec.* **2013**, *2370*, 1–9. [CrossRef]
16. Mandal, T.; Hanz, A.; Bahia, H. Challenges in using the Disc-Shaped Compact Tension (DCT) test to determine role of asphalt mix design variables in cracking resistance at low temperatures. *Int. J. Pavement Eng.* **2019**, *20*, 1275–1284. [CrossRef]
17. Xiang, S.; Bao, H.; Dragon, V. Evaluation of cracking resistance of recycled asphalt mixture using semi-circular bending test. In Proceedings of the GeoShanghai 2010, Paving Materials and Pavement Analysis, Shanghai, China, 3–5 June 2010; pp. 58–65.
18. Fu, Z.; Soohyok, I.; Li, S.; Tom, S. Development of an IDEAL cracking test for asphalt mix design and QC/QA. *Road Mater. Pavement Des.* **2016**, *18*, 405–427.
19. Jing, H.; Cong, Y.; Zhang, Y.; Song, L.; Romanovich, M. Mix ratio optimization design method for hot in-place recycled asphalt mixtures. *DYNA* **2020**, *95*, 553–560. [CrossRef] [PubMed]

20. Moon, K.; Falchetto, A.; Marasteanu, M.; Turos, M. Using recycled asphalt materials as an alternative material source in asphalt pavements. *KSCE J. Civ. Eng.* **2014**, *18*, 149–159. [CrossRef]
21. Ming, L.; Li, L.; Wei, H.; Hua, W. Study on the mixing process improvement for hot recycled asphalt mixture. *Constr. Build. Mater.* **2023**, *365*, 130068.
22. Seyedeh, G.H.; Kordani, A.; Mohammad, Z. Effect of recycled additives on pure mode I fracture resistance and moisture susceptibility of hot mix asphalt (HMA): An experimental study using semicircular bending (SCB) and indirect tensile strength (ITS) tests. *Theor. Appl. Fract. Mech.* **2023**, *128*, 104168.
23. Dai, L.; Marie, E.; Yeong, J.; Filippo, G. Future recyclability of hot mix asphalt containing recycled plastics. *Constr. Build. Mater.* **2023**, *368*, 130396.
24. Guo, Q.; Chen, Z.; Liu, P.; Yi, L.; Hu, J.; Gao, Y.; Li, X. Influence of basalt fiber on mode I and II fracture properties of asphalt mixture at medium and low temperatures. *Theor. Appl. Fract. Mech.* **2021**, *112*, 102884. [CrossRef]
25. Pirmohammad, S.; Amani, B.; Shokorlou, Y.M. The effect of basalt fibres on fracture toughness of asphalt mixture. *Fatigue Fract. Eng. M* **2020**, *43*, 1446–1460. [CrossRef]
26. Celauro, C.; Praticò, F.G. Asphalt mixtures modified with basalt fibres for surface courses. *Constr. Build. Mater.* **2018**, *170*, 245–253. [CrossRef]
27. Çetín, A.; Oral, G. Performance evaluation of porous asphalt mixtures modified with basalt fiber. *Rev. Constr.* **2022**, *21*, 93–104. [CrossRef]
28. Ying, H.; Guang, M.; Peng, X.; Qin, T.; Fang, H.; Ai, K.; Zheng, W. Recent advances in basalt fiber reinforced asphalt mixture for pavement applications. *Materials* **2022**, *15*, 6826. [CrossRef]
29. *JTG 5142-2019*; Technical Specifications for Maintenance of Highway Asphalt Pavement. Ministry of Transport of the People's Republic of China: Beijing, China, 2019.
30. *JTG E20-2011*; Standard Test Methods of Bitumen and Bituminous Mixtures for Highway Engineering. Ministry of Transport of the People's Republic of China: Beijing, China, 2011.
31. *JTG/T 5521-2019*; Technical Specifications for Highway Asphalt Pavement Recycling. Ministry of Transport of the People's Republic of China: Beijing, China, 2019.
32. *JTG E42-2005*; Test Methods of Aggregatate for Highway Engineering. Ministry of Transport of the People's Republic of China: Beijing, China, 2005.
33. *GB/T 7690.5*; Reinforcements. Test Method for Yarns. Part 5:Determination of Fiber Diameter for Glass Fiber. General Administration of Quality Supervision, Inspection and Quarantine of the People's Republic of China: Beijing, China, 2013.
34. *JT/T 776.1-2010*; Basalt Fiber and Product for Highway Engineering—Part 1: Basalt Fiber Chopped Strand. Ministry of Transport of the People's Republic of China: Beijing, China, 2010.
35. *GB/T 20310-2006*; Textile Glass-Rovings-Manufacture of Test Specimens and Determination of Tensile Strength of Impregnated Rovings. General Administration of Quality Supervision, Inspection and Quarantine of the People's Republic of China: Beijing, China, 2006.
36. *GB/T 1549-2008*; Chemical Analysis Methods of Glass, Glass Marble and Vitreous Fiber. General Administration of Quality Supervision, Inspection and Quarantine of the People's Republic of China: Beijing, China, 2008.
37. *AASHTO TP 124-16*; Determining the Fracture Potential of Asphalt Mixtures Using Semicircular Bend Geometry (SCB) at Intermediate Temperature. American Association of State Highway and Transportation Officials: Washington, DC, USA, 2016.
38. Pei, Z.; Xu, M.; Cao, J.; Feng, D.; Hu, W.; Ren, J.; Jing, R.; Yi, J. Analysis of the microcharacteristics of different kinds of asphalt based on different aging conditions. *Mater. Struct.* **2022**, *55*, 250. [CrossRef]
39. Ke, L.; Peng, X.; Ai, K.; Zheng, W.; Peng, L. Suitability of fiber lengths for hot mix asphalt with different nominal maximum aggregate size: A pilot experimental investigation. *Materials* **2020**, *13*, 3685. [CrossRef]
40. Bo, L.; Yang, Z.; Ai, K.; Ke, L.; Qian, G. Research on fracture behavior of fiber-asphalt mixtures using digital image correlation technology. *Materials* **2023**, *16*, 6825. [CrossRef] [PubMed]
41. Wu, B.W.; Meng, W.J.; Xia, J.; Xiao, P. Influence of basalt fibers on the crack resistance of asphalt mixtures and mechanism analysis. *Materials* **2022**, *15*, 744. [CrossRef]
42. Martin, D.; Zuzana, P.; Michal, J.; Maria, K. Evaluation of the effect of average annual temperatures in Slovakia between 1971 and 2020 on stresses in rigid pavements. *Land* **2022**, *11*, 764. [CrossRef]

materials

MDPI

Article

Effect of Brick Aggregate Content on Performance of Recycled Construction-Solid-Waste Aggregate

Xuan Zhu [1,2,*], Le Ding [1], Yuexing Wu [1], Xinzhong Wang [1] and Xianliang Tan [1]

[1] School of Civil Engineering, Hunan City University, Yiyang 413000, China; dingle@hncu.edu.cn (L.D.); wuyuexing@hncu.edu.cn (Y.W.); wxz811@hncu.edu.cn (X.W.); tanxianliang@hncu.edu.cn (X.T.)
[2] National Engineering Research Center for Highway Maintenance Technology, Changsha University of Science & Technology, Changsha 410114, China
* Correspondence: zhuxuan@hncu.edu.cn; Tel.: +86-133-4131-5163

Abstract: In road engineering, road construction requires a large amount of natural aggregate; its substitution with recycled construction-solid-waste aggregate not only saves resources but also reduces the burden on the environment. The main components of construction solid waste are concrete blocks and brick slag; the breakability of the latter can affect the performance of mixed recycled aggregate, which hinders the use of construction solid waste in road engineering applications. To analyze the applicability of recycled construction-solid-waste aggregate containing brick slag aggregate in the subgrade layer, the effect of brick aggregate content on the CBR (California bearing ratio) and crushing value of mixed recycled aggregates was evaluated based on laboratory tests, and the field compaction quality of the recycled aggregates was analyzed. The results show that the 9.5–19 mm mixed recycled aggregate samples were crushed to a higher degree during the compaction process. A brick aggregate content less than 40% had little effect on the performance of mixed recycled construction-solid-waste aggregate. It is recommended to use a 22 t road roller for five passes (two weak vibrations + two strong vibrations + one weak vibration) at a speed of 3 km/h in the main compaction stage of the subgrade filling.

Keywords: construction solid waste; recycled aggregate; brick slag content; subgrade

Citation: Zhu, X.; Ding, L.; Wu, Y.; Wang, X.; Tan, X. Effect of Brick Aggregate Content on Performance of Recycled Construction-Solid-Waste Aggregate. *Materials* **2024**, *17*, 2616. https://doi.org/10.3390/ma17112616

Academic Editor: Francisco Agrela

Received: 16 April 2024
Revised: 24 May 2024
Accepted: 27 May 2024
Published: 29 May 2024

1. Introduction

1.1. Research Background

With the development of urbanization, the production of construction solid waste has increased rapidly, which has put enormous pressure on the environment [1]. The traditional method for construction-solid-waste treatment is landfill, which not only occupies a large amount of land but is also harmful to the environment and human health [2,3]. In contrast, the utilization of renewable resources is an important process in clean production, and the utilization of construction waste could achieve sustainable development and environmental protection in the construction industry [4,5].

Construction solid waste is difficult to use directly due to its large volume. Therefore, in order to improve its utilization efficiency, it is generally crushed and recycled into aggregate [6,7]. In structural engineering, there are strict requirements for mechanical performance; as the strength of subgrade materials is relatively low [8–10], the application of recycled construction-solid-waste aggregates for subgrade filling is one of the ways to efficiently utilize waste-based resources [11].

Recycled construction-solid-waste aggregate mainly includes recycled concrete aggregate and brick slag aggregate [12]. There are bottlenecks in the development of technology for the separation of the two; therefore, the recycling and utilization of brick–concrete mixed aggregates are particularly important [13,14]. The hardness of recycled brick aggregate from broken brick slag is low, which affects the strength and stability of mixed recycled construction-solid-waste aggregate [15,16]. Previous research studies have shown that brick aggregate content has a significant impact on the volume index of mixed recycled aggregate [17]; specifically, with the increase in brick aggregate content, the apparent density increases, and the maximum dry density decreases [18,19]. However, the effect of brick aggregate content on the mechanical performance of mixed recycled aggregates is still unclear. Furthermore, researchers have investigated the performance of mixed recycled aggregate [20,21], finding that the resilient modulus and deflection of construction-solid-waste material containing a small amount of recycled brick aggregate can meet the technical standard requirements [22,23].

At present, research on recycled aggregate mainly focuses on the evaluation of mechanical properties based on laboratory tests [24,25]. Although laboratory experiments can be easily performed, they cannot simulate field conditions well, which causes a discrepancy between laboratory design and subgrade construction [26]. The field compaction method for construction solid waste, the main cause of uneven settlement of the subgrade, is determined by the on-site engineers according to their experience [27]. Accurate control and scientific evaluation of compaction quality are effective ways to solve the above problems. The portable falling weight deflectometer (PFWD) is widely used in road engineering due to its advantages of being fast, accurate, lightweight, and environmentally friendly; the subgrade deflection and resilient modulus under different material compositions can be rapidly evaluated based on the pressure and displacement sensors of a PFWD [28].

Therefore, in order to evaluate the effect of brick aggregate content on the performance of recycled construction-solid-waste aggregate, a research method that innovatively combines laboratory experiments and field investigation was used in this study. The effect of brick aggregate content on the mechanical properties of recycled aggregate was analyzed through laboratory tests, and the optimal brick aggregate content and construction method were determined based on field measurement parameters. Our research results can provide theoretical support for the application of recycled brick–concrete-solid-waste aggregates in subgrade filling.

1.2. Research Plan

The purpose of this study was to evaluate the effect of brick aggregate content on the performance of recycled mixed construction-solid-waste aggregate and to select the optimal content and field compaction method. The research plan was executed as follows: Firstly, the properties of recycled mixed aggregate were tested to investigate the differences between brick aggregate and recycled concrete aggregate. Secondly, the effects of brick aggregate content on the gradation, CBR, and crushing value of mixed recycled aggregate were studied based on laboratory experiments. Finally, recycled construction-solid-waste aggregates with different brick aggregate contents were used for subgrade filling in field tests. The deflection and resilient modulus were tested by using a PFWD to evaluate the applicability of brick construction-solid-waste aggregate in subgrade filling, and the optimal brick aggregate content and compaction method were determined. The research plan is shown in Figure 1.

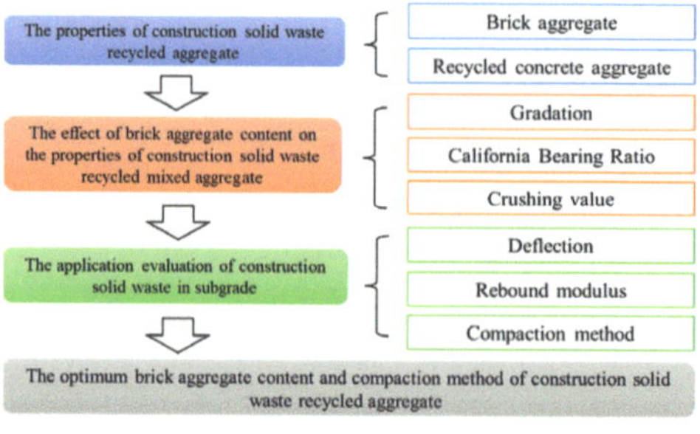

Figure 1. The research plan.

2. Materials and Methods

2.1. Properties of Construction-Solid-Waste Materials

The jaw breaker produced by Yinhe Analytical Instrument Chemical Co., Ltd. (Hebi, China) was used to crush brick slag and waste concrete blocks and form recycled aggregates with a particle size range of 1.18–37.5 mm. To remove soil from the construction solid waste, the recycled aggregate was washed and dried for 3–4 h at a temperature of 100 °C, as shown in Figure 2.

Figure 2. Recycled construction-solid-waste aggregate.

Performance tests were conducted on the brick aggregate and the recycled concrete aggregate for recycled construction-solid-waste aggregate according to technical standard JTG E42-2005 [29], and a natural aggregate was used for comparison. As shown in Table 1, the water absorption rate of the recycled concrete aggregate was 5.30%, and the crushing value was 18.6%. The brick aggregate had low apparent density, and its water absorption rate and crushing value were higher than those of the recycled concrete aggregate. The results also indicate that the strength of the recycled construction-solid-waste aggregate was poor.

Table 1. The properties of the recycled construction-solid-waste aggregate.

Material	Apparent Density ($kg{\cdot}m^{-3}$)	Water Absorption Rate (%)	Crushing Value (%)	Moisture Content (%)
Brick aggregate	2038	12.70	42.3	4.03
Concrete aggregate	2580	5.30	18.6	2.50
Natural aggregate	2863	1.01	13.2	1.23

The brick slag was subjected to mechanical cutting and vibration during the crushing process. Due to the lower hardness of the brick slag, its edges and texture were more easily polished to produce brick slag powder with a particle size smaller than 1.18 mm. To analyze the effect of moisture content on the mechanical state of the brick slag powder, the plasticity characteristics were evaluated by using the liquid and plastic limit indices [30]. As shown in Table 2, the liquid limit (Wl) of the brick slag powder was 36.5%, which indicates a limit moisture content of 36.5% for the transition from the flowing state to the plastic state. The plastic limit (Wp) of the brick slag powder was 19.9%, which indicates a limit moisture content of 19.9% for the transition from the plastic state to the semi-solid state. The results also show that the plasticity index (Ip) of the brick slag powder was 17.5%, which indicates good plasticity.

Table 2. Results of liquid and plastic limit tests.

Material	Liquid Limit (W_l) (%)	Plastic Limit (W_p) (%)	Plasticity Index (I_p) (%)
Brick slag powder	36.5	19.9	16.6

2.2. Research Method

2.2.1. Laboratory Tests

The recycled concrete aggregate and brick aggregate in the dry state were selected, and mixed aggregate samples were prepared according to the following six different contents of brick aggregate: 0%, 20%, 40%, 60%, 80%, and 100%. Two laboratory experiments were carried out to determine the samples' bearing capacity and mechanical strength.

The California bearing ratio (CBR) index was used to evaluate the bearing capacity of the recycled mixed aggregates. The samples were prepared by using heavy compaction, according to technical standard JTG E40-2007 [31]. We considered three layers of compaction, and each layer was compacted 98 times, where 45 cm was selected as the falling height of the heavy compaction hammer. A 2.5 mm/min loading rate was applied until the sample settled at 25.4 mm [31]. The force and settlement data during the test were recorded to calculate the CBR.

In order to evaluate the firmness and wear resistance of the mixed recycled construction-solid-waste aggregate, the mechanical properties at the six brick aggregate content values were studied based on the crushing value index. The 9.5–13.2 mm aggregate samples were screened according to technical standard JTG E42-2005 [29]. Different proportions of brick aggregate and concrete recycled aggregate were prepared, and the crushing value test was carried out. In the test, we adopted the uniform-loading mode; once the total load reached 400 kN at 10 min, the pressure was stabilized for 5 s; then, the load was removed. Finally, an aggregate sample with a particle size smaller than 2.36 mm was employed to calculate the crushing value of the recycled mixed aggregate.

2.2.2. Field Investigation

A test section of construction-solid-waste subgrade was constructed in Guangdong Province, China. Six different proportions of brick aggregate and recycled concrete aggregate were used to fill in the first part of the test section, and the deflection and resilient modulus of the field were measured by using a PFWD to evaluate the construction quality of mixed recycled aggregate with different proportions of construction solid waste. The optimal proportion of brick aggregate was used for field compaction in the second part of the test section, and the settlement characteristics of the subgrade were evaluated based on the thickness evolution characteristics of the fill layer.

3. Effect of Brick Aggregate Content on Performance of Mixed Recycled Aggregate

3.1. Effect of Brick Aggregate Content on CBR

To analyze the stability of mixed recycled aggregate from construction solid waste during the preparation process of CBR specimens, screening tests were conducted on recycled aggregate with different brick aggregate contents, and the impact of brick aggregate crushing characteristics on the gradation of the mixed recycled aggregates during compaction was evaluated. Figure 3 shows the grading curves before and after sample preparation. It can be seen that, when the brick aggregate content was less than 80%, it had little influence on the gradation of the mixed recycled solid-waste aggregate before sample preparation, but, when it was greater than 80%, the pass rate of the 9.5 mm sieve was relatively high, which indicated that its gradation was relatively fine. It was not difficult to find that the gradation change was not evident in the compaction process when the brick aggregate content was 0. With an increase in brick aggregate content, the passing percent of the 9.5 mm sieve and the degree of change in gradation increased. When the brick aggregate content was 100%, the passing percentage of the 9.5 mm sieve increased by around 10%.

Each mixed recycled aggregate sample was divided into four parts according to particle size, i.e., between 19 and 37.5 mm, 9.5–19 mm, 1.18–9.5 mm, and 0.075–1.18 mm, and the residual percentage variation was analyzed. As shown in Figure 4, when the brick aggregate content was less than 40%, the residual percentage variation of aggregates with size greater than 19 mm was relatively small and that of aggregates with size less than 9.5 mm was relatively stable. On the contrary, with the increase in brick aggregate content, the residual percentage variation of aggregates with size less than 9.5 mm increased gradually and that of 9.5–19 mm recycled aggregate was reduced, which indicated that the aggregate was crushed to a higher degree during the compaction process. The results show that a greater content of brick aggregate led to a greater change in its residual percentage.

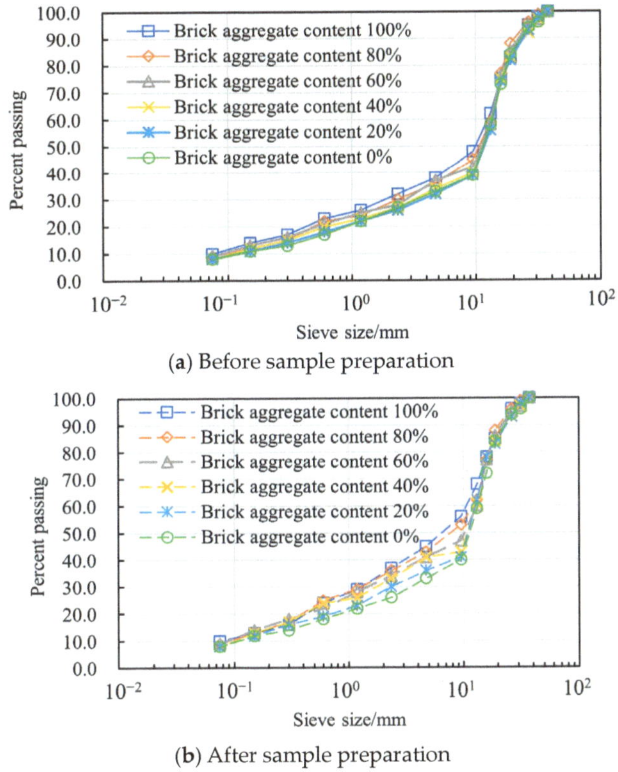

Figure 3. Gradation curves of recycled construction-solid-waste aggregate.

As a performance index, the CBR directly reflects the strength of mixed construction-solid-waste aggregate and has clear physical and mechanical significance. The load at a penetration depth of 5 mm was used to calculate the CBR. Four specimens for each brick aggregate content were prepared, and Table 3 shows the CBR results. It can be seen that the CBR continuously decreased with the increase in brick aggregate content, which indicates that the bearing capacity of the recycled aggregate was reduced by the brick aggregate. The CBR of the mixed recycled aggregate samples was also far higher than the requirement of 8% to meet the technology standard [32] when the content of brick aggregate was 100%. When the brick aggregate content was 0–40%, the reduction rate of the CBR was lower than that found for 40–100% brick aggregate content. The main reason for this finding is that, when the content of brick aggregate was less than 40%, there was enough high-strength concrete aggregate in the mixed recycled aggregate to form a stable skeleton structure. Therefore, with the increase in brick aggregate content, the strength of the mixed aggregate decreased slowly. Due to the stronger water absorption and expansion ability of brick aggregates, the water absorption rate and expansion amount of mixed recycled construction-solid-waste aggregates gradually increased with the increase in brick aggregate content. To ensure the bearing capacity of mixed recycled construction-solid-waste aggregate, the optimum content of brick aggregate is recommended to be 40%.

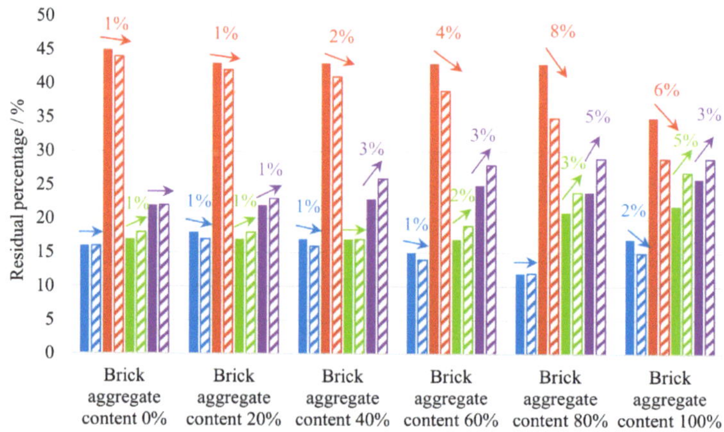

Figure 4. Aggregate content changes of samples with different particle sizes before and after compaction. Note: The blue, red, green, and purple columns represent the 19–37.5 mm, 9.5–19 mm, 1.18–9.5 mm, and 0.075–1.18 mm aggregates, respectively. Additionally, solid-filled columns indicate the residual percentages of the aggregate before compaction, and the diagonal-filled columns represent after compaction.

Table 3. CBR test results of recycled aggregate for construction solid waste.

Content of Brick Aggregate (%)	CBR (%)	Requirement of CBR (%)	Expansion Ratio (%)	Water Absorption Rate (%)
0	36.1		0.017	7.6
20	34.2		0.017	10.0
40	32.8		0.021	11.8
60	28.4	8	0.031	10.6
80	24.7		0.038	12.8
100	18.4		0.044	13.1

3.2. *Effect of Brick Aggregate Content on Crushing Value*

The crushing value was used to evaluate the resistance of aggregates to crushing under gradually increasing loads. A larger crushing value corresponded to more crushing of the aggregates under loading and a lower strength. The crushing performance of the mixed recycled aggregates with different brick aggregate contents was evaluated. As shown in Figure 5, the crushing value of the mixed recycled aggregate increased with the increase in brick aggregate content, and the greater the amount of brick aggregate, the greater the dispersion of the test results. When the brick aggregate content was 100%, the crushing value was 31.84%, which was 1.68 times the value found when the brick aggregate content was 0. The recycled aggregate's crushing value sharply increased with an increase in brick aggregate content above 40%, and the crushing resistance decreased noticeably, which indicates that brick aggregate content above 40% has a greater impact on the overall strength of mixed recycled aggregate.

In previous studies, the water absorption characteristics of recycled aggregate have been evaluated [33,34], but the influence of water immersion on the mechanical properties of recycled aggregate has not. In this study, the crushing value test was conducted on the recycled aggregates after immersion. Because the properties of brick aggregate are easily affected by water, the soaking crushing value of the brick aggregate was analyzed after 72 h. As shown in Figure 5, when the mixed recycled aggregate did not contain brick aggregate, the soaking crushing value only increased by around 1%. Since the strength of brick aggregate is low after water absorption, the strength decrease in the samples was more evident when the content of brick aggregate was increased. Therefore, when the brick aggregate content was 100%, the soaking crushing value increased by around 7%, which indicates that a greater brick aggregate content corresponded with a greater influence of soaking on the crushing value. Measures should be taken to improve the moisture resistance of subgrade recycled aggregate. Consistent with the change characteristics of the crushing value in the dry state, when the content of brick aggregate was 40%, was that the crushing value had a noticeable inflection point. To ensure mixed recycled aggregate compressive strength, it is recommended that the content of brick aggregate not exceed 40%.

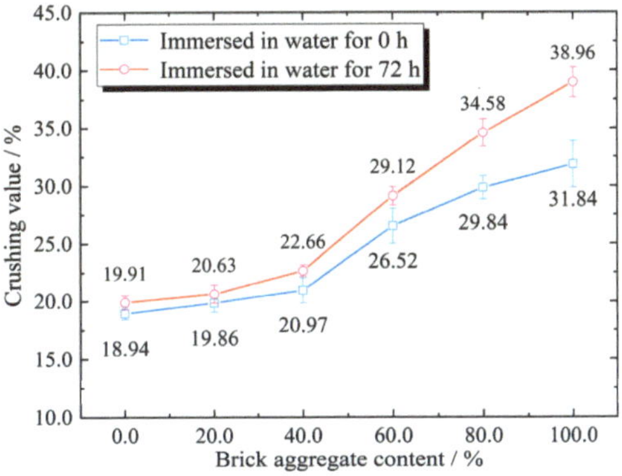

Figure 5. The crushing values of mixed recycled aggregates with different brick aggregate contents.

4. Field Compaction of Recycled Aggregate

4.1. Evaluation of Field Compaction Quality

In previous studies, laboratory tests and numerical simulations of recycled aggregates have been evaluated [35,36], but the influence of field parameters on pavement design has not been considered. In order to obtain the field parameters of recycled aggregates and analyze the applicability of recycled aggregates containing brick aggregate in subgrade filling, a test section of construction-solid-waste subgrade was constructed in Guangdong, China. The total length of the test section was 600 m, with a width of 28.5 m. The first 120 m of the test section was used to investigate the field deflection and resilient modulus of different brick aggregate contents, and the last 480 m was used to evaluate the compaction characteristics of the subgrade under different compaction methods. In order to reduce the interference of other factors during the experiments, the first part of the test section was divided into six different areas, each of which was 20 m in length and had a field compaction degree of 95%. The brick aggregate contents of the six areas were 0%, 20%, 40%, 60%, 80%, and 100%.

Subgrade construction quality was evaluated according to the index of deflection and the resilient modulus. At present, there are many methods for assessing deflection, among which the PFWD can simulate dynamic loads well and scientifically evaluate the dynamic deflection and resilient modulus generated under dynamic loads. In addition, the data measured by the sensor can be processed directly by a computer, which is faster and more accurate than human processing.

A PFWD was adopted for the field test, and a working plane with uniform and flat characteristics was selected, as shown in Figure 6. The test location was determined according to the method of uniform distribution. To reduce the boundary effect, starting from 2.85 m away from the edge of the subgrade, the test was conducted every 5.7 m in the horizontal direction and every 2 m in the vertical direction. A total of 300 points were considered in the test section.

Figure 6. Recycled construction-solid-waste aggregate used for subgrade filling.

The resilient modulus and deflection of the test section with different brick slag contents were obtained, and the results are shown in Figure 7. It was found that, with the increase in brick slag content, the resilient modulus decreased, and the deflection increased. The reason for this finding is that the brick aggregate was crushed under the action of the dynamic load, and the greater its content, the higher the degree of crushing. In the crushing process, the relative displacement between the particles was large, which led to the increase in deflection and the decrease in resilient modulus.

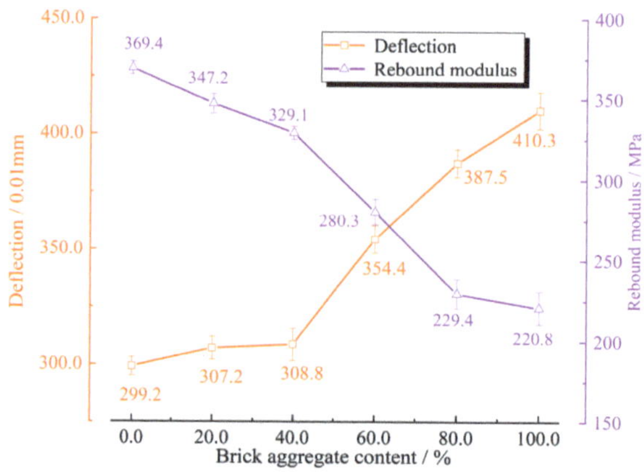

Figure 7. The results of the field test.

There was discreteness in the field detection data of the test section, which gradually increased with the increase in brick slag content. In addition, it could also be seen that the deflection curve had a noticeable inflection point corresponding to the slag content of 40%, while the resilient modulus curve was smoother. When the content of brick slag was 40%, the measured deflection was 3.088 mm, and the resilient modulus was 329.1 MPa.

In order to analyze the construction uniformity of the mixed recycled aggregates, we obtained the deflection and resilient modulus at different transverse locations at distances of 2.85 m, 8.55 m, 14.25 m, 19.95 m, and 25.65 m from the subgrade edge, representing the leftmost, slightly left, center, slightly right, and rightmost portions of the subgrade, respectively. As shown in Figure 8a, compared with the subgrade center, the deflection on both sides of the subgrade was slightly greater when the brick aggregate content was less than 20% and smaller when the brick aggregate content was greater than 60%; finally, when the brick aggregate content was 40%, the deflection at different distances from the subgrade edge was nearly the same, which indicates that the construction uniformity was high. The measured deflection value at the edge of the subgrade did not increase strictly with an increase in brick aggregate content, which indicates that the construction uniformity at the subgrade edge was more discrete.

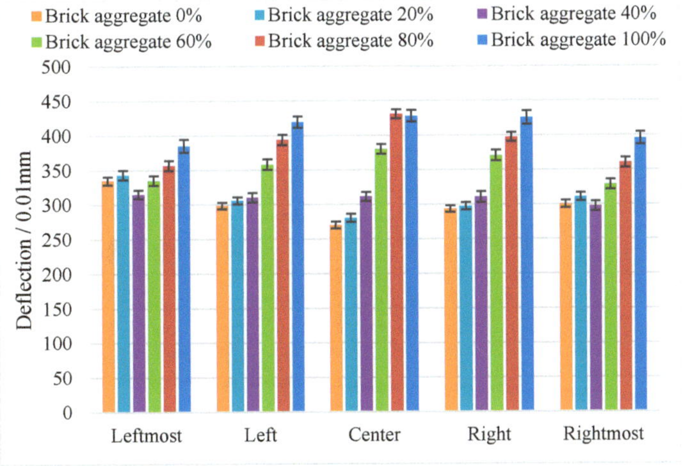

(**a**) Deflection

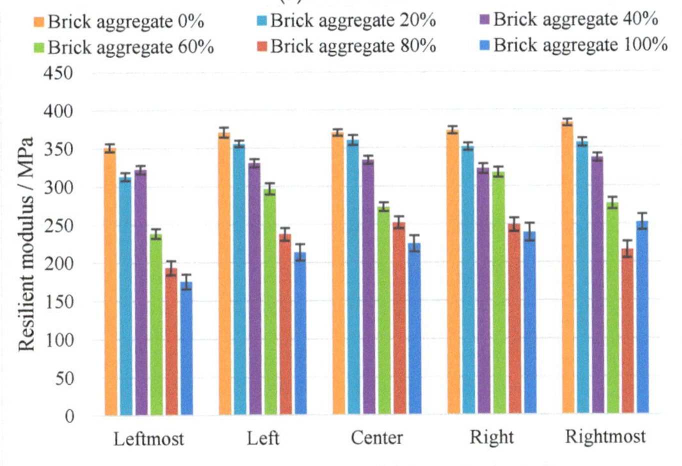

(**b**) Resilient modulus

Figure 8. The results for different transverse locations.

The results of the resilient modulus at different transverse locations are shown in Figure 8b. It can be seen that, when the brick aggregate content was less than 20%, the resilient modulus was mostly consistent, except for the leftmost side of the subgrade. When the brick aggregate content was 40%, the uniformity of the subgrade was better. When the content of brick aggregate was greater than 60%, the resilient modulus at different positions was different. Finally, the resilient modulus in the center of the subgrade decreased with the increase in brick aggregate content, which indicates that the construction uniformity in the center of the subgrade was higher.

4.2. Compaction Parameters of Construction-Solid-Waste Subgrade

The compaction degree is a phenomenological volume parameter that does not directly reflect the stress and deformation characteristics of mixed recycled aggregate during the compaction process. To evaluate the effect of the compaction method on the thickness variation of the mixed recycled aggregates during the subgrade construction process, the second part of the test section was constructed with the mixed recycled aggregate with 40%

brick aggregate content determined from the first experiment. The length was 480 m, and the filling thickness of the subgrade was 40 cm. A grid of 5.7 m × 5.87 m was delineated within the fill area, and mixed recycled aggregate (approximately 12 m^3) was poured into each grid to prevent its segregation. Finally, the mixed recycled aggregate materials in different grids were pressed together to avoid large voids as much as possible.

A compacting machine, a Xugong 22 t vibratory roller produced by XCMG Construction Machinery Co., Ltd. (Xuzhou, China), was used for compaction from the edge of the subgrade to the center, and the speed of the roller was set to 3 km/h. The initial compaction stage included three passes of the compaction roller, followed by static pressure, weak vibration, and strong vibration. The main compaction stage was from the fourth to the eighth pass, and the ninth pass represented the final compaction stage, after which static pressure was applied. Four different methods were used for field compaction during the main compaction stage, as shown in Table 4, where each was employed in a test section length of 120 m. The LyCRA TCR402 total station produced by Eurasia Measuring System Equipment Co., Ltd. (Shanghai, China) was used to detect the height of the subgrade after each roller pass during the main compaction stage. The thickness of the fill layer was calculated based on the height difference, and the changes in fill thickness with different compaction methods were analyzed.

Table 4. Construction methods in the repressing stage.

Mode	A	B	C	D
Compaction method	2 strong vibrations + 3 weak vibrations	3 weak vibrations + 2 strong vibrations	1 strong vibration + 3 weak vibrations + 1 strong vibration	2 weak vibrations + 2 strong vibrations + 1 weak vibration

The results of fill layer thickness under different compaction methods are shown in Figure 9a. It can be seen that the fill thickness continued to decrease with the increase in roller passes in the main compaction stage. The significant difference in the thickness variation under different compaction methods indicates that the compaction method had an impact on fill thickness during the compaction process. Although the final thickness of the fill layer was around 0.35 m after the main compaction stage, the variation in thickness under different compaction methods at the eighth pass was inconsistent, which indicates that the stability of the final subgrade was different.

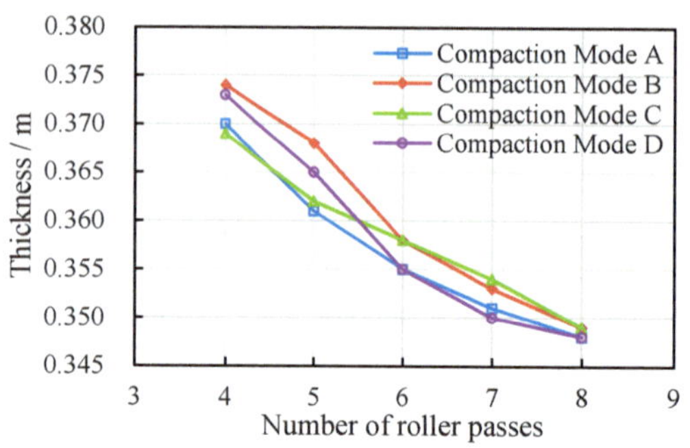

(**a**) Thickness

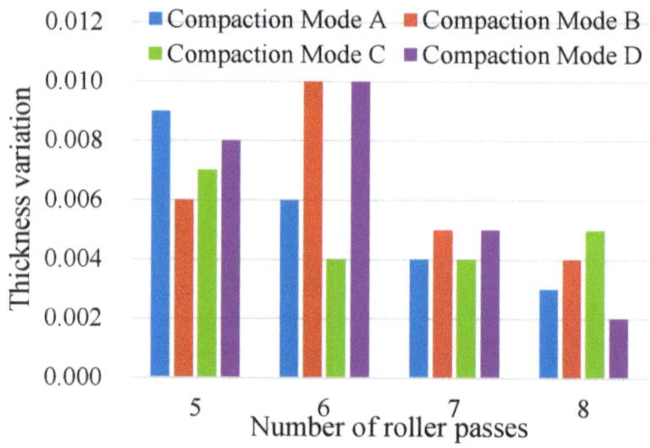

(**b**) Thickness variation

Figure 9. Thickness results under different compaction methods.

The variation in fill thickness after each pass was calculated. As shown in Figure 9b, when Mode A was applied in the process of compaction, the variation in layer thickness decreased with the increase in the number of passes, which indicates that it became more and more stable. When Modes B and D were applied, the layer thickness first increased and then decreased with the increase in passes, and the layer thickness changed the most after the sixth pass. The layer thickness variation under Modes D and B was inconsistent after the fifth and eighth passes, where Mode D caused a greater change in layer thickness after the fifth pass but a smaller one after the eighth pass of only 2 mm. The layer thickness under Mode C showed the smallest change after the sixth pass and then did not decrease with the increase in passes. The results of the thickness change show that the construction effects of Modes A and D were better.

The layer thickness variation at different transverse locations during the main compaction stage was analyzed. As shown in Figure 10, due to the difference in the properties of the recycled aggregates, there were a few void structures in the materials, so the measured data at each location were not completely consistent. The test results obtained under Modes B and C differ greatly at different transverse positions, which indicates that uniformity was poor for both compaction methods. There was a significant difference in thickness between the edge and the middle of the fill layer under Mode A after the seventh and eighth passes. The results for five different transverse locations obtained under Mode D are mostly consistent. Moreover, the layer thickness varied less after the eighth pass, which indicates that the compaction quality obtained with Mode D was high. Therefore, it is recommended to use a 22 t roller for five passes (two weak vibrations + two strong vibrations + one weak vibration) at a speed of 3 km/h as the compaction method for recycled construction-solid-waste aggregate subgrade filling.

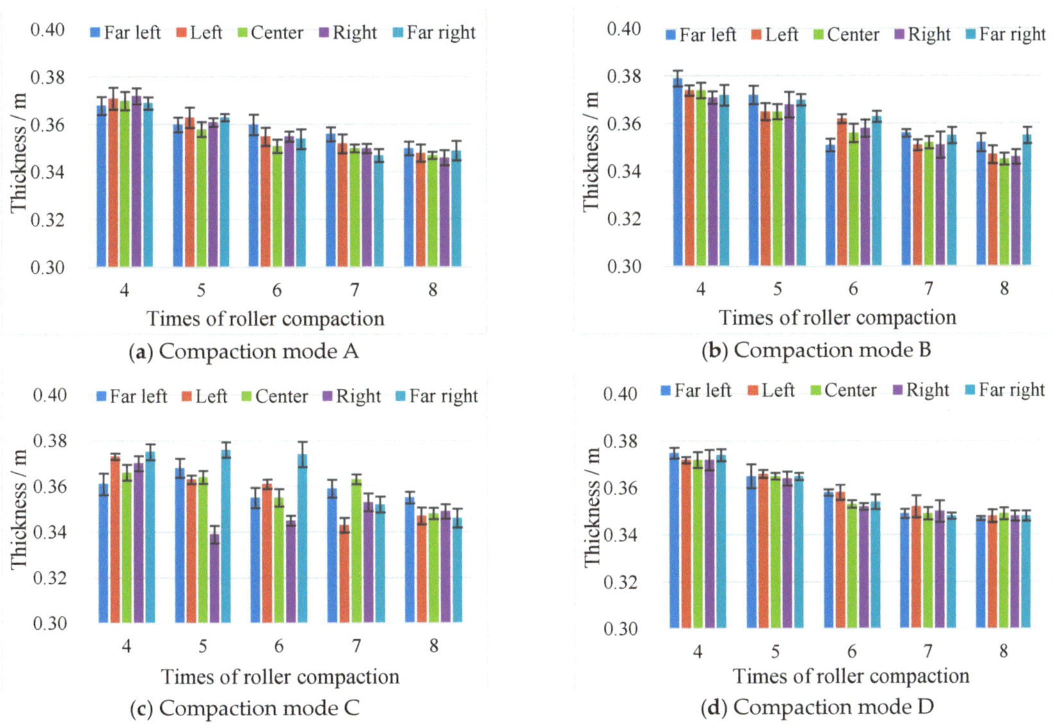

Figure 10. Thickness results for different transverse locations (The error bars represent standard deviation).

5. Conclusions

(1) A brick aggregate content less than 80% had little effect on the gradation of mixed recycled aggregate. During the compaction process, the recycled aggregate with a

particle size of 9.5 mm to 19 mm was crushed to a higher degree, and a larger amount of brick aggregate led to a higher degree of crushing.

(2) With an increase in brick aggregate content, the CBR decreased continuously, and when the content was 0–40%; the reduction rate of the CBR was smaller than that found for 40–100% content. However, even when the content of brick aggregate was 100%, the CBR was much higher than the technology standard requirement of 8%.

(3) The crushing value of the mixed recycled aggregates increased with an increase in brick aggregate content. The larger the value of the latter, the greater the dispersion of the test results and the greater the influence of the soaking crushing value. When the brick aggregate content was greater than 40%, the crushing value of the recycled aggregates increased sharply, and the crushing resistance decreased noticeably.

(4) With an increase in brick slag content, the resilience modulus decreased, the deflection increased, and the brick aggregate content affected the uniformity of compaction. When the brick aggregate content was 40%, the construction uniformity was good.

(5) The recommended ratio of brick aggregate to recycled concrete aggregate for mixed recycled construction-solid-waste aggregate is 40% to 60%. The results show that the subgrade filling construction method, whereby a 22 t roller is used for compaction for five passes (two weak vibrations + two strong vibrations + one weak vibration) at a speed of 3 km/h in the main compaction stage, achieves better construction quality.

(6) Due to limitations in engineering practices, only recycled aggregates with six different brick aggregate contents were selected for performance evaluation in this study. Future research could continue to analyze the influence of subgrade structural parameters on the performance of recycled aggregates from construction solid waste based on further field testing.

Author Contributions: X.Z.: conceptualization, writing—original draft, methodology. L.D.: investigation, data curation. Y.W.: data curation, validation. X.W.: software. X.T.: supervision. All authors have read and agreed to the published version of the manuscript.

Funding: This research was supported by the Natural Science Foundation Project of Hunan Provincial (2024JJ6111 and 2024JJ7077); and Key Project of Hunan Provincial Department of Education (Grant Number: 23A0559 and 22A0561); and Excellent Youth Project of Hunan Provincial Department of Education (Grant Number: 23B0749, 23B0736 and 23B0732), and Open Fund of National Engineering Research Center of Highway Maintenance Technology (Changsha University of Science & Technology) (Grant Number. kfj230204).

Institutional Review Board Statement: Not applicable.

Informed Consent Statement: Not applicable.

Data Availability Statement: The original contributions presented in the study are included in the article, further inquiries can be directed to the corresponding author.

Conflicts of Interest: The authors declare no conflict of interest.

References

1. Yu, H.; Ge, J.; Qian, G.; Zhang, C.; Dai, W.; Li, P. Evaluation on the rejuvenation and diffusion characteristics of waste cooking oil on aged SBS asphalt based on molecular dynamics method. *J. Clean. Prod.* **2023**, *406*, 136998. [CrossRef]
2. Chen, Y.; Pei, Z.; Mao, X.; Fan, L.; Xu, M.; Li, Y.; Wang, D.; Yi, J. Evaluation of electromagnetic scattering characteristics of construction solid waste—A theoretical study of solid waste identification. *Sci. Total Environ.* **2024**, *927*, 172312. [CrossRef] [PubMed]
3. Cong, H.; Meng, H.; Chen, M.; Song, W.; Xing, H. Co-processing paths of agricultural and rural solid wastes for a circular economy based on the construction concept of "zero-waste city" in China. *Circ. Econ.* **2023**, *2*, 100065. [CrossRef]
4. He, Z.; Xie, T.; Yu, H.; Ge, J.; Dai, W. Evaluation of quantum dot composite graphene /Titanium oxide enhanced UV aging resistance modified asphalt. *Constr. Build. Mater.* **2023**, *408*, 133732. [CrossRef]
5. Peng, X.; Xie, N.; Xia, C.; Zhou, X.; Zhao, P.; Ma, S.; Zhang, C.; Lv, S. Laboratory evaluation of different bio-oil recycled aged asphalts: Conventional performances and microscopic characteristics. *J. Clean. Prod.* **2023**, *428*, 139442. [CrossRef]
6. Wang, P.; Xu, C.; Li, Q.; Wang, L.; Guo, Y. Experimental study on the preparation of wood aggregate recycled concrete using waste wood and recycled fine aggregate from construction and demolition wastes. *J. Build. Eng.* **2024**, *90*, 109471. [CrossRef]

7. Khan, Z.A.; Balunaini, U.; Costa, S. Environmental feasibility and implications in using recycled construction and demolition waste aggregates in road construction based on leaching and life cycle assessment—A state-of-the-art review. *Clean. Mater.* **2024**, *12*, 100239. [CrossRef]
8. Zhang, C.; Yu, H.; Zhu, X.; Yao, D.; Peng, X.; Fan, X. Unified Characterization of Rubber Asphalt Mixture Strength under Different Stress Loading Paths. *J. Mater. Civ. Eng.* **2024**, *36*, 04023498. [CrossRef]
9. Zhu, X.; Qian, G.; Yu, H.; Yao, D.; Shi, C.; Zhang, C. Evaluation of coarse aggregate movement and contact unbalanced force during asphalt mixture compaction process based on discrete element method. *Constr. Build. Mater.* **2022**, *328*, 127004. [CrossRef]
10. Zhu, X.; Yu, H.; Qian, G.; Yao, D.; Dai, W.; Zhang, H.; Li, J.; Zhong, H. Evaluation of asphalt mixture micromechanical behavior evolution in the failure process based on Discrete Element Method. *Case Stud. Constr. Mater.* **2023**, *18*, e01773. [CrossRef]
11. Zhang, Z.; Wang, Q.; Zhang, K.; Fang, J.; Li, S.; Ge, A.; Huang, H. Study on the physical and mechanical properties of recycled weathered rock materials in expressway subgrade in permafrost areas. *Constr. Build. Mater.* **2024**, *430*, 136494. [CrossRef]
12. Zhu, L.; Lei, S.; Li, T. Investigation on thermal performance of recycled brick aggregate concrete structural walls. *Constr. Build. Mater.* **2024**, *425*, 136083. [CrossRef]
13. Dadd, L.; Xie, T.; Bennett, B.; Visintin, P. Exploring the physical and mechanical characteristics of multi-generation recycled aggregate concrete at equivalent compressive strengths. *J. Clean. Prod.* **2024**, *451*, 141965. [CrossRef]
14. Xu, Y.; Chen, H.; Liang, Y.; Shen, J.; Yang, H. Study on fracture characteristics and fracture mechanism of fully recycled aggregate concrete using AE and DIC techniques. *Constr. Build. Mater.* **2024**, *419*, 135540. [CrossRef]
15. Chen, W.-Z.; Chen, X.-F. Experimental Study and Mathematical Modeling of Mechanical Properties of Basalt Fiber-Reinforced Recycled Concrete Containing a High Content of Construction Waste. *Constr. Mater.* **2023**, *3*, 462–473. [CrossRef]
16. Gong, S.; Bai, L.; Tan, Z.; Xu, L.; Bai, X.; Huang, Z. Mechanical Properties of Polypropylene Fiber Recycled Brick Aggregate Concrete and Its Influencing Factors by Gray Correlation Analysis. *Sustainability* **2023**, *15*, 11135. [CrossRef]
17. Khalife, E.; Sabouri, M.; Kaveh, M.; Szymanek, M. Recent Advances in the Application of Agricultural Waste in Construction. *Appl. Sci.* **2024**, *14*, 2355. [CrossRef]
18. Xia, P.; Wang, S.; Gong, F.; Cao, W.; Zhao, Y. Rapid recognition method of red brick content in recycled brick-concrete aggregates and powder based on color segmentation. *J. Build. Eng.* **2024**, *84*, 108633. [CrossRef]
19. Xue, K.; Su, M.; Yang, X.; Qi, J.; Liu, M.; Ju, C.; Peng, X.; Wu, Y. Investigation of the water resistance of non-sintered wrap-shell lightweight aggregates with high content solid waste and its application in water permeable bricks. *Constr. Build. Mater.* **2023**, *400*, 132612. [CrossRef]
20. Zhang, R.; Jiao, J.; He, M.; Tao, Z.; He, P. Design, implementation and performance prediction of profiled steel sheet-mixed aggregate recycled concrete hollow composite slab. *J. Build. Eng.* **2023**, *79*, 107839. [CrossRef]
21. Tafreshi, S.M.; Khanjani, A.; Dawson, A.; Faramarzi, A. Performance of recycled waste aggregate mixed with crushed glass over a weak subgrade. *Constr. Build. Mater.* **2023**, *402*, 133002. [CrossRef]
22. Abbey, S.J.; Amakye, S.Y.O.; Eyo, E.U.; Booth, C.A.; Jeremiah, J.J. Wet–Dry Cycles and Microstructural Characteristics of Expansive Subgrade Treated with Sustainable Cementitious Waste Materials. *Materials* **2023**, *16*, 3124. [CrossRef] [PubMed]
23. Zhi, X.; Yang, T.; Zhang, X.; Ren, Y.; Deng, P.; Chen, Y.; Xiao, Y. Experimental Study on the Mechanical Properties and Permeability of Cement-Stabilized Permeable Recycle Aggregate Materials. *Sustainability* **2023**, *15*, 14063. [CrossRef]
24. Belabbas, O.; Bouziadi, F.; Boulekbache, B.; Hamrat, M.; Haddi, A.; Amziane, S. Mechanical properties of multi-recycled coarse aggregate concrete, with particular emphasis on experimental and numerical assessment of shrinkage at different curing temperatures. *J. Build. Eng.* **2024**, *89*, 109333. [CrossRef]
25. Htet, P.; Chen, W.; Huang, Z.; Hao, H. Physical and mechanical properties of deflection-hardening hybrid fibre-reinforced recycled aggregate concrete. *J. Build. Eng.* **2024**, *90*, 109510. [CrossRef]
26. Hameed, R.; Tahir, M.; Abbas, S.; Sheikh, H.U.; Kazmi, S.M.S.; Munir, M.J. Mechanical and Durability Characterization of Hybrid Recycled Aggregate Concrete. *Materials* **2024**, *17*, 1571. [CrossRef] [PubMed]
27. Deng, M.; Xie, X.; Zhuo, J.; He, Y.; Wang, K. Experimental Study on the Strength and Hydration Products of Cement Mortar with Hybrid Recycled Powders Based Industrial-Construction Residue Cement Stabilization of Crushed Aggregate. *Materials* **2023**, *16*, 4233. [CrossRef] [PubMed]
28. Rasheed, A.H.; Ahmed, B.A. Dynamic monitoring of saturated stiff clay soil foundation structure by falling weight deflectometer system under impact loads sensors effect. *Meas. Sens.* **2023**, *25*, 100673. [CrossRef]
29. *JTG E42-2005*; Test Methods of Aggregate for Highway Engineering 2005. Ministry of Transport of the People's Republic of China: Beijing, China, 2005.
30. Jianqiao, L.; Xiaodong, Z.; Meng, Z.; Hao, L. Soil Liquid Limit and Plastic Limit Treating System Based on Analytic Method. *Procedia Earth Planet. Sci.* **2012**, *5*, 175–179. [CrossRef]
31. *JTG E40-2007*; Test Methods of Soils for Highway Engineering. Ministry of Transport of the People's Republic of China: Beijing, China, 2007.
32. *JTG/T 3610-2019*; Technical Specifications for Construction of Highway Asphalt Subgrades. Ministry of Transport of the People's Republic of China: Beijing, China, 2019.
33. Thérèné, F.; Keita, E.; Naël-Redolfi, J.; Boustingorry, P.; Bonafous, L.; Roussel, N. Water absorption of recycled aggregates: Measurements, influence of temperature and practical consequences. *Cem. Concr. Res.* **2020**, *137*, 106196. [CrossRef]

34. Liang, K.; Hou, Y.; Sun, J.; Li, X.; Bai, J.; Tian, W.; Liu, Y. Theoretical analysis of water absorption kinetics of recycled aggregates immersed in water. *Constr. Build. Mater.* **2021**, *302*, 124156. [CrossRef]
35. Jin, L.; Wang, Z.; Wu, T.; Liu, P.; Zhou, P.; Zhu, D.; Wang, X. Mesoscale-based mechanical parameters determination and compressive properties of fully recycled coarse aggregate concrete. *J. Build. Eng.* **2024**, *90*, 109366. [CrossRef]
36. Fanara, A.; Courard, L.; Collin, F.; Hubert, J. Transfer properties in recycled aggregates concrete: Experimental and numerical approaches. *Constr. Build. Mater.* **2022**, *326*, 126778. [CrossRef]

materials

MDPI

Article

Preparation and Performance Evaluation of Castor Oil-Based Asphalt Regeneration Agent

Pan Pan [1,2], Yibo Chen [1], Xinhe Hu [3], Bingquan Dai [1], Xiaodi Hu [1,2,*] and Ning Wang [4]

1 School of Civil Engineering and Architecture, Wuhan Institute of Technology, Wuhan 430205, China; panp@wit.edu.cn (P.P.)
2 Hubei Provincial Engineering Research Center for Green Civil Engineering Materials and Structures, Wuhan 430073, China
3 The College of Post and Telecommunication of Wuhan Institute of Technology, Wuhan 430073, China
4 China Three Gorges Construction Engineering Corporation, Chengdu 610095, China
* Correspondence: 11010101@wit.edu.cn; Tel.: +86-27-8719-4823

Abstract: Regeneration agents play a critical role in modifying the mechanical properties and durability of RAP asphalt mixtures. This paper aimed to develop a castor oil-based asphalt regeneration agent. The effects of this regeneration agent on the pavement performance of laboratory-aged asphalt and an RAP asphalt mixture were comparatively studied by a series of laboratory tests. For the developed castor oil-based asphalt regeneration agent, the weight ratio of the castor oil to dibutyl phthalate was determined as 1:4. Moreover, the regeneration effectiveness of the castor oil-based regeneration agent was tested on three laboratory-aged asphalt binders and an RAP asphalt binder; the penetration, softening point and ductility of the RAP asphalt binder recovered to 83 dmm, 50.3 °C, and more than 100 cm, respectively. The optimum content of the regeneration agent was 5% by the weight of the aged asphalt binder. Furthermore, the castor oil-based regeneration agent could effectively restore the pavement performance of an RAP asphalt mixture. In this study, the RAP percentage can reach up to 60% by the weight of the HMA mixture using the castor oil-based asphalt regeneration agent according to the Chinese specification.

Keywords: regeneration agent; castor oil; laboratory aging; asphalt binder; RAP asphalt mixture

Citation: Pan, P.; Chen, Y.; Hu, X.; Dai, B.; Hu, X.; Wang, N. Preparation and Performance Evaluation of Castor Oil-Based Asphalt Regeneration Agent. *Materials* **2024**, *17*, 2078. https://doi.org/10.3390/ma17092078

Academic Editor: Simon Hesp

Received: 11 April 2024
Revised: 23 April 2024
Accepted: 24 April 2024
Published: 28 April 2024

1. Introduction

The maintenance of asphalt pavement not only produces a large amount of asphalt waste, but also consumes a considerable quantity of asphalt binder and mineral aggregate [1–3]. In China, the amount of RAP (reclaimed asphalt pavement) is very huge and tends to increase annually [4,5]. Undoubtedly, application of the RAP in new asphalt mixtures can reduce the accumulation of RAP material and the demand for the non-renewable pavement materials, e.g., mineral aggregate and asphalt binder [6,7]. It has been shown that the material cost would be reduced by 14~34% when the content of RAP ranges from 20% to 50% by the total weight of asphalt mixture [8]. The Ministry of Transport of China proposed that the RAP utilization rate should be over 80% by 2025, aiming to promote green transportation development [9]. Therefore, recycling the RAP for the maintenance or construction of asphalt pavement is highly significant from the policy perspective.

According to the construction temperature, there are three methods for recycling the RAP, including hot recycling, cold recycling, and warm recycling technologies [10,11]. There are many factors affecting the pavement performance of a recycled asphalt mixture, e.g., the new asphalt, aggregate gradation, the regeneration agent, the aged asphalt of the RAP, the particle characteristics of the RAP, the amount of RAP material, and the production and compaction temperatures during the construction stage [12–14]. Considering both the pavement performance and durability, the RAP percentages range from 20% to 30% by the weight of the asphalt mixture, and it is commonly used in the middle layer or bottom layer of asphalt pavement in actual pavement engineering. Hence, it has become a hot topic to increase the RAP percentage in asphalt mixtures without sacrificing the pavement performance.

In recent decades, a regeneration agent is usually used to recover the aged asphalt binder and improve the bonding performance between the RAP particles and the new aggregate particles to enhance the mechanical properties and durability of an RAP asphalt mixture [15]. The most common asphalt regeneration agent is a mineral oil-based regeneration agent according to the base material source. In addition, different types of additives are commonly added and mixed with the base material to modify the regeneration performance of the regeneration agent [16,17]. However, mineral oil is non-renewable and its production process results in environmental pollution [18]. Therefore, some researchers all over the world have prepared asphalt regeneration agents with vegetable oil as the base material.

In their published papers, the regeneration effects of different vegetable oil-based and mineral oil-based regeneration agents on aged asphalt were compared, and the results showed that the vegetable oil outperforms the mineral oil-based regeneration agents [19–21]. Zheng et al. studied the regeneration effect of vegetable oil from the aspects of permeability and molecular level. They found that vegetable oil shows better compatibility with the aged asphalt binder, owing to its greater permeability and stronger molecular activity [22]. For the same aged asphalt binder, the amount of vegetable oil was only one-third that of the mineral oil-based regeneration agent to obtain the same regeneration effect [23]. Since plant resources are abundant and renewable, it is significant to develop the vegetable oil-based regeneration agent.

In addition, a few studies were conducted on the regeneration properties of fresh vegetable oil [24,25]. Zhong found that the "amphipathic structure" in most vegetable oils has excellent dissolution and dispersion ability for asphaltene. Among them, castor oil was considered as one of the most promising vegetable oil-based materials to prepare regeneration agents due to its good thermal stability, storage stability, and regeneration effect. However, a low dosage of castor oil cannot obtain a satisfactory regeneration effect while a high dosage might increase the risk of moisture damage and permanent deformation of the RAP pavement [26,27]. Moreover, the content of RAP in an asphalt mixture is limited in practical engineering to below 30% by the weight of the mixture. Therefore, a modified castor oil-based regeneration agent is needed for improving the RAP percentage without sacrificing the pavement performance of an asphalt mixture.

This paper aimed to develop a new castor oil-based asphalt regeneration agent. The effects of the regeneration agent on the pavement performance of laboratory-aged asphalt and an RAP asphalt mixture were comparatively studied. Figure 1 shows a technical diagram of the present study.

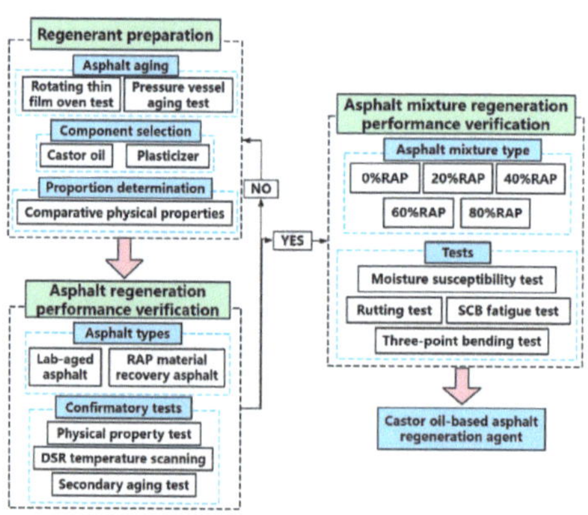

Figure 1. A technical diagram of this study.

2. Materials and Methods

2.1. Raw Materials

In this paper, all the raw materials came from Wuhan, China. Three kinds of 70# paving asphalt were selected, including JL AH-70, SL AH-70, and Shell AH-70. Table 1 shows their physical properties.

Table 1. Physical properties of original asphalts.

Physical Properties	JL AH-70	SL AH-70	Shell AH-70
Penetration (dmm)	60.9	73.1	60.3
Softening point (°C)	47.3	46.6	50.8
Ductility (cm)	>150	>150	>150

The gneiss aggregate had a density of 2.649 g/cm^3 and a particle size less than 26.5 mm. The mineral filler was limestone filler with a density of 2.745 g/cm^3. Their physical and mechanical properties met the requirements of the Chinese specification JTG F40. The reclaimed asphalt pavement (RAP) material, obtained from a pavement project in Macheng City, Hubei, had an asphalt content of 4.6% and a particle size less than 19 mm.

The raw materials of the asphalt regeneration agent included castor oil, Plasticizer ESO, Plasticizer TC, and Plasticizer DP. Since all the raw materials are liquid at 25 °C, during the mixing of the raw materials and the application of the regeneration agent, a low-speed shear device was used to shear at a speed of 500 r/min for 15 min. Table 2 shows the performance of the raw materials for preparing the asphalt regeneration agent.

Table 2. Performance of raw materials for preparing asphalt regeneration agent.

Performance	Castor Oil	Plasticizer ESO	Plasticizer TC	Plasticizer DP
Viscosity (mPa·s)	84	325	32	96
Flash Point (°C)	322	310	368	172
Density (g/mL)	0.963	0.985	1.042	1.045
Toxicity	non-toxic	non-toxic	low toxicity	slightly toxic

2.2. Sample Preparation

2.2.1. Preparation of Aged Asphalt

The aged samples were prepared according to the Chinese specification of JTG E20 [28]. Firstly, the short-term-aged asphalt binders were obtained by a rotating thin-film oven test (RTFOT, JTG E20 T0610) at 163 °C for 85 min. Then, the short-term-aged asphalt binders were subjected to the pressure vessel aging tests (PAV, JTG E20 T0630) at 100 °C for 20 h under 2.1 MPa to obtain the long-term-aged asphalt binders.

In addition, actual aged asphalt binder was extracted from RAP to verify the regeneration performance of the castor oil-based regeneration agent. Firstly, the RAP mixtures were cleaned by an extractor to obtain a trichloroethene bitumen concentrate. Then, the mixed solution was centrifuged at a speed of 3000 r/min for 5 min to remove the sediment. The solution after centrifugation was subjected to the rotary evaporation test according to JTG E20 T0727 [28]. Table 3 shows the physical properties of the aged asphalt binders.

Table 3. Physical properties of aged asphalt.

Physical Properties	Long-Term Aging			Actual Aged Asphalt
	JL AH-70	SL AH-70	Shell AH-70	
Penetration (dmm)	23.7	29.1	21.7	29.2
Softening point (°C)	63.6	60.7	68.9	64.3
Ductility (cm)	1.0	2.2	0.9	14.2

2.2.2. Asphalt Mixture Design

Figure 2 shows the gradation curve of the asphalt mixture with the nominal maximum size of 19 mm. The upper and lower limits of gradation followed the Chinese specification of JTG F40-2004 [29]. The content of the mineral filler was 4% by weight of the aggregate and the optimum asphalt content was 4.4% by weight of the asphalt mixture.

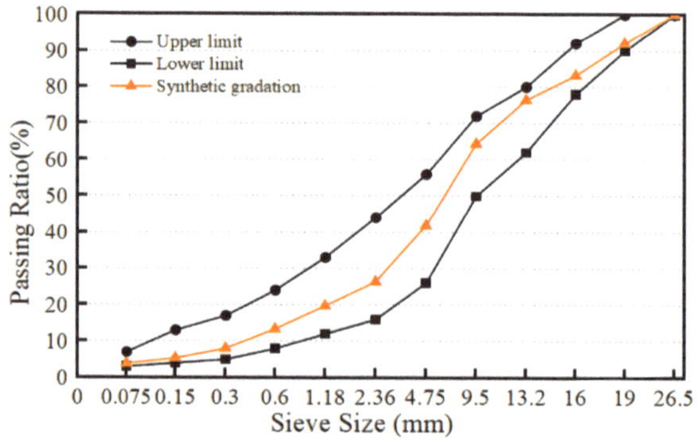

Figure 2. Chart of aggregate gradation.

2.3. *Test Method*

2.3.1. Physical Properties Test of Asphalt

The physical properties of the asphalt binder including penetration (25 °C), ductility (15 °C), and softening point were studied in accordance with Chinese specifications JTG E20 T0604, T0605, and T0606, respectively [28].

2.3.2. DSR Temperature Scanning Test of Asphalt

The DSR temperature scanning test was carried out at high temperature (40~80 °C) and at low temperature (0~40 °C) with a temperature interval of 5 °C. The cracking factor ($G^*\cos^2\delta/\sin\delta$) and rutting factor ($G^*/\sin\delta$) of the asphalt were calculated by using the complex shear modulus G^* and phase angle δ.

2.3.3. Pavement Performance Test of Asphalt Mixture

According to the Chinese specification JTG E20, the high-temperature performance and the low-temperature performance of the asphalt mixtures were evaluated by wheel tracking test (T0719) and the three-point bending test (T0715), while the moisture susceptibility of the asphalt mixtures was determined by the Marshall stability test (T0709) and the freeze–thaw split test (T0729).

2.3.4. SCB Fatigue Test

The SCB fatigue test was carried out to evaluate the fatigue performance of the aged and rejuvenated asphalt mixtures. The stress control model was adopted and the loading pressure of each asphalt mixture was 0.7 times its maximum load. Four test repetitions were prepared for each mixture and the average value was used. The frequency of cyclic loading was 10 Hz and the test temperature was 25 °C. The loading cycles were defined as fatigue life when the samples was damaged.

3. Preparation of Asphalt Regeneration Agent

3.1. *Selection of Asphalt Regeneration Agent Materials*

3.1.1. Effect of Castor Oil on Physical Properties of Asphalt Binder

Castor oil was adopted as the base material for preparing the asphalt regeneration agent. Castor oil with five different dosages (6%, 7%, 8%, 9%, 10%) was added to the long-term-aged JL AH-70. Figure 3 illustrates the effects of the aging process and the castor oil on the physical properties of the asphalt binders. The penetration increased with the increase in castor oil content. When the castor oil content increased from 0% to 10%, the penetration of the aged asphalt binder increased from 24 dmm to 72 dmm. Since penetration is related to the stiffness of asphalt, asphalt binder with low penetration has high stiffness. The result implied that castor oil can efficiently soften and recover aged asphalt binder.

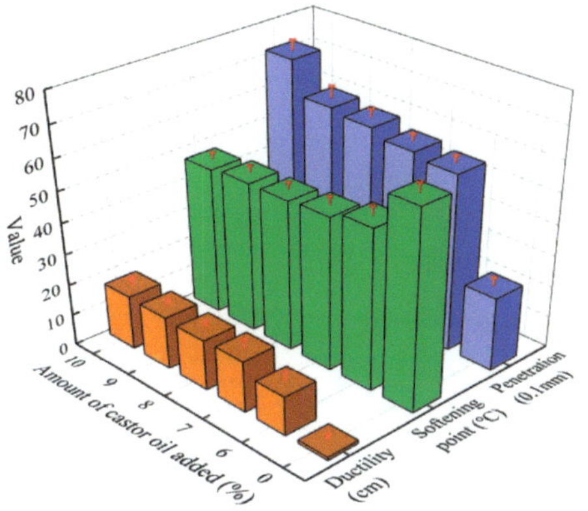

Figure 3. Physical properties of aged asphalt mixed with castor oil.

Since asphalt binder would become harder after aging and the softening point increase due to the aging effect, it can be assumed that an aged asphalt mixture is less prone to permanent deformation at high temperature. The softening point showed a clear decline trend with the castor oil. The softening point of aged asphalt with 10% castor oil was reduced by 25% compared with the original asphalt. This indicated that castor oil softened the aged asphalt as well, which was in accordance with the penetration results.

The test results for the physical properties confirmed that castor oil can restore the performance of asphalt binder. The reason might be that it contains a large number of aromatic components and increases the content of light components in asphalt binders. Compared with the penetration and softening point, the improvement of ductility was not significant. When the castor oil content was 10%, the ductility of aged asphalt was only 17 cm and could not meet the requirement of the JTG E40-2004 specification in China.

Ductility is generally used to evaluate the low-temperature cracking performance of asphalt binder. Asphalt binder with low ductility would increase the risk of the cracking of an asphalt pavement at low temperature. Therefore, plasticizers were adopted as a modifier to improve the low-temperature performance of the regenerated asphalt binder. The penetration and softening point of the aged asphalt binder with 8% castor oil was similar to the original asphalt. On this basis, the effect of plasticizers on the ductility of aged asphalt binders was investigated in follow-up research.

3.1.2. Effect of Plasticizers on Physical Properties of Asphalt Binder

Three common plasticizers, including epoxy soybean oil (ESO), tributyl citrate (TC), and dibutyl phthalate (DP), were used in this study. On the basis of adding 8% castor oil into the asphalt binder, plasticizers with 5% (high content) and 1% (low content) by the weight of the asphalt binder were added as well. The purpose of the low content was to find out whether the plasticizer can function effectively at lower content; the purpose of the high content was to find out whether the plasticizer would have saturation effect or performance degradation at higher content. The physical properties of the asphalt binders were tested to compare the effects of the different plasticizers on the performance of aged asphalt binders. At the same time, the approximate range of plasticizer content which can restore the ductility to the standard value was estimated. The test results are shown in

Figures 4–6. Since castor oil showed good recovery effects on the penetration and softening point of aged asphalt binders, more attention was paid to the effect of plasticizers on improving the ductility.

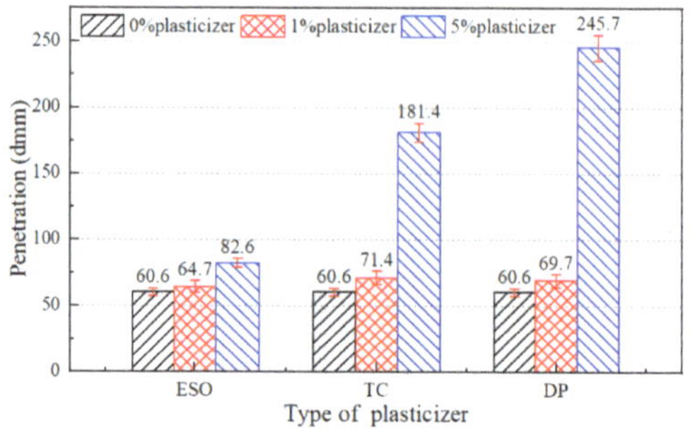

Figure 4. Penetration of 8% castor oil and plasticizer asphalt binders.

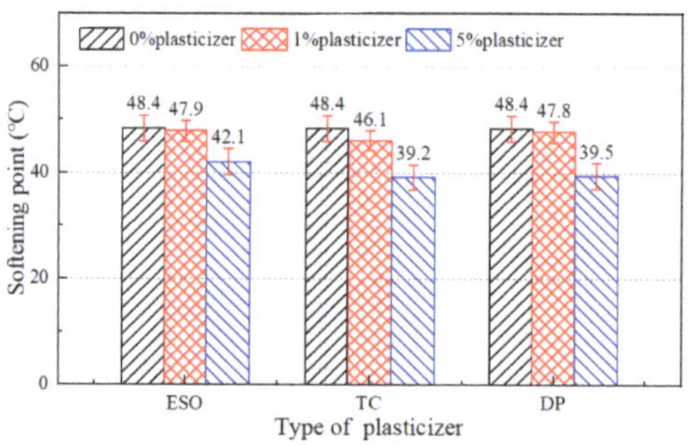

Figure 5. Softening point of 8% castor oil and plasticizer asphalt binders.

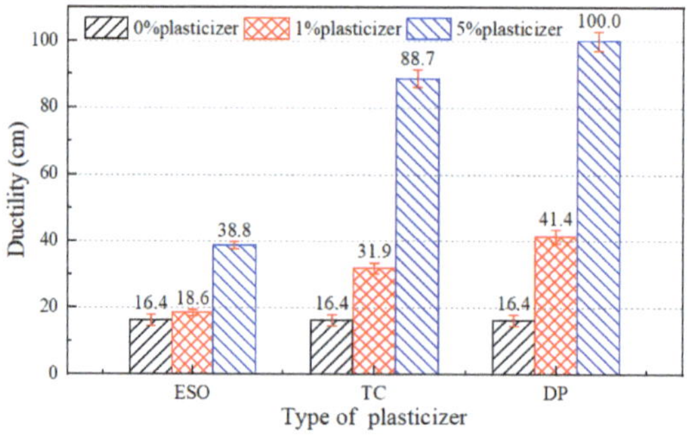

Figure 6. Ductility of 8% castor oil and plasticizer asphalt binders.

As shown in Figure 6, the ductility improvement of aged asphalt binders with dibutyl phthalate was greater than for those with tributyl citrate and epoxy soybean oil. Of all the test results, only aged asphalt binder containing 5% dibutyl phthalate met the ductility requirements of the Chinese specification. It could be concluded that dibutyl phthalate showed the best recovery effect on ductility and thus it was selected as another component of the asphalt regeneration agent.

Moreover, the penetration of aged asphalt binders increased and the softening point decreased with the addition of plasticizers. After adding 5% dibutyl phthalate, the softening point and penetration were 39.5 °C and 246 dmm, respectively. Too low a softening point might cause the risk of permanent deformation of the pavement. For instance, the softening point of asphalt binder is required to be higher than 46 °C in China. Therefore, it was necessary to determine the optimum ratio of castor oil and plasticizer for obtaining a regeneration asphalt mixture with sufficient stability to avoid cracking and permanent deformation.

3.2. Asphalt Regeneration Agent Proportion Design

Under the combined effects of castor oil and plasticizer, the penetration of aged asphalt binder with 8% castor oil and 5% plasticizer was too high and the softening point could not meet the Chinese standard. Since the castor oil has a significant softening effect on the aged asphalt binder, the content of castor oil was decreased, which ranged from 1% to 5% in further study. In addition, the content of plasticizer ranged from 2% to 5%. The physical properties of asphalt binders were tested for all combinations of the two additives with different contents. The test results and their comparison with the Chinese specification are shown in Figures 7–9.

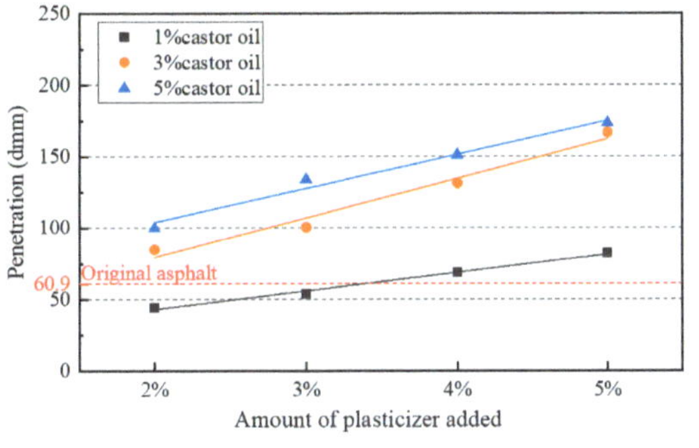

Figure 7. Penetration of aged asphalt with different amounts of additives.

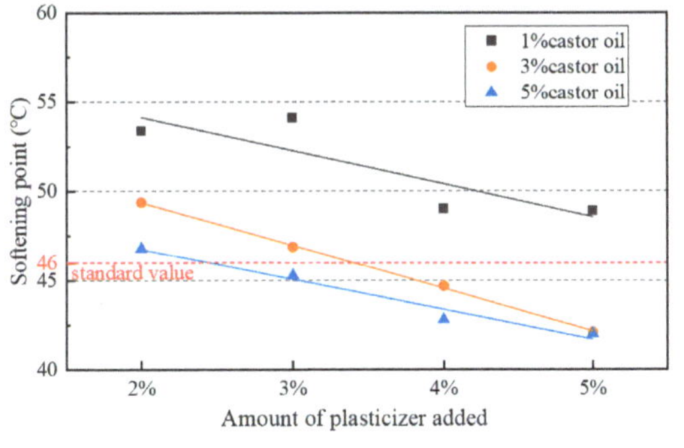

Figure 8. Softening point of aged asphalt with different amounts of additives.

The softening point and ductility needed to be more than 46 °C and 100 cm according to the specification, respectively. So, only five combinations of dosages met the requirements. Considering both the physical properties and the cost, it was determined that the ratio of castor oil to dibutyl phthalate was 1:4 and the amount of asphalt regeneration agent was 5% of asphalt. The penetration and the softening point of the asphalt were 69 dmm and 48.9 °C, respectively, and the ductility was more than 100 cm. In order to avoid accidental phenomena, the castor oil-based asphalt regeneration agent was added to different kinds of aged asphalt binders and asphalt mixtures to verify its regeneration performance.

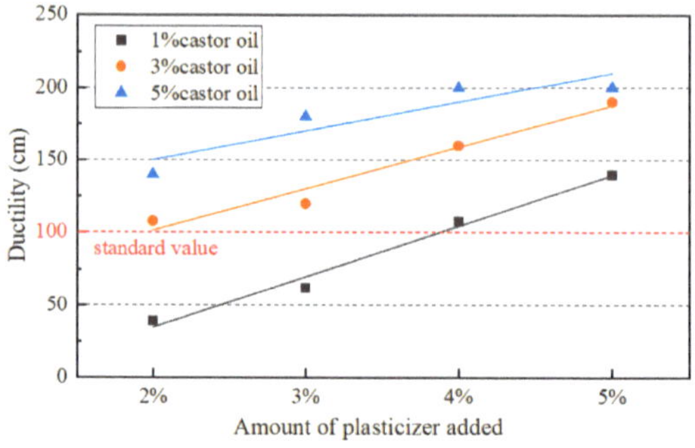

Figure 9. Ductility of aged asphalt with different amounts of additives.

3.3. *Validation of Regeneration Performance*

3.3.1. Effect on Physical Properties of Aged Asphalt Binder

DSR temperature scanning test results of the original asphalt, the aged asphalt, and the regeneration asphalt are shown in Figures 10 and 11. A greater cracking factor means worse low-temperature performance while a greater rutting factor means better high-temperature performance. The high-temperature performance was improved and the low-temperature anti-cracking performance was degraded after aging. The DSR temperature scanning test results of the regeneration asphalt were similar to those of the original asphalt. The performance of the aged asphalt could be restored to the original asphalt level after adding the regeneration agent. This implied that the castor oil-based agent has good regeneration performance.

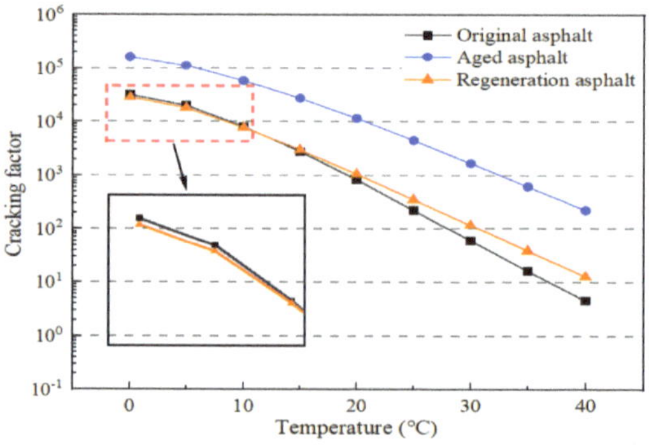

Figure 10. Low-temperature scanning of asphalt DSR test.

In addition, there were three different asphalt binders to test the regeneration effect of the developed regeneration agent. The SL and Shell aged asphalt binders were obtained by laboratory aging tests and the actual aged asphalt (MC) was reclaimed from RAP. The physical properties of the asphalt binders were tested after adding asphalt regenerating agent and the results are shown in Figures 12–14.

The results showed that the castor oil-based regeneration agent could restore the physical properties of the three aged asphalt binders. For actual aged asphalt, its penetration, softening point, and ductility could be recovered to 83 dmm, 52.3 °C, and exceeding 100 cm, respectively. For laboratory-aged asphalt, the physical properties of the aged asphalt binders when adding castor oil-based regeneration agent were even better than the original asphalt binder. Therefore, an accidental phenomenon could be ruled out and it was assumed that the castor oil-based asphalt regeneration agent can be applied to most aged asphalt binders.

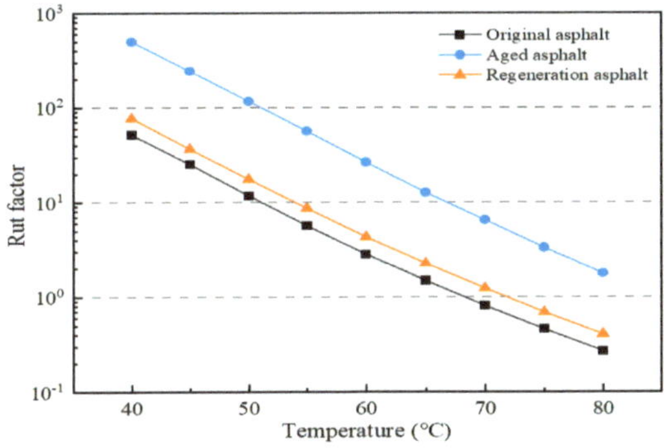

Figure 11. High-temperature scanning of asphalt DSR test.

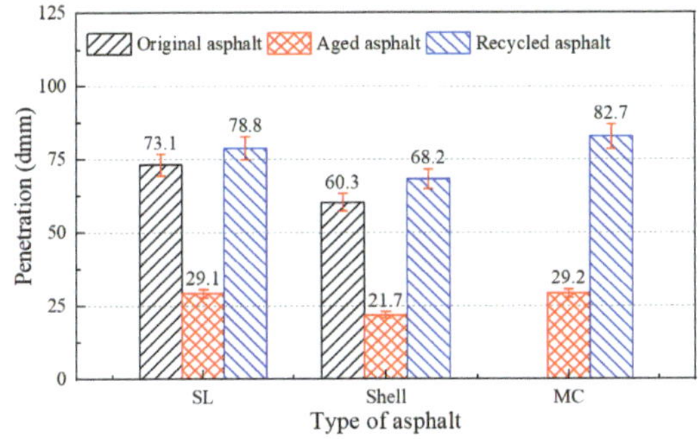

Figure 12. Penetration of aged asphalt and asphalt with regeneration agent.

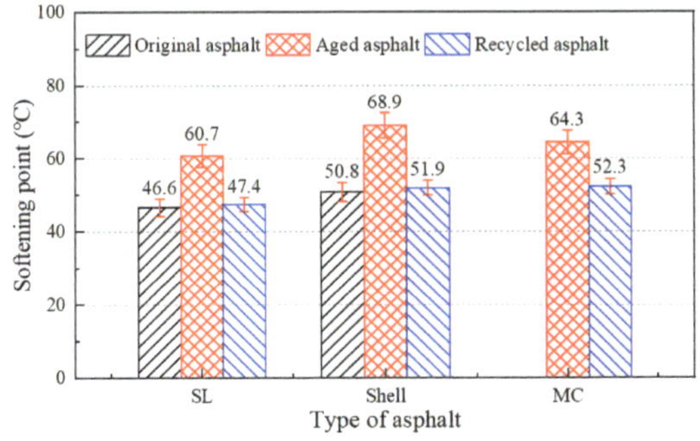

Figure 13. Softening point of aged asphalt and asphalt with regeneration agent.

Due to the high cost of asphalt pavement renovation, it is necessary to ensure the anti-aging performance of regeneration asphalt. The JL regeneration asphalt was re-aged to study the anti-aging performance of the regeneration agent. The physical properties of the re-aged asphalt binder were tested and the results obtained are shown in Figure 15. Under the same aging conditions, the physical properties were better than those of the JL aged asphalt. This showed that the anti-aging performance of the regeneration asphalt is better than that of the original asphalt from the repeated aging aspect.

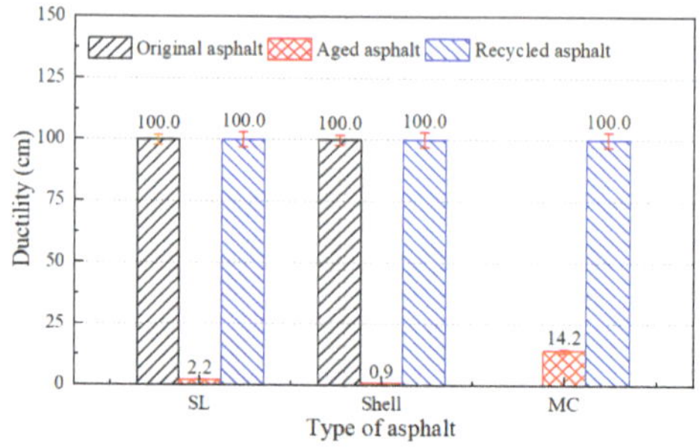

Figure 14. Ductility of aged asphalt and asphalt with regeneration agent.

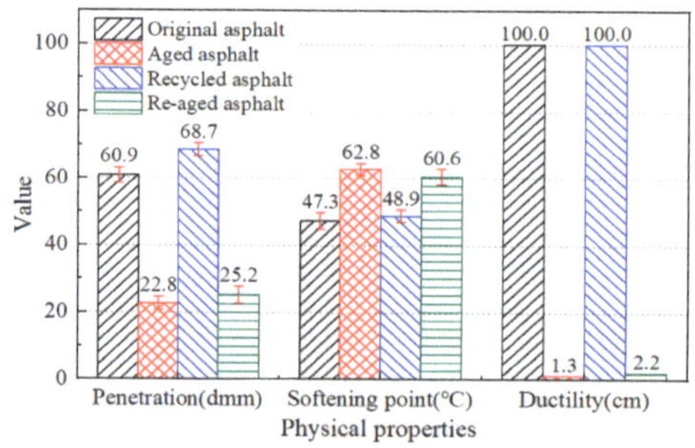

Figure 15. Physical properties of re-aged asphalt and control group asphalt.

Through the above tests, it can be concluded that the castor oil-based asphalt regeneration agent can effectively regenerate aged asphalt and cause its physical properties to return to the level of the original asphalt. In order to compare the regeneration effect of the castor oil-based asphalt regeneration agent, two different kinds of traditional mineral oil-based asphalt regeneration agents, including Regeneration Agent A and Regeneration Agent B, were selected to regenerate aged asphalt by adding 5% of each of them, respectively. Figure 16 shows that all three asphalt reclaiming agents can restore the physical properties of aged asphalt. Compared with the castor oil-based regeneration agent, the Regeneration Agents A and B showed a worse effect on the regeneration of aged asphalt. The softening point was restored to 59.0 °C and 57.2 °C, respectively, and the penetration and ductility did not meet the requirements of the specification. The physical properties of the regeneration asphalt after adding the 5% castor oil-based asphalt regeneration agent were all better than those of the original asphalt. It can be concluded that the developed agent had a better regeneration effect than the traditional mineral oil-based asphalt regeneration agents.

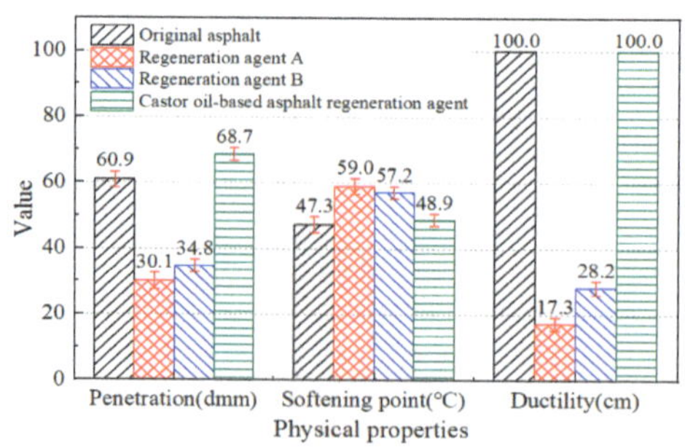

Figure 16. Physical properties of aged asphalt mixed with different asphalt regeneration agents.

3.3.2. Effect of Regeneration Agent on Pavement Performance of RAP Mixture

The above results indicated that the castor oil-based asphalt regeneration agent had superior improvement effects on the physical properties of aged asphalt binders. Since the effect of regeneration agent on aged asphalt binders and on asphalt mixtures is quite different, it was necessary to add castor oil-based regeneration agent in asphalt mixtures to verify its regeneration performance. In this study, 20%, 40%, 60%, and 80% RAP were used to replace the original aggregate and asphalt binder. The following experiments were carried out on asphalt mixtures with different RAP content.

(1) Moisture Susceptibility Performance

Figure 17 shows the Marshall stability results of the original asphalt mixture, the RAP asphalt mixtures (20%RAP, 40%RAP, 60%RAP, 80%RAP), and the recycled asphalt mixtures (R20%, R40%, R60%, R80%). Based on the results in Section 3.2, the amount of regeneration agent was determined as 5% by weight of the aged asphalt binder in the recycled asphalt mixture. The Marshall stability ratios of the asphalt mixtures with 20%, 40%, 60%, and 80% RAP were 92.8%, 90.0%, 86.0%, and 81.3%, respectively. The Marshall stability ratio gradually decreased with the increase in RAP content in the RAP asphalt mixtures. The Chinese specification requires that the Marshall stability ratio of hot mix asphalt be greater than 80%. Although the Marshall stability ratios of RAP asphalt mixtures which added 40%, 60%, and 80% RAP were greater than 80%, these were still lower than that of the original asphalt mixture which was 91.8%. This implies that the addition of RAP will make the asphalt mixture more susceptible to moisture effects and reduce the strength of the asphalt mixture. With the addition of the asphalt regeneration agent, the Marshall stability ratio of the recycled asphalt mixture was up about 6% and even higher than that of the original asphalt mixture. It can be concluded that the castor oil-based regeneration agent improved the resistance of RAP asphalt mixtures to moisture based on the results obtained from the Marshall stability test.

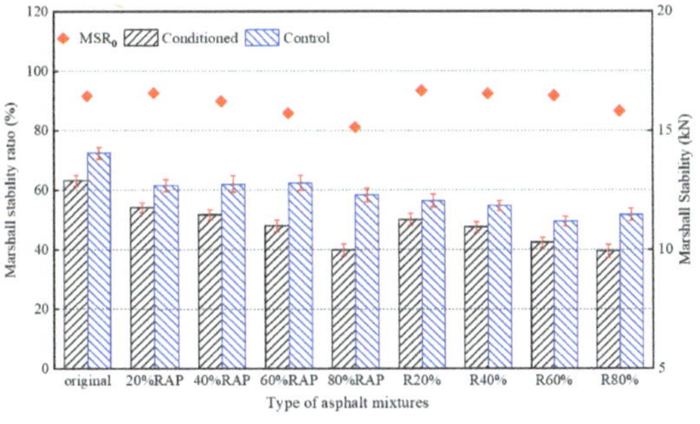

Figure 17. Marshall stability test results of asphalt mixtures.

In addition, we carried out the freeze–thaw split test to evaluate the moisture susceptibility of the asphalt mixtures and the results are shown in Figure 18. The tensile strength ratios of asphalt mixtures decreased with the increase in RAP content. And the TSR values of the 60% and 80% RAP samples were 74.3% and 72.8%, which were lower than the performance requirement of 75% for hot mix asphalt according to the Chinese specification of JTG F40-2004. However, the TSR values of the regenerated samples were all higher than 80% and met the specification. This confirmed the effectiveness of the asphalt regeneration agent on moisture susceptibility.

It is worth noting that the MSR and TSR values of the 20% RAP asphalt mixture were increased compared with those of the original asphalt mixture. The reason might be that the aggregate of the original asphalt mixture was acidic gneiss and the RAP aggregate was alkaline limestone, but the asphalt was a weak acid material. Although aged asphalt in RAP has poor adhesion to aggregates, exposed limestone surfaces during pavement milling have better adhesion to new asphalt than gneiss. The combination effect improved the water stability of the asphalt mixture. The content of new asphalt decreased with the increase in RAP content and consequently the new asphalt adhering to limestone decreased. When the content of RAP was beyond 20%, the negative effect of the RAP content played a dominant role and hence the water stability of asphalt mixture was degraded.

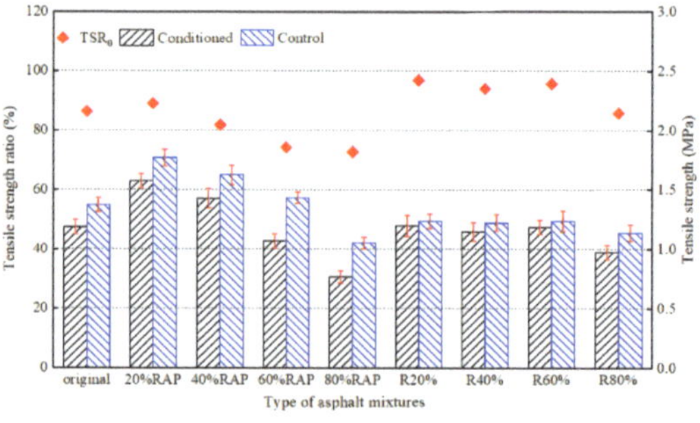

Figure 18. Freeze–thaw split test results of asphalt mixtures.

(2) High-Temperature Performance

Figure 19 shows the wheel tracking test results of the original asphalt mixtures, the RAP asphalt mixtures (20%RAP, 40%RAP, 60%RAP, 80%RAP), and the regenerated asphalt mixtures (R20%, R40%, R60%, R80%). The dynamic stability of the original asphalt mixture and that of the 80% RAP asphalt mixture were the minimum and maximum, respectively. The dynamic stability of the asphalt mixture with 80% RAP content was almost six times higher than that of the original asphalt mixture. Due to the aging effect, the asphalt mixtures with higher RAP content showed greater resistance to permanent deformation under the same repeated loading. Since the regeneration agent softens the RAP in the asphalt mixture, the dynamic stability of the regenerated asphalt mixtures decreased. It can be concluded that the softening effect of the castor oil-based regeneration agent degrades the high-temperature stability of the asphalt mixture, which was in accordance with the results of the study of asphalt.

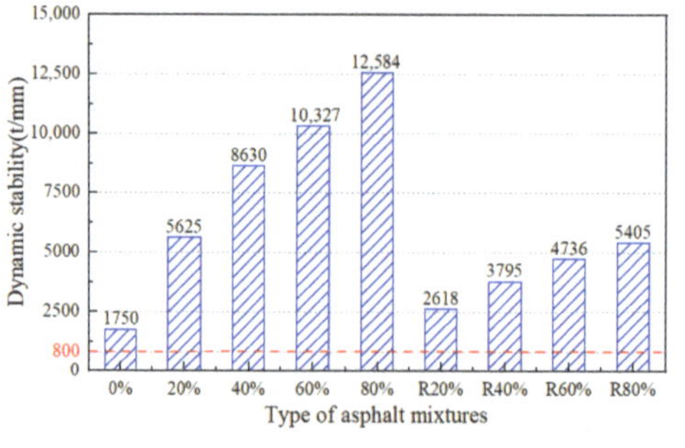

Figure 19. Wheel tracking test results of asphalt mixtures.

(3) Low-Temperature Performance

Figure 20 illustrates the flexural strain of asphalt mixtures with different RAP content and recycled asphalt mixtures. With the increase in RAP content, the flexural strain of the RAP asphalt mixtures decreased gradually and the flexural strain of none of them met the Chinese standard of 2000 $\mu\varepsilon$. The maximum bending strain of the RAP asphalt mixtures was 1752 $\mu\varepsilon$ and 1235 $\mu\varepsilon$, respectively, when the content of RAP was 20% and 80%, which was decreased by 21.9% and 44.9% compared with the original asphalt mixture.

The reason was that the RAP material degraded the low-temperature performance of the asphalt mixture and increased the risk of cracking of the asphalt pavement. Compared with the RAP asphalt mixtures, the flexural strain of recycled asphalt mixtures was increased due to the effect of the regeneration agent. It can be concluded that the castor oil-based regeneration agent can improve the low-temperature performance of asphalt mixtures.

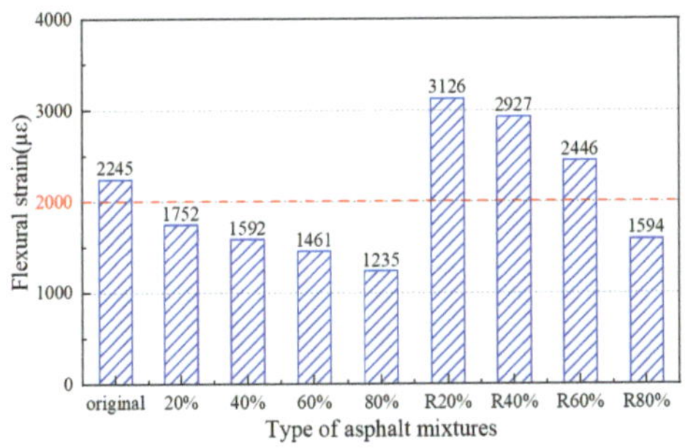

Figure 20. Three-point bending test results of asphalt mixtures.

The recovery rate of the regenerated mixture was 78% when the RAP content was 20%. However, the recovery rate of the regenerated mixture decreased to 29% when the RAP content was 80%. And the flexural strain could not meet the standard requirements in China. Therefore, the RAP content could be determined as 60% according to the low-temperature performance of the asphalt mixtures.

(4) Fatigue Performance

Figure 21 illustrates the fatigue lives with different RAP content and recycled asphalt mixtures. The fatigue life of the asphalt mixture decreases with the RAP materials. The fatigue life decreased by 50% when the RAP content was 80%. The lower the fatigue life is, the easier it is for a crack to occur under repeated loading. By comparing the fatigue lives of RAP asphalt mixtures and recycled asphalt mixtures with the same content, it was seen that the fatigue lives of asphalt mixtures were improved after adding a regeneration agent. The fatigue life of the asphalt mixture with 80% RAP recovered by 47.5%. The results showed that the recovery effect of the castor oil-based regeneration agent on fatigue life was strong. Although the performance of the original asphalt mixture cannot be achieved, it can restore the fatigue life of an RAP asphalt mixture to a similar level.

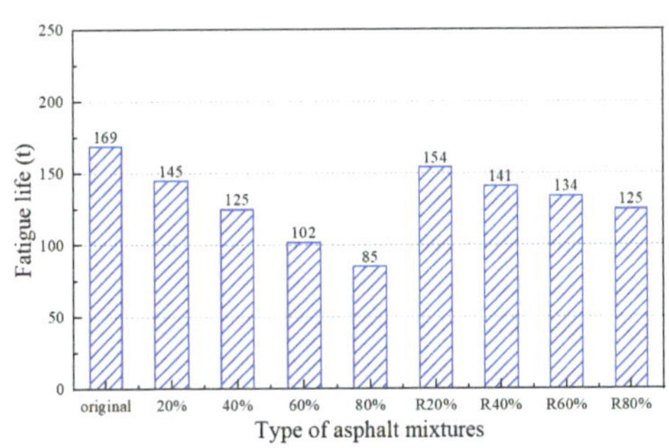

Figure 21. Fatigue life of asphalt mixtures.

(5) Anti-aging Performance

When the content of RAP was 60%, the pavement performance of regenerated asphalt mixtures met the Chinese standard. Therefore, asphalt mixtures containing 60% RAP were prepared to investigate the anti-aging performance. First, the recycled asphalt mixture was heated in an oven at 135 °C ± 3 °C for 4 h ± 5 min (short-term aging), and then in an oven at 85 °C ± 3 °C for 120 h ± 0.5 h (long-term aging). After aging the recycled asphalt mixture in the laboratory, the pavement performance of the original asphalt mixture (0% RAP), the aged asphalt mixture (60% RAP), the recycled asphalt mixture (R60% RAP), and the re-aged asphalt mixture (2–60% RAP) was tested to study the anti-aging performance of recycled asphalt mixtures.

Figure 22 shows the dynamic stability of the asphalt mixtures measured by the rutting test. The dynamic stability of the re-aged asphalt mixture was 7247 t/mm, which was 53% higher than that of the recycled asphalt mixture and four times that of the original asphalt mixture, while the result of the aged asphalt mixture was six times that of the original asphalt mixture. A greater increase in the dynamic stability of an asphalt mixture means the aging degree of asphalt is deeper. This shows that the anti-aging performance of the recycled asphalt mixture is strong.

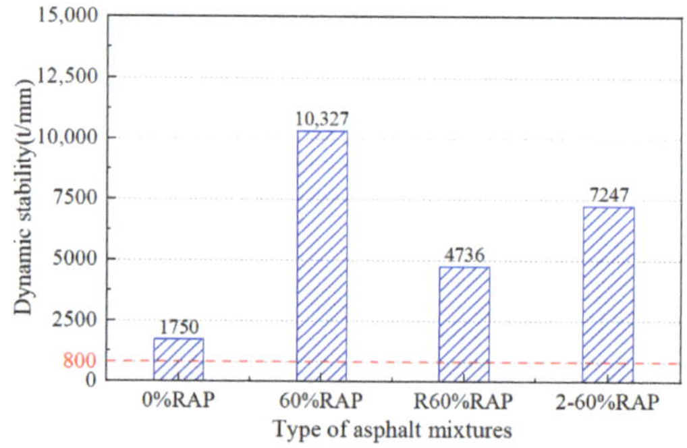

Figure 22. Wheel tracking test results of re-aged asphalt mixtures.

As shown in Figure 23, the maximum flexural tensile strain of the re-aged asphalt mixture and the aged asphalt mixture was 1757 με and 1461 με, respectively, which were 22% and 35% lower than that of the original asphalt mixture. Since the maximum flexural strain of the re-aged asphalt mixture was greater than that of the aged asphalt mixture, this indicates that the anti-aging performance of the recycled asphalt mixture was excellent as well.

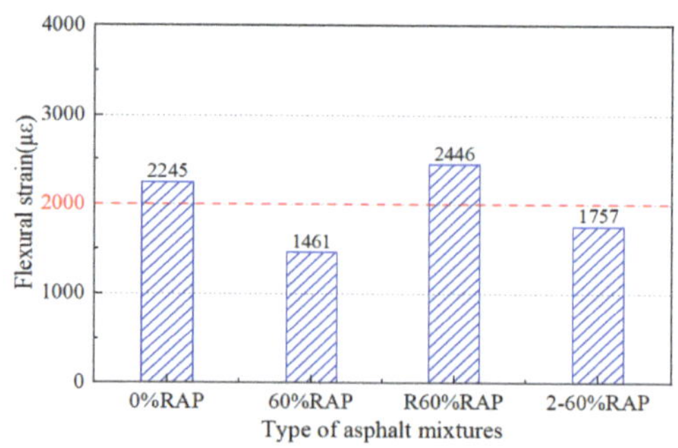

Figure 23. Three-point bending test results of re-aged asphalt mixtures.

Figures 24 and 25 show that the Marshall stability ratio and the tensile strength ratio of the re-aged asphalt mixture were 87.5% and 78.1%, respectively, which were higher than those of the aged asphalt mixture. This implies that the water stability of the re-aged asphalt mixture was better. It is well-known that the aging effect will reduce the Marshall stability ratio and tensile strength ratio of an asphalt mixture. Compared with the original asphalt mixture, the Marshall stability ratio and tensile strength ratio of the aged asphalt mixture decreased by 6.2% and 12.3%, respectively, while the re-aged asphalt mixture only by 4.7% and 9.6%. All the pavement performance tests showed that the anti-aging performance of the recycled asphalt mixture is desirable. Therefore, the recycled asphalt mixture can be expected to be used in an asphalt pavement for a long time.

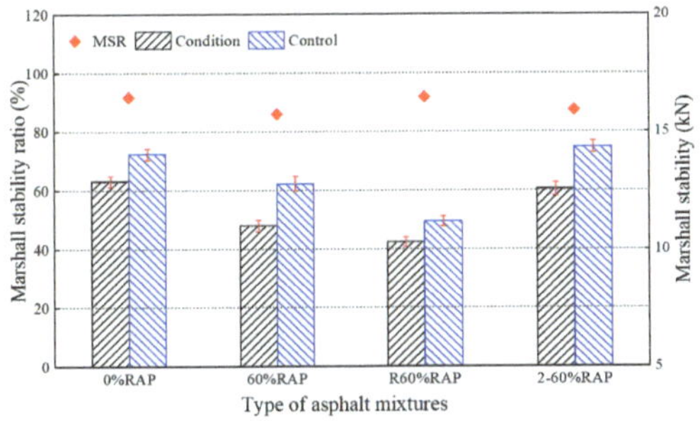

Figure 24. Marshall stability test results of re-aged asphalt mixtures.

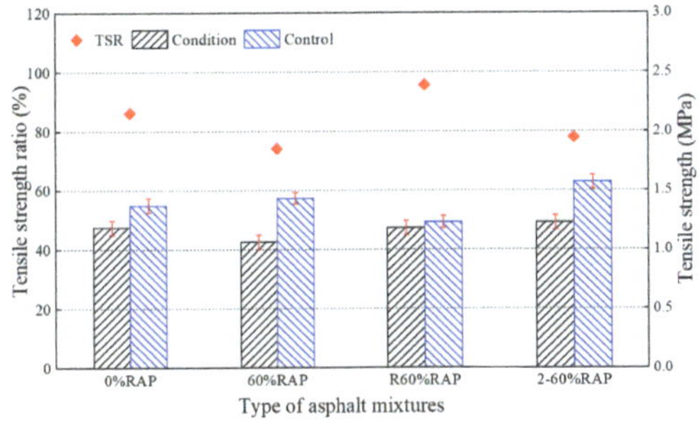

Figure 25. Freeze–thaw split test results of re-aged asphalt mixtures.

4. Conclusions

In this study, a new castor oil-based asphalt regeneration agent was developed and its effects on aged asphalt binders and RAP materials were comparatively investigated by a series of laboratory tests. According to the test results, conclusions could be drawn as follows.

1. The castor oil-based asphalt regeneration agent was developed by a series of tests on laboratory-aged asphalt. It consists of castor oil and dibutyl phthalate at a weight ratio of 1:4. The optimum content of the regeneration agent was 5% by the weight of the aged asphalt binder in this study.
2. The regeneration effectiveness of the castor oil-based regeneration agent was tested with three laboratory-aged asphalt binders and an actual aged asphalt binder, and the anti-aging performance of the recycled asphalt was good. The regeneration effect of the castor oil-based asphalt regeneration agent was better than that of the two traditional mineral oil-based asphalt regeneration agents.
3. The castor oil-based regeneration agent can effectively restore the pavement performance of an RAP asphalt mixture, especially the low-temperature anti-cracking performance. The content of the RAP can reach 60% by the weight of the asphalt mixture with the castor oil-based regeneration agent. The anti-aging performance of the recycled asphalt mixture was better than that of the original asphalt mixture, so it can be used in an asphalt pavement for a long time.

Overall, this paper provides a reference for the material design and performance evaluation of a castor oil-based regeneration agent. However, there are still some problems that can be explored in future research, including the following aspects:

(1) The universality of the castor oil-based asphalt regeneration agent. The asphalt binders and RAP materials used in this study were limited. In future research, asphalts from different sources and RAP materials of different gradation and different years could be added to verify the regeneration performance of the castor oil-based asphalt regeneration agent.
(2) The regeneration mechanism of the castor oil-based asphalt regeneration agent. Combined with advanced material testing technology, determining the effect mechanism of the castor oil-based asphalt regeneration agent on aged asphalt in RAP, including composition adjustment, interface fusion, etc., would provide a theoretical basis and technical support for the engineering application of the castor oil-based asphalt regeneration agent.
(3) It is strongly recommended to pave a pavement test section to investigate the pavement performance and durability of the RAP asphalt mixture with the castor oil-based regeneration agent.

Author Contributions: Conceptualization, writing—original draft preparation, project administration, P.P.; Methodology, investigation, Y.C.; Methodology, writing—original draft preparation, X.H. (Xinhe Hu); Methodology, investigation, B.D.; writing—review and editing, supervision, X.H. (Xiaodi Hu); Methodology, investigation, N.W. All authors have read and agreed to the published version of the manuscript.

Funding: This research was funded by the Technology Project of the Department of Transportation of Hubei Province (No. 2022-11-1-2) and the National Natural Science Foundation of China (No. 52178248).

Institutional Review Board Statement: Not applicable.

Informed Consent Statement: Not applicable.

Data Availability Statement: All the data in the tests of this study have been listed in the paper.

Acknowledgments: We would like to express our gratitude to the editors and the reviewers for their constructive and helpful review comments. The authors also wish to thank Zhan Pang (Wuhan Institute of Technology) for his help in experiments.

Conflicts of Interest: Author Ning Wang was employed by the company China Three Gorges Construction Engineering Corporation. The remaining authors declare that the research was conducted in the absence of any commercial or financial relationships that could be construed as a potential conflict of interest.

References

1. Pan, P.; Kuang, Y.; Hu, X.; Zhang, X. A Comprehensive Evaluation of Rejuvenator on Mechanical Properties, Durability, and Dynamic Characteristics of Artificially Aged Asphalt Mixture. *Materials* **2018**, *11*, 1554. [CrossRef] [PubMed]
2. Sengoz, B.; Isikyakar, G. Evaluation of the properties and microstructure of SBS and EVA polymer modified bitumen. *Constr. Build. Mater.* **2007**, *22*, 1897–1905. [CrossRef]
3. Shenoy, A. Prediction of high temperature rheological properties of aged asphalts from the flow data of the original unaged samples. *Constr. Build. Mater.* **2002**, *16*, 509–517. [CrossRef]
4. Holleran, G.; Wieringa, T.; Tailby, T. *Rejuvenation Treatments for Aged Pavements*; Transit New Zealand and New Zealand Institute of Highway Technology (NZIHT): Auckland, New Zealand, 2006.
5. Lin, J.; Hong, J.; Xiao, Y. Dynamic characteristics of 100% cold recycled asphalt mixture using asphalt emulsion and cement. *J. Clean. Prod.* **2017**, *156*, 337–344. [CrossRef]
6. Wang, H.; Zhang, R.; Chen, Y.; You, Z.; Fang, J. Study on microstructure of rubberized recycled hot mix asphalt based X-ray CT technology. *Constr. Build. Mater.* **2016**, *121*, 177–184. [CrossRef]
7. Behnood, A. Application of rejuvenators to improve the rheological and mechanical properties of asphalt binders and mixtures: A review. *J. Clean. Prod.* **2019**, *231*, 171–182. [CrossRef]
8. Kandhal, P.S.; Mallick, R.B. *Pavement Recycling Guidelines for State and Local Governments: Participant's Reference Book*; Final Report; U.S. Department of Transportation/Federal Highway Administration: Washington, DC, USA, 1997.

9. *The "14th Five Year Plan" for Green Transportation Development*; Ministry of Transport of the People's Republic of China: Beijing, China, 2022.
10. Kim, Y.; Im, S.; Lee, H. Impacts of Curing Time and Moisture Content on Engineering Properties of Cold In-Place Recycling Mixtures Using Foamed or Emulsified Asphalt. *J. Mater. Civ. Eng.* **2011**, *23*, 542–553. [CrossRef]
11. Xing, C.; Li, M.; Liu, L.; Lu, R.; Liu, N.; Wu, W.; Yuan, D. A comprehensive review on the blending condition between virgin and RAP asphalt binders in hot recycled asphalt mixtures: Mechanisms, evaluation methods, and influencing factors. *J. Clean. Prod.* **2023**, *398*, 136515. [CrossRef]
12. Dinis-Almeida, M.; Castro-Gomes, J.; Sangiorgi, C.; Zoorob, S.E.; Afonso, M.L. Performance of Warm Mix Recycled Asphalt containing up to 100% RAP. *Constr. Build. Mater.* **2016**, *112*, 1–6. [CrossRef]
13. Presti, L.D.; Carrión BD, J.A.; Airey, G.; Hajj, E. Towards 100% recycling of reclaimed asphalt in road surface courses: Binder design methodology and case studies. *J. Clean. Prod.* **2016**, *131*, 43–51. [CrossRef]
14. Zhu, Y.; Ma, T.; Xu, G.; Fan, J.; Zhang, Y.; Wu, M. Study of the Mixing between Asphalt and Rejuvenator in Hot In-Place Recycled Layer. *J. Transp. Eng. Part B Pavements* **2023**, *149*, 04023005. [CrossRef]
15. Cao, Z.; Chen, M.; Han, X.; Wang, R.; Yu, J.; Xu, X.; Xue, L. Influence of characteristics of recycling agent on the early and long-term performance of regenerated SBS modified bitumen. *Constr. Build. Mater.* **2020**, *237*, 117631. [CrossRef]
16. El-Shorbagy, M.A.; El-Badawy, M.S.; Gabr, R.A. Investigation of waste oils as rejuvenators of aged bitumen for sustainable pavement. *Constr. Build. Mater.* **2019**, *220*, 228–237. [CrossRef]
17. Jiao, Y.; Xie, Y.; Liu, Z. Optimization design and performance evaluation of a novel asphalt rejuvenator. *Front. Mater.* **2022**, *9*, 1081858.
18. Felipe, S.B.; Leandro, A.F.; Sandra SA, D. Effect of Soy-Based Asphalt Rejuvenators on Asphalt Binders of Different Performance Grades. *J. Transp. Eng. Part B Pavements* **2021**, *147*, 04020090.
19. Davies, G. Release Agent Formulas and Methods. U.S. Patent 20050132925 A1, 7 June 2005.
20. Leng, B.; Chen, M.; Wu, S. Effect of Waste Edible Vegetable Oil on High Temperature Properties of Different Aged Asphalts. Key. *Eng. Mater.* **2014**, *3019*, 135–140.
21. Hohmann, A.D.; Forrester, M.J.; Staver, M.; Kuehl, B.W.; Hernández, N.; Williams, R.C.; Cochran, E.W. Chemically mediated asphalt rejuvenation via epoxidized vegetable oil derivatives for sustainable pavements. *Fuel* **2024**, *355*, 129374. [CrossRef]
22. Zheng, X.; Xu, W.; Ji, W.; Cao, K. Study on the Wetting and Permeation Properties of Bio-Oil as Bitumen Rejuvenator. *Int. J. Mol. Sci.* **2023**, *24*, 6521. [CrossRef]
23. Somé, S.C.; Gaudefroy, V.; Delaunay, D. Effect of vegetable oil additives on binder and mix properties: Laboratory and field investigation. *Mater. Struct.* **2016**, *49*, 2197–2208. [CrossRef]
24. Elkashef, M.; Williams, C.R.; Cochran, W.E. Thermal and cold flow properties of bio-derived rejuvenators and their impact on the properties of rejuvenated asphalt binders. *Thermochim. Acta* **2018**, *671*, 48–53. [CrossRef]
25. Elkashef, M.; Podolsky, J.; Williams, R.C.; Cochran, E.W. Introducing a soybean oil-derived material as a potential rejuvenator of asphalt through rheology, mix characterisation and Fourier Transform Infrared analysis. *Road Mater. Pavement Des.* **2018**, *19*, 1750–1770. [CrossRef]
26. Cao, Z.; Chen, M.; Liu, Z.; He, B.; Yu, J.; Xue, L. Effect of different rejuvenators on the rheological properties of aged SBS modified bitumen in long term aging. *Constr. Build. Mater.* **2019**, *215*, 709–717. [CrossRef]
27. Cao, X.; Wang, H.; Cao, X.; Sun, W.; Zhu, H.; Tang, B. Investigation of rheological and chemical properties asphalt binder rejuvenated with waste vegetable oil. *Constr. Build. Mater.* **2018**, *180*, 455–463. [CrossRef]
28. *JTG E20*; Standard Test Methods of Bitumen and Bituminous Mixtures for Highway Engineering. Ministry of Transport of the People's Republic of China: Beijing, China, 2011.
29. *JTG F40*; Technical Specifications for Construction Highway Asphalt Pavements. Ministry of Transport of the People's Republic of China: Beijing, China, 2004.

materials

Article

Influence of Basalt Fiber Morphology on the Properties of Asphalt Binders and Mixtures

Chenhao Cai [1], Keke Lou [1,*], Fuxin Qian [2] and Peng Xiao [1]

1 College of Civil Science and Engineering, Yangzhou University, Yangzhou 225127, China; 221404602@stu.yzu.edu.cn (C.C.); pengxiao@yzu.edu.cn (P.X.)
2 Administration of Urban Facilities, Yangzhou 225003, China; 18936272588@163.com
* Correspondence: lkkyzu@163.com

Abstract: Basalt fiber (BF) has been proven to be an effective additive for improving the properties of asphalt mixtures. However, the influence of basalt fiber morphology on the properties of asphalt binders and mixtures remains inadequately explored. In this study, chopped basalt fiber (CBF) and flocculent basalt fiber (FBF) were selected to make samples for testing the influence of the two types of basalt fibers on asphalt materials. Fluorescence microscopy was used to obtain the dispersion of fiber in asphalt binders. Then, a temperature sweep test and a multiple stress creep recovery (MSCR) test were carried out to appraise the rheological characteristics of the binder. Moreover, the performance of the fiber-reinforced asphalt mixture was evaluated by a wheel tracking test, a uniaxial penetration test, an indirect tensile asphalt cracking test (IDEAL-CT), a low-temperature bending test, a water-immersion stability test, and a freeze–thaw splitting test. The results indicate that the rheological behavior of asphalt binders could be enhanced by both types of fibers. Notably, FBFs exhibit a larger contact area with asphalt mortar compared to CBFs, resulting in improved resistance to deformation under identical shear conditions. Meanwhile, the performance of the asphalt mixture underwent different levels of enhancement with the incorporation of two morphologies of basalt fiber. Specifically, as for the road property indices with FBFs, the enhancement extent of DS in the wheel tracking test, that of R_T in the uniaxial penetration test, that of the CT_{index} in the IDEAL-CT test, and that of ε_B in the low-temperature trabecular bending test was 3.1%, 6.8%, 15.1%, and 6.5%, respectively, when compared to the CBF-reinforced mixtures. Compared with CBFs, FBFs significantly enhanced the elasticity and deformation recovery ability of asphalt mixtures, demonstrating greater resistance to high-temperature deformation and a more pronounced effect in delaying the onset of middle- and low-temperature cracking. Additionally, the volume of the air void for asphalt mixtures containing FBFs was lower than that containing CBFs, thereby reducing the likelihood of water damage due to excessive voids. Consequently, the moisture susceptibility enhancement of CBFs to asphalt mixture was not obvious, while FBFs could improve moisture susceptibility by more than 20%. Overall, the impact of basalt fibers with different morphologies on the properties of asphalt pavement materials varies significantly, and the research results may provide reference values for the choice of engineering fibers.

Keywords: basalt fiber; asphalt binder; asphalt mixture; rheological behavior; pavement performance

Citation: Cai, C.; Lou, K.; Qian, F.; Xiao, P. Influence of Basalt Fiber Morphology on the Properties of Asphalt Binders and Mixtures. *Materials* **2024**, *17*, 5358. https://doi.org/10.3390/ma17215358

Academic Editor: Carlos Leiva

Received: 11 October 2024
Revised: 23 October 2024
Accepted: 24 October 2024
Published: 1 November 2024

1. Introduction

Continuous basalt fiber, a kind of amorphous inorganic nonmetallic fiber, is made from basalt ore as the raw material by melting at 1450~1500 °C and rapid drawing of the platinum–rhodium alloy leakage plate [1,2]. For different purposes, continuous basalt fiber is reprocessed into different products to meet diverse needs [3,4]. There are two main types of basalt fibers: chopped basalt fiber (CBF) and flocculent basalt fiber (FBF) [5–10]. CBFs, as shown in Figure 1a,b, are made from continuous basalt fibers by cutting them to a certain length and dimension, whereas FBFs (shown in Figure 1c,d) are produced from continuous basalt fibers by dispersion, shearing, and granulation after the centrifugal or blowing-out process, leading them to have a rougher surface and a smaller diameter.

Figure 1. Basalt fibers: (**a**) CBFs; (**b**) microstructure of CBFs; (**c**) FBFs; (**d**) microstructure of FBFs.

As a kind of durable pavement material with excellent mechanical properties and low economic cost, basalt-fiber-reinforced asphalt mixture has been widely used [11]. At present, scholars mainly focus on three aspects of CBF on the performance and damage characteristics of asphalt binders or asphalt mixtures: content, length, and diameter [12,13]. The investigation of Lou [14] on the macro-performance of ultra-thin wearing course with different fiber types proved that the basalt fiber with an optimum content of 0.2~0.3% exhibited excellent performances at high or low temperatures. Sun [15] pointed out that when the content of basalt fiber in asphalt mixture reaches 0.4% by weight, the ultimate tensile stress of asphalt mixture reaches a peak value. Zhang [16] added different types of fibers with contents of 0.15%, 0.3%, and 0.45% into the open-graded friction course (OGFC) asphalt mixture, and focused on evaluating the drainage performance, rutting resistance, stiffness, and fatigue life of the mixture. Some scholars have evaluated the adaptability of fiber length to hot-mixed asphalt mixtures with different nominal maximum particle sizes of aggregates; then, an optimum length for CBFs for asphalt mixtures with different gradations and maximum nominal grain sizes was proposed [17–20]. In addition, relevant studies have shown that [17,21,22] a stone-matrix asphalt (SMA) mixture has a skeleton structure with large spacing between coarse aggregates, so longer basalt fibers can play a better bonding role in the mixture. Meanwhile, research on the influence of basalt fiber diameter on the performance of asphalt mixtures has been performed recently. Wu et al. [23] found that the bond strength between the three kinds of short-cut basalt fiber with a diameter in the range of 13–16 μm and asphalt is greater than that of polyester fiber or polyacrylonitrile fiber under the condition of 25 °C. Pei [13] selected basalt fibers with three different diameters to conduct comparative analyses on the crack resistance of asphalt mixture and evaluated the crack inhibition effect of fiber diameters on asphalt mixtures.

However, FBFs have attained the attention of researchers slightly late, and research into the application of FBFs in pavement engineering has gradually emerged only recently. In Shi's and Kou's studies, chopped basalt fiber, flocculated basalt fiber, polyester fiber, and lignin fiber were adopted to compare and analyze the reinforcement effects on the asphalt binder [24,25]. The results revealed that flocculated basalt fiber could heighten the high-/low-temperature properties of asphalt mastic, which showed better improvement in elasticity and deformation recovery. Wu and Gu [23,26] compared and measured the moisture absorption, pH value, and asphalt adsorption capacity of three kinds of CBFs, coated with different treating compounds, and one kind of FBF; then, they discussed the rheological properties and micro-morphologies of the asphalt binders. Research on FBFs is limited at present, and most studies focus on their effects on asphalt mortar. Furthermore, the properties of the two fibers are very different, and there is a lack of research into comparative analyses between FBFs and CBFs in relation to their influence on the properties of asphalt mixtures.

Therefore, the preliminary evaluation and comparison of an asphalt–binder mixture blended with basalt fibers were conducted to explore the effect of different morphologies of basalt fibers on the overall performance of asphalt pavement materials in this study. The temperature sweep and MSCR tests were carried out on fiber–asphalt binder samples to evaluate the rheological properties. The original asphalt mixture, the CBF-reinforced asphalt mixture and the FBF-reinforced asphalt mixture were prepared to test the high-temperature, medium-temperature and low-temperature performance and moisture susceptibility via a wheel tracking test, a uniaxial penetration test, an IDEAL-CT test, a low-temperature trabecular bending test, a water immersion stability test as well as a freeze–thaw splitting test. This study analyzes the interaction between basalt fibers and asphalt, elucidating the enhancing mechanisms of basalt fibers in asphalt materials. Furthermore, the impact of basalt fiber on the performance of asphalt pavement is examined. The findings of this research provide valuable insights for road management professionals, enabling them to select basalt-fiber-reinforced materials that exhibit superior performance and are more appropriate for specific engineering scenarios. Ultimately, this can lead to an extension in pavement service life, a reduction in maintenance frequency and costs and an enhancement in the overall road service quality.

2. Materials and Test Methods

2.1. Raw Materials and Sample Preparation

2.1.1. Fiber

CBFs and FBFs, produced by Jiangsu Tianlong Basalt Continuous Fiber Co., Ltd., Yangzhou, China, were adopted. CBFs are golden brown, straight and free of impurities, while FBFs are dark gray cotton-like particles with different sizes. The technical specifications of the two fibers are summarized in Table 1.

Table 1. Basic properties of different basalt fibers.

Fiber Types	Density (g/cm^{-3})	Length (mm)	Diameter (μm)	Tensile Strength (MPa)	Water Content (%)	Oil Absorption Rate (%)
CBFs	2.762	6	13–16	2831	<0.1	104
FBFs	2.817	4–7	6–8	2804	<0.1	230

2.1.2. Asphalt

Styrene-butadiene-styrene block copolymer (SBS)-modified asphalt (PG 76-22) was acquired from Nantong Tongsha Asphalt Technology Co., Ltd. (Nantong, China). The related property indicators were tested in strict accordance with the specifications [27]. The softening point of asphalt was 84 °C and the viscosity at 135 °C was 1.8 Pa·s, respectively.

2.1.3. Aggregate and Filler

The nominal maximum sieve size of aggregates is 13.2 mm in this experiment, and the filler is the limestone mineral powder, with a moisture content of 0.3% and a proportion of 92.2%. The particle size is less than 0.075 mm. The aggregate performance tests, etc. were carried out according to JTG E42-2005 [28], and all the indices met the requirements of the specification.

2.1.4. Fiber–Asphalt Binder

Generally, relevant studies have shown that the weight content of the basalt fiber in asphalt mixture varies from 0.1% to 0.4%. Therefore, the fiber weight content in the

SBS-modified asphalt was selected from 1% to 4%, setting 1% as the interval [29,30], and the asphalt without fiber was treated as a control group. Figure 2 illustrates the mixing process of the fiber–asphalt binder samples. The detailed steps are as follows. First, the fibers were placed in an oven at 120 °C for 3 h to avoid the influence of moisture on the test results. Next, the asphalt was placed on a temperature control plate and heated to 175 °C. Then, the fibers were divided into 3 parts and were slowly added to the asphalt in batches. The fibers were then stirred at 1000 RPM for 30 min. The temperature of the fiber–asphalt binder was maintained at 175 °C ± 5 °C throughout the preparation process. In addition, the SBS original asphalt binder would also undergo the above manufacturing process to eliminate the effects of aging.

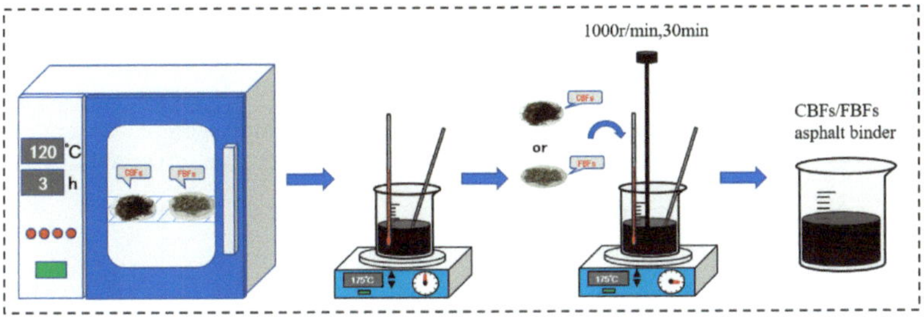

Figure 2. Mixing process of fiber–asphalt binders.

2.1.5. Gradation Design

The AC-13 graded asphalt mixture was designed by the Marshall design method. Generally, 2.36 mm is the key control sieve size, and its passing rate is lower than the median value of the target sieve size range when the mixture gradation is the coarse gradation. Owing to the better skeleton structure and high temperature stability, the coarse gradation of AC-13 was selected in this paper. After several trials, the final design grading curve is reported in Figure 3.

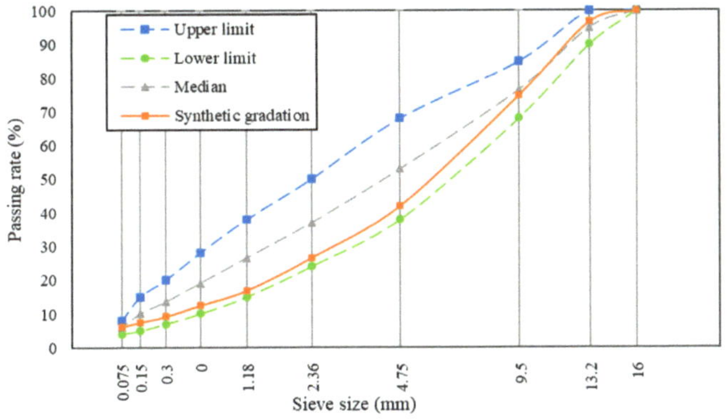

Figure 3. The design grading curve of AC-13.

According to the differences between the additives, three types of asphalt mixtures were fabricated, namely neat asphalt mixture, CBF-reinforced asphalt mixture, and FBF-reinforced asphalt mixture, respectively. Based on the engineering practice and the previous research results of the research group [12,19,31], the proposed weight content of basalt fiber in asphalt mixture was selected as 0.3%. Therefore, the 0.3% content of basalt fiber was used to prepare asphalt mixture samples. The optimum asphalt content (OAC) was determined by the Marshall test [27]. The composition materials of different asphalt mixtures and Marshall test results are listed in Table 2. Among them, the volume parameter data of air void (VV), void in mineral aggregate (VMA) and void filled with asphalt (VFA) were kept at similar levels, which could help to better evaluate the influence of other factors on the performance of the asphalt mixtures.

Table 2. Components of asphalt mixtures and Marshall test results.

Fiber Stabilizers	Fiber Contents (%)	OAC (%)	VV (%)	VMA (%)	VFA (%)	Marshall Stability (kN)	Flow Values (cm)
Neat	/	4.9	3.73	13.84	72.26	10.02	3.1
CBFs	0.3	5.1	3.99	14.72	72.87	12.03	3.2
FBFs	0.3	5.2	3.87	14.61	73.52	12.56	3.3

It was found from Table 2 that in the asphalt mixtures, the mixture with FBFs has the largest OAC value, followed by that with CBFs. This is because basalt fiber has a certain oil absorption ability. Under the same weight content of fiber, FBFs have wider spatial distribution and greater total contact area with asphalt.

2.2. Test Methods

2.2.1. Asphalt Binder Tests

Fluorescence Microscopy

Under certain fluorescent irradiation conditions, the basalt fiber coated with the infiltrator and SBS modifier will show a yellow (bright) color; by contrast, the asphalt will show a black (dark) color. Therefore, the LSM 700 3D measurement laser microscope, as shown in Figure 4, produced by Carl Zeiss in Germany was used to capture the distribution photo of fibers in the asphalt binders. A 405 nm solid-state laser was adopted to acquire the fluorescence microscopic images of fiber–asphalt binders at 100 μm magnification. The steps to prepare the observed samples were as follows:

(1) First, a metal container with a diameter of 5–10 mm and a height of 15–20 mm was placed on a horizontal table;
(2) Second, the different asphalt binders were uniformly poured into metal containers and kept at room temperature for half an hour;
(3) Next, the samples were demoulded and then placed in a refrigerator at −24 °C for 4 h;
(4) Finally, the frozen samples were taken out and cut immediately, so as to obtain a relatively flat observation surface.

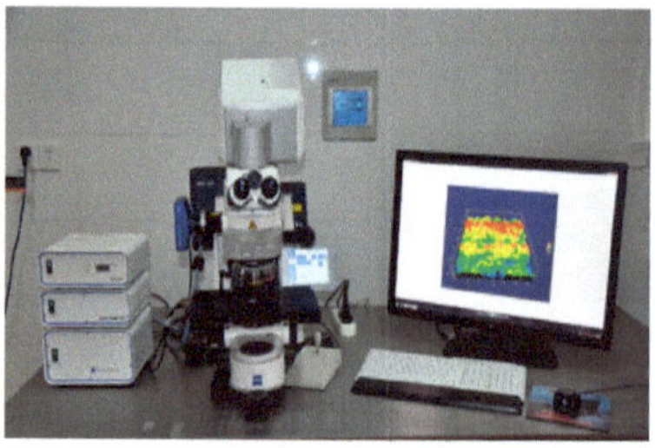

Figure 4. LSM 700 3D measurement laser microscope.

The Temperature Sweep Test

According to AASHTO T315 [32], the temperature sweep test was conducted to investigate the impact of fiber morphologies on the rheological properties. The test equipment is the DHR-2 rheometer produced by TA Instrument of Waters Company, New Castle, DE, USA. The test method is suitable for the determination of phase angle (δ), complex shear modulus (G*) and rutting factor (G*/sinδ) of the asphalt binders. The diameter of the sample is 25 mm and the thickness is 1 mm. A sinusoidal vibration load with an angular frequency of 10 rad/s was applied to the testing samples. And the test temperature was from 52 °C to 82 °C, with an increment temperature of 6 °C.

Multi-Stress Creep Recovery (MSCR) Test

The test temperature of the MSCR test was set as 64 °C, and the test equipment was the same as the temperature sweep test. According to AASHTO T350 [33], the test samples were loaded 10 times under two stress levels (0.1 kPa and 3.2 kPa). During each loading cycle, the creep loading time is 1 second and the recovering time is 9 seconds. The samples were placed between two parallel stainless steel plates with a diameter of 25 mm, and the gap value was set as 1000 μm. Nonrecoverable creep compliance (J_{nr}), elastic recovery rate (R) and relative difference in nonrecoverable creep compliance ($J_{nr\text{-}diff}$) were used to analyze the rutting resistance of asphalt binders.

2.2.2. Asphalt Mixture Tests

Wheel Tracking Test

The wheel tracking test was conducted in accordance with the JTG E20-2011 specification [27] to assess the high-temperature deformation resistance of asphalt mixtures. The samples with the size of 300 mm × 300 mm × 50 mm were prepared, and the temperature of the instrument chamber was preheated to 60 °C before loading. During the test, the loading speed was set to 42 cycles per minute and the loading stress was set to 0.7 MPa. The dynamic stability (DS) was measured to appraise the high-temperature properties of the samples. Usually, a higher DS value indicates a better resistance of the asphalt mixture against rutting damage.

Uniaxial Penetration Test

Uniaxial penetration tests can not only measure the shear strength of asphalt mixtures but also reflect the actual stress state of the asphalt materials in the pavement structure [34]. According to the JTG D50-2017 specification [35], cylindrical samples with a diameter of 150 mm and a thickness of 100 mm were prepared by the rotating compactor. The samples were placed into the chamber of the UTM-25 testing machine, and a metal indenter with a diameter of 42 mm and a thickness of 50 mm was placed above the samples. The test loading rate and temperature were set as 1 mm per minute and 60 °C, respectively. The loading diagram of the uniaxial penetration test is shown in Figure 5. The penetration strength (R_T) is calculated in compliance with Equations (1) and (2) [36].

$$\sigma_P = \frac{P}{A} \tag{1}$$

$$R_T = f_T \sigma_P \tag{2}$$

where P is the maximum load, N; A is the contact area, mm^2; f_T is the coefficient of penetration stress with a value of 0.35.

Figure 5. Loading diagram of uniaxial penetration test.

Indirect Tensile Asphalt Cracking Test (IDEAL-CT)

The IDEAL-CT test was adopted to test the cracking resistance of asphalt mixtures at medium temperatures according to ASTM D8225-19 [37]. Figure 6 presents the load–displacement curve of the experiment. It can be divided into two stages: the left part before the 100% peak load exists is called the crack initiation stage, while the right part is called the crack propagation stage [38]. Therefore, this test focuses on capturing the load value and the corresponding vertical displacement. CT_{index} is determined by analyzing the load–displacement curve, and the larger CT_{index} value means better anti-cracking ability. In this study, samples with a diameter of 150 mm and a thickness of 62 mm were formed by the rotary compactor. The compression load was applied to the sample at 25 °C under a constant rate of 50 mm/min. It was worth mentioning that the cylindrical asphalt mixture samples were subjected to indirect tension along the diameter plane direction.

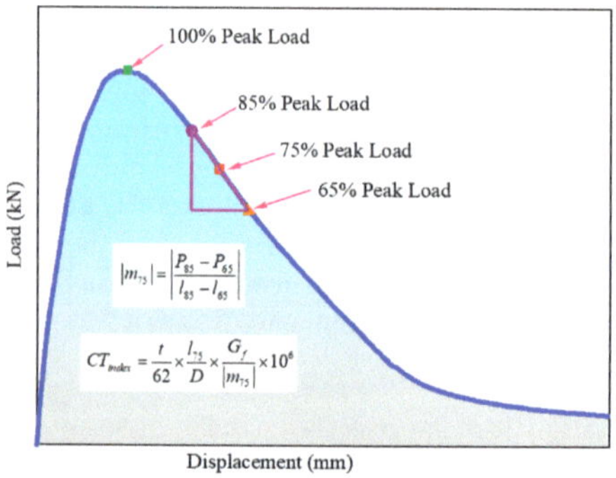

Figure 6. The load–displacement curve of IDEAL-CT test.

Low-Temperature Bending Test

According to the JTG E20-2011 specification [27], a low-temperature bending test was also conducted to evaluate the cracking resistance of three asphalt mixtures. The samples with a length of 250 mm, a width of 30 mm and a thickness of 35 mm were placed in a fixed temperature container at −10 °C for 4 h. Then, a point load was applied to the middle span of trabecular samples with a speed of 50 mm/min.

Water Immersion Stability Test and Freeze–Thaw Splitting Test

The asphalt mixtures were subjected to a combination of a water immersion stability test and a freeze–thaw splitting test, which could better reflect the anti-moisture damage of asphalt mixtures [39]. Residual stability (MS_0) was defined by the ratio of the Marshall stability value of the asphalt samples immersed in hot water (60 °C) for 48 h and for 0.5 h. Meanwhile, the freeze–thaw splitting tensile strength ratio (TSR) was the splitting strength ratio of the samples after and before the freeze–thaw process. The two indices were adopted to appraise the moisture susceptibility of samples.

3. Experimental Results and Analysis

3.1. Properties Evaluation of Fiber-Reinforced Asphalt Binders

3.1.1. Fluorescence Microscopy Test

At the microscopic scale, fluorescence microscopy can help to reveal the actual state of existence of fibers in asphalt binders to some extent. It can be seen from Figure 7 that both types of fibers have good dispersibility in asphalt binder and that no clumping phenomenon occurred. The CBFs still maintain their straight status, which could form a good bridging effect in the mixture. Nevertheless, the FBFs are cotton-like fibers formed by the curved basalt fiber monofilaments, which can help to form a three-dimensional network structure. Under this situation, the asphalt is adsorbed on the surface of the flocculent fiber, and the interface between the asphalt and aggregates is more stable, thus increasing the thickness of the asphalt film and the bond strength between asphalt and aggregates. At the same time, the microscopic picture of the fiber–asphalt binder could verify the view that there are clear boundary areas between fibers and asphalt, which can effectively increase the contact surface between fibers and the asphalt. Then, it can produce an anchoring effect when the fiber and asphalt interact so as to further improve the adhesion of fibers and asphalt.

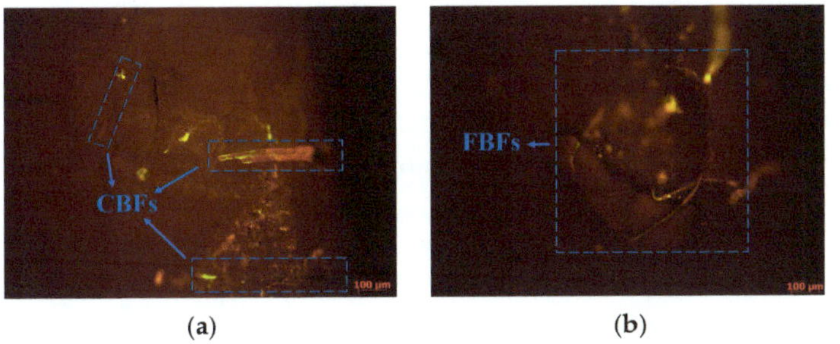

Figure 7. Fluorescent microscopic images: (**a**) CBFs asphalt binder; (**b**) FBFs asphalt binder.

3.1.2. The Temperature Sweep Test

Comparative analyses of the δ, G* and G*/sinδ values of the three types of asphalt binders are depicted in Figure 8. When the temperature increases, the δ, G* and G*/sinδ values decrease.

After the addition of fiber, the phase angle of asphalt binder changes significantly but also retains the curve characteristics of asphalt binder. As to CBFs, when the fiber dosage is 1%, 2% and 3%, the phase angle curve at high temperature is not obviously different from that of SBS original asphalt, while at 4%, the phase angle curve has a dramatic decline. A possible reason for this phenomenon may be that enough CBFs can form a three-dimensional grid structure in the asphalt binder, which enhances the elasticity of the asphalt material and limits its fluidity under high-temperature conditions. For FBFs, the phase angle of asphalt binder under the condition of low dosage is also significantly lower than that of SBS original modified asphalt, and the phase angle decreases more significantly with the increase in fiber content. By looking back at Figures 1 and 7, and comparing the appearance technical indicators of the two basalt fibers, it is clear that the specific surface area and gross volume of the flocculent basalt fibers with the same mass are larger. The flocculent basalt fiber with the same dosage makes it easier to stabilize the asphalt binders.

The results of complex shear modulus and rutting factor before and after adding basalt fibers are illustrated in Figure 8b,c. The results of G* and G*/sinδ values of asphalt binders decrease sharply with the increase in temperature, indicating that temperature is an important factor affecting the modulus values of asphalt binders, which also verifies the previous findings [23,25,40]. The curves between temperature and the complex modulus/rutting factor of fiber–asphalt binders are higher than those of SBS original asphalt, which indicates that fiber–asphalt binders have better deformation resistance at high temperatures. The curves of asphalt binders containing flocculent basalt fibers remain at the top all the time, and the index of asphalt binders with 4% flocculent basalt fiber increases remarkably, mainly because the high flexibility of flocculent basalt fiber can reduce the overall stiffness of asphalt binder.

Overall, it is obvious that the basalt fibers can minimize the influence of high temperature on the rheological properties of asphalt binders, and FBFs reveal more excellent ability in this respect, showing their better temperature sensitivity.

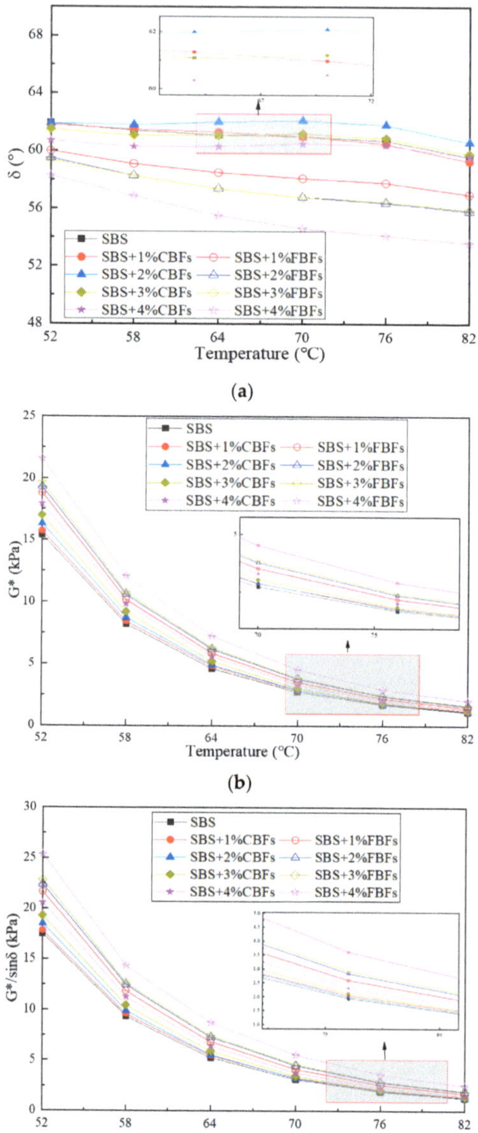

Figure 8. The results of the temperature sweep test: (**a**) δ; (**b**) G*; (**c**) G*/sinδ.

3.1.3. MSCR Test

Previous studies [13,23,41] have shown that Jnr is used to reflect the nonlinear rheological response of asphalt under large stress, and it has a good correlation with the rutting resistance of asphalt mixture. When the stress increases, the R values of asphalt binders decrease hastily and the Jnr values increase drastically to a different degree, which indicates that the effect of "heavy load" on the resistance to permanent deformation of asphalt binders bears a resemblance to the "high temperature". In other words, the stress and temperature have some kind of equivalence. Under the shear stress of 0.1 kPa at

64 °C, the rheological properties of asphalt binders are basically in the linear range. With the increase in stress level, the differences in R and Jnr between fiber–asphalt binders and the control group become more obvious. Moreover, the rheological properties of asphalt binders are likely to have entered into the nonlinear range, which should be able to simulate the situation of asphalt pavement under large loads. More specifically, the elastic recovery rates of asphalt binders are improved and non-recoverable creep compliances decrease after adding fibers. The improvement amplitude tends to be stable with the increase in the dosage. The R/Jnr curves of the chopped basalt fiber asphalt binder tend to be flat when the fiber dosage is 3%, whereas it tends to flatten out with a 2% fiber dosage for flocculent basalt fiber.

For different fiber properties, under 3.2 kPa load, the elastic recovery rate of the FBFs asphalt binder is always slightly higher than that of the CBF asphalt binder. And the nonrecoverable creep compliance is inferior to that of the CBF asphalt binder, showing excellent high-temperature deformation resistance. By observing the stress-sensitive index Jnr-diff in Figure 9c, it is not difficult to find that the Jnr-diff values increase by 46.1%, 75.2%, 75.7% and 104.8%, respectively, with the increase in the dosage of CBFs. Meanwhile, as for FBFs, the Jnr-diff values increase by 31.6%, 84.4%, 101.8% and 133.6%, respectively. This yields increasingly good results on the data. The stress sensitivity of asphalt binder shows an increasing trend with the increase in fiber dosage, indicating that fiber is conducive to improving the stress sensitivity of asphalt binder. Meanwhile, superior results are seen for FBF asphalt binders, which is broadly in line with the results of the temperature sweep test.

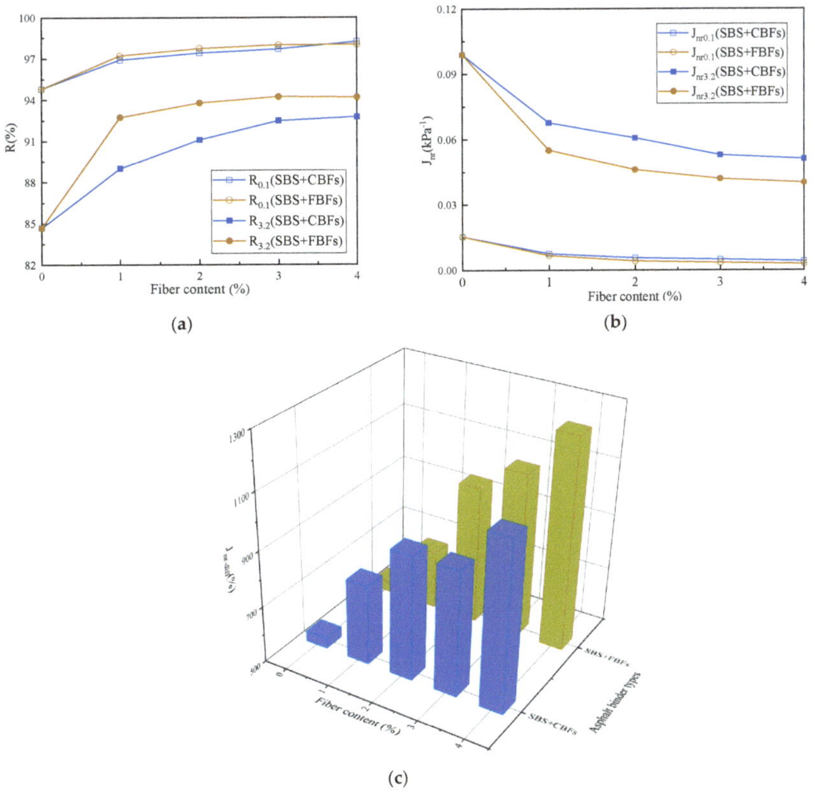

Figure 9. The MSCR results of different asphalt binders: (**a**) elastic recovery rate; (**b**) nonrecoverable creep compliance; (**c**) relative difference in nonrecoverable creep compliance.

3.2. Properties Evaluation of Fiber Reinforced Asphalt Mixtures

3.2.1. Wheel Tracking Test

The DS values of different asphalt mixtures are shown in Figure 10. It was obvious that the DS values increased after adding basalt fibers. Compared with the control group, the DS value of asphalt mixture with CBF increases from 4050 to 5175 times·mm^{-1}, and the increasing extent is 28%. The DS value of asphalt mixture with FBFs increases from 4050 to 5335 times·mm^{-1}, with a 32% increasing extent. Some scholars believe that fibers can better bond the asphalt and aggregate together, which could help to better resist the rolling loads [17,42,43]. Additionally, the highest DS values belong to the samples with FBFs, which show a slight advantage on the high temperature rutting resistance. This is because CBFs are dense and smooth, resulting in low fiber oil absorption rates. While the structure of the FBF is more complex, it can contribute to absorbing more asphalt. This phenomenon agrees with the results of the SEM test [23]. The high fiber oil absorption rate is conducive to preventing the segregation and overflow of the asphalt binder under high-temperature conditions.

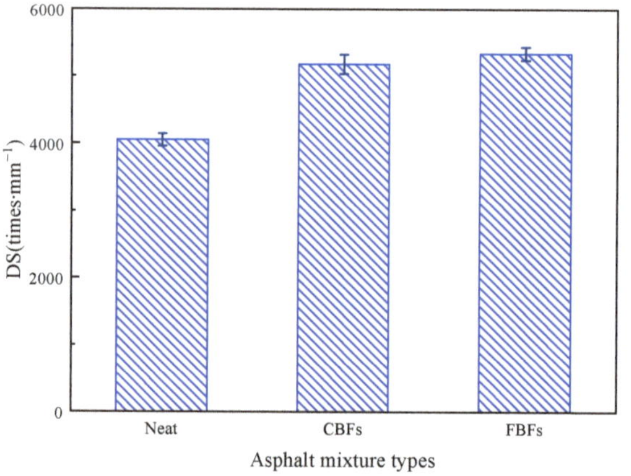

Figure 10. The dynamic stability results from the wheel tracking test of different samples.

3.2.2. Uniaxial Penetration Test

Figure 11 illustrates the effects of the two basalt fibers on the uniaxial penetration test of the examined asphalt mixture samples. Compared with the control group, the P values of asphalt mixtures increase to 3.558 kN for mixtures containing CBFs and increase to 3.802 kN for mixtures containing FBFs. Simultaneously, the penetration strength value of the mixture increases by 23.2% for the mixture containing CBFs, and it increases by 31.6% for mixtures containing FBFs. The reason is mainly because the basalt fibers can absorb free asphalt in the mixture, which can be transformed into structural asphalt. Then, the aggregates are bonded to form a frame support in the mixture, thereby enhancing the shear deformation resistance of asphalt pavements.

Since the oil absorption rate of the FBFs is higher than that of the CBFs, the FBFs can overlap with each other and form a network structure in the mixture. This indicates that FBFs can better transfer the load and further reduce the structural damage caused by the stress concentration inside the asphalt pavements. This concurs well with the results of the wheel tracking test, which further proves the effect of basalt fibers on the high-temperature performance of the mixture and the effect of fiber morphologies on the mixtures.

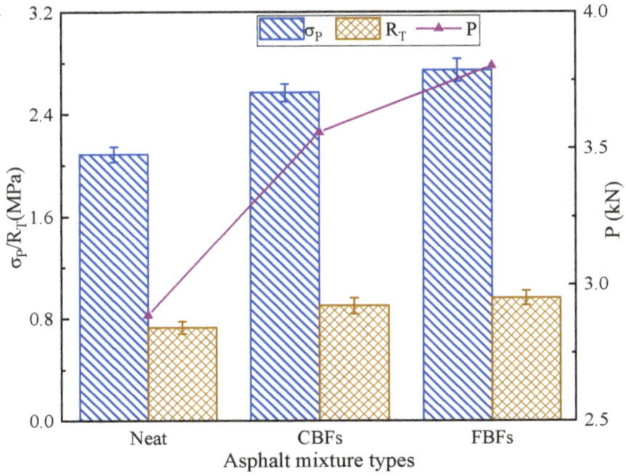

Figure 11. The uniaxial penetration test results of different samples.

3.2.3. IDEAL-CT Test

As is seen in Figure 12a, the displacement increases with the increase in the compressive load. Under the same load (before 100% peak load), the displacement of the samples with fibers is smaller, and it is more significant in the mixes with FBFs. In the whole process of the test, the initiation energy (G0), fracture energy and CT_{index} of the asphalt mixtures containing fibers are greatly improved when compared with the ordinary non-fiber asphalt mixture. After adding CBFs and FBFs, the initiation energy values of asphalt mixtures increase by 9.78% and 26.4%, the fracture energy values of mixtures increase by 15.1% and 32.5%, and the cracking index CT_{index} is elevated by 81.7% and 109.1%, respectively. Therefore, there is an effective enhancing effect on the initial cracking resistance of asphalt mixtures. This can be largely explained by the reason that basalt fiber is a strong inorganic material with high modulus, and the fibers can form an interconnected three-dimensional network structure. In the process of crack formation, the fibers help to transfer the internal stress of the mixture and increase the stiffness of the suspended dense skeleton structure. Moreover, the combination of basalt fibers and asphalt helps to delay the propagation speed of cracks [13,19,44].

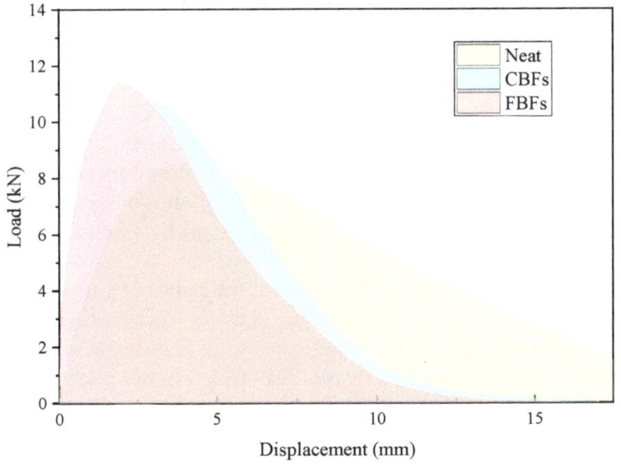

(a)

Figure 12. *Cont.*

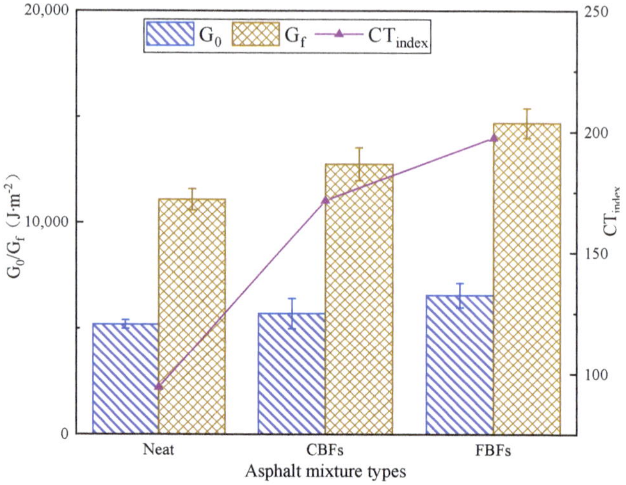

(**b**)

Figure 12. The IDEAL-CT results of different samples: (**a**) displacement–load curves; (**b**) initiation energy (G_0), fracture energy (Gf) and CT_{index}.

In addition, from the perspective of composite materials, when additives with specific strength are incorporated, the mechanical strength of the composite will increase accordingly. Compared with the relevant test results of the asphalt mixture with fibers, the initiation energy, fracture energy and CT_{index} of the asphalt mixture with FBFs are all bigger than those of CBFs, which means that the crack growth rate of FBF-reinforced asphalt mixture is slower than that of CBFs reinforced asphalt mixture. This is because FBFs can help to absorb excess asphalt inside the asphalt mixture and can make the structural asphalt on the aggregates surface more tightly connected to each other. Since FBFs are randomly distributed in the asphalt mixture, it is helpful to connect fibers with the structural asphalt on the aggregate surface. Therefore, the cracking resistance of the asphalt mixture is better.

3.2.4. Low-Temperature Bending Test

As is shown in Figure 13a, at −10 °C, the samples with FBFs have the maximum damage load, while the samples with CBFs have the minimal mid-span deflection. On the contrary, the control group has minimal damage load and maximum mid-span deflection. This demonstrates that the basalt-fiber-modified asphalt mixtures have stronger toughness and better crack resistance at low temperatures. The explanation may be that the asphalt becomes harder and more brittle under low-temperature conditions, which makes the adhesion between the fiber and the asphalt worse, resulting in the insignificant bridging effect. As shown in Figure 13b–d, after adding CBFs or FBFs, the flexural tensile strength (RB) values of asphalt mixtures increase by 2.36% and 4.72%, the failure strain (εB) values enlarge to 3067 με and 3267 με, and the bending stiffness modulus (SB) of asphalt mixture reduces by 4.37% and 8.55%, respectively. As mentioned before, the oil absorption rate of FBFs is relatively high, which promotes the flexibility of the asphalt mixture and forms more asphalt films. As a result, FBFs play a better role in filling and healing the micro-cracks between the aggregates.

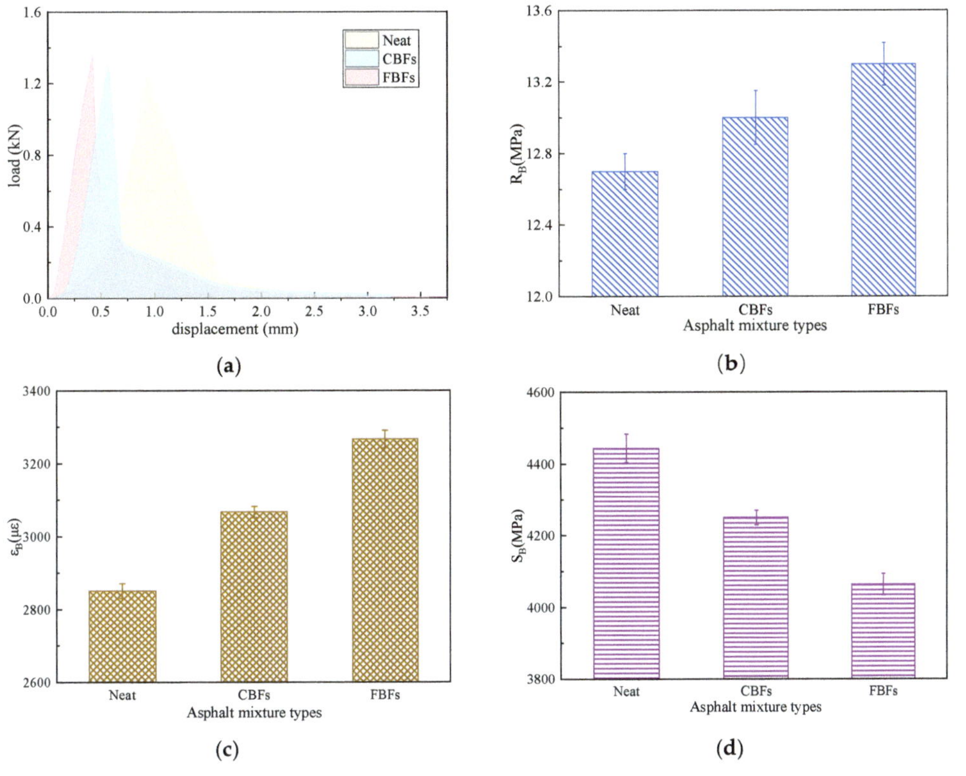

Figure 13. The low-temperature trabecular bending test results of different asphalt mixtures: (**a**) displacement–load curves; (**b**) flexural tensile strength; (**c**) failure strain; (**d**) bending stiffness modulus.

3.2.5. Water Immersion Stability Test and Freeze–Thaw Splitting Test

The shear force transferred from the road surface under traffic load accelerates the stripping and spalling of aggregates, leading to the aggravation of water damage [45]. Water immersion stability test results have been presented in Figure 14a. Compared with the control group, the Marshall stability values of the asphalt mixture with CBFs before and after the immersion process are slightly increased and the MS0 values are almost unchanged. However, the relevant indexes of the asphalt mixture after the adding FBFs are improved evidently, and the MS0 value increases by about 20.0%. The explanation may be that the presence of fibers plays a crucial role in increasing the consistency of the asphalt binder and improving the binding effect with aggregates. As mentioned before, FBFs can better absorb the light components in asphalt, increasing the thickness of asphalt film. In addition. It could then improve the bond strength as well as prevent water from entering the interface between asphalt and aggregates effectively.

According to the results of freeze–thaw splitting test (Figure 14b), the TSR value of the asphalt mixture with CBFs is 88.54%, which increases by 1.04% more than that of the non-fiber asphalt mixture. Conversely, the TRS value of the mixture with FBFs reaches 90.65%, which is about 3% higher than that of the control group. It is evident that the TSR value of the asphalt mixture with FBFs increases more obviously than that with CBFs, which shows the same pattern as the result of the immersion Marshall test. On the one hand, the oil absorption rate of FBFs is higher than that of CBF, and the higher OAC can make the water less likely to be immersed in the asphalt mixture. On the other hand, FBFs maintain high strength characteristics and show a better reinforcement effect than the ordinary dense graded asphalt mixture. In a word, FBFs can help to control the spread of asphalt pavement cracks and reduce the occurrence of water damage by absorbing part of the fracture energy during the service of asphalt pavement.

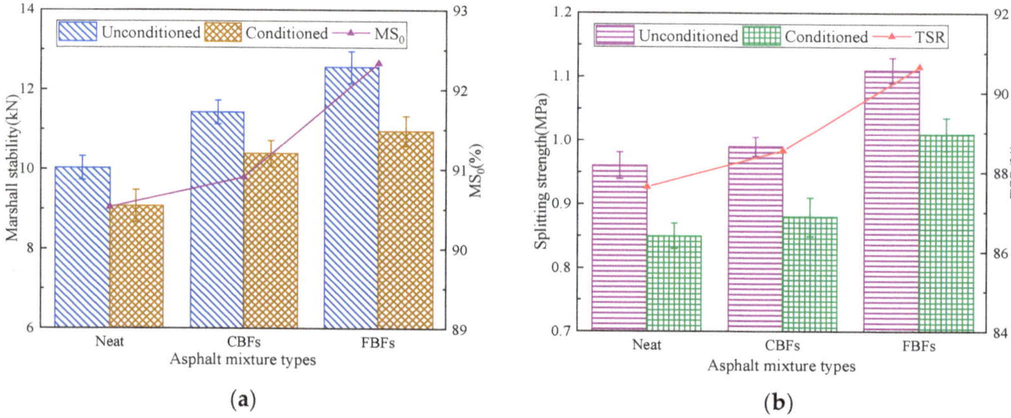

Figure 14. Moisture susceptibility test results: (**a**) water immersion stability test; (**b**) freeze–thaw splitting test.

3.3. Comprehensive Analysis

The rheological properties of different asphalt binders and road performance of asphalt mixtures are multidimensionally expressed by radar maps, as shown in Figure 15. In terms of asphalt binders, δ, G*, G*/sinδ (at 64 °C), and R, Jnr, Jnr-diff (under 3.2 kPa at 64 °C) were chosen to represent the rheological properties. For the asphalt mixtures, DS and R_T are considered to represent the high-temperature stabilities, CT_{index}/ε_B is selected to stand for the crack resistance at medium/low temperature, and MS0 as well as TSR are considered to represent the moisture susceptibilities.

The improvement effects of different basalt fiber morphologies on the asphalt binders or asphalt mixtures are various. In Figure 15a, the rheological properties of asphalt binders mixed with CBFs are slightly higher than that of ordinary SBS-modified asphalt, while the elasticity of asphalt binders mixed with FBFs is distinctly increased, and the deformation resistance at high temperatures is significantly improved. As seen from the radar chart of asphalt mixture pavement performance (Figure 15b), FBFs have a more favorable influence on the comprehensive performance of asphalt mixture, which could prolong the service life of asphalt pavement. However, the impact of CBFs is slightly inferior to that of FBFs. Therefore, in addition to the traditional parameters such as fiber dosage, length, diameter, etc., the morphology of the fibers also has a crucial effect on the properties of asphalt binders or asphalt mixtures.

In order to evaluate the effect of different morphologies of basalt fiber on the properties of asphalt mixtures more directly, the normalization method was used for quantitative analysis. The performance index of the mixture without basalt fiber is used as the reference material, which was marked as "1". Then, the increase or decrease in the properties of the other samples relative to the baseline was obtained, where the increase was positive and the decrease was negative, respectively. The selection of asphalt mixture indicators is similar to that of radar maps. It should be noted in particular that the average changes in DS and RT are considered as the changes in the high-temperature deformation resistance, and the average changes in MS_0 and TSR are considered as the changes in moisture stability. Finally, the sum of normalized values of mixtures was calculated and depicted in Figure 16.

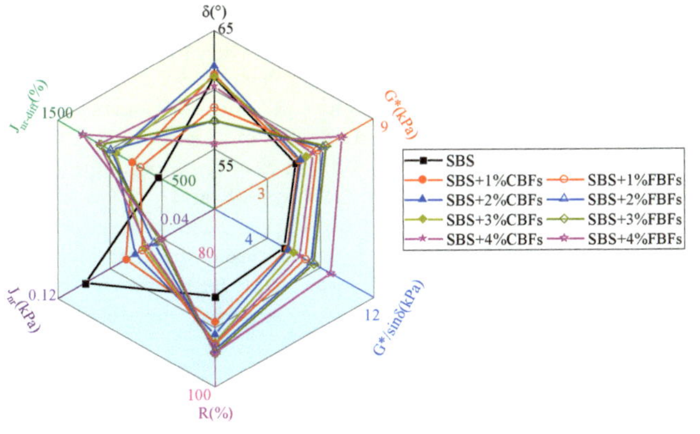

(**a**)

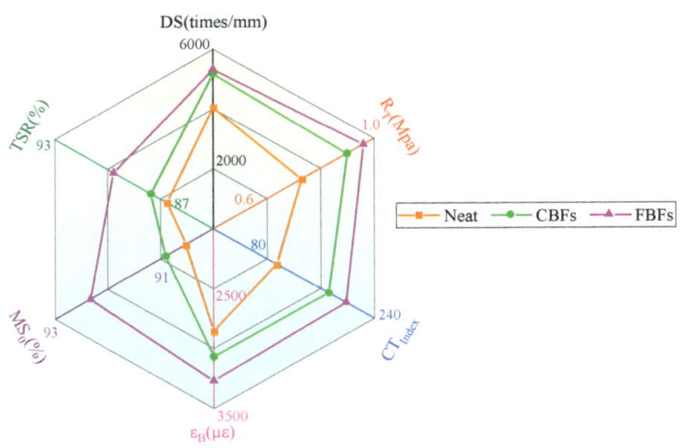

(**b**)

Figure 15. Radar maps: (**a**) rheological behavior of different asphalt binders with 1% fiber dosage; (**b**) pavement performance of asphalt mixtures.

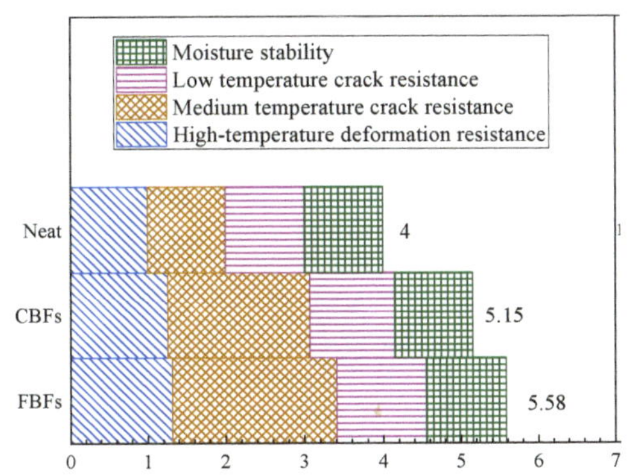

Figure 16. The sum results of normalized values.

The results verify that the two different morphologies of basalt fibers can significantly improve the overall performance of the asphalt mixture. The normalized analysis values of high-temperature deformation resistance and medium-temperature crack resistance are between 1.255 and 1.316 and 1.817 and 2.091, respectively. This indicates that basalt fiber can greatly enhance the medium- and high-temperature performance of asphalt mixture. Meanwhile, the normalized values of the low-temperature crack resistance and moisture susceptibility range from 1.076 to 1.146 and 1.007 to 1.027, respectively. This phenomenon suggests that basalt fiber can improve the corresponding properties of asphalt mixtures, although the amplitude is not obvious. Simultaneously, in terms of the normalized sum results, mixtures with FBFs show better enhancement capability. Consequently, morphology should also be considered when selecting fibers in engineering practice.

4. Conclusions

In this paper, two basalt fibers (CBFs and FBFs) were blended with SBS-modified asphalt to prepare the fiber-reinforced asphalt binder. The microscopic morphology and the rheological properties of fiber–asphalt binders were studied by a fluorescence microscopy test, a temperature sweep test and an MSCR test. Subsequently, two kinds of fibers were mixed into asphalt mixtures by the dry method, and the mixture samples were tested to investigate the pavement performance. The main conclusions can be summarized as follows:

(1) Adding basalt fibers with different morphologies makes the volume parameters of the asphalt mixture change to some extent. With the same amount of fiber in asphalt mixtures, FBFs can absorb more structural asphalt than CBFs, and the OAC values and Marshall stability values of asphalt mixture with FBFs are slightly higher than those with CBFs.

(2) Both basalt fibers can be well dispersed in the asphalt binder and form a three-dimensional grid structure. Compared with CBFs, the larger specific surface area of FBFs with the same weight content makes the asphalt binder structure more stable. And the additives of fibers could increase the elasticity, decrease the viscosity and improve the temperature sensitivity of asphalt materials. An asphalt binder containing FBFs shows better high-temperature performance and elastic recovery ability, followed by CBFs.

(3) The enhancement effect of FBFs on the medium/high-temperature properties and moisture susceptibility of the asphalt mixture is obviously better than that of CBFs. The small volume of air void in the FBF asphalt mixture also ensures its resistance to water damage. However, FBFs only have a slight advantage in terms of low-temperature cracking resistance. Therefore, if the fiber is used in areas with high requirements for low-temperature crack resistance, the fibers with lower costs could be chosen.

(4) FBFs have a superior impact on the rheological properties of asphalt binders and the comprehensive pavement performance of asphalt mixtures. The morphology of the fiber should be taken into account, especially in the areas having middle or high temperatures.

Author Contributions: Conceptualization, K.L.; methodology, K.L. and P.X.; software, C.C.; formal analysis, C.C.; data curation, F.Q.; writing—original draft preparation, C.C. and F.Q.; writing—review and editing, K.L. and P.X.; project administration, P.X.; funding acquisition, K.L. and C.C. All authors have read and agreed to the published version of the manuscript.

Funding: This research was funded by the National Natural Science Foundation of China, grant number 52408498, Innovative and Entrepreneurial Training Program for College Students of Jiangsu Province, grant number 202411117128Y and University-Industry Cooperation project of Jiangsu Province, grant number BY20230505.

Institutional Review Board Statement: Not applicable.

Informed Consent Statement: Not applicable.

Data Availability Statement: The original contributions presented in the study are included in the article, further inquiries can be directed to the corresponding author.

Conflicts of Interest: The authors declare no conflicts of interest.

References

1. Kim, J.S.; Lim, J.H.; Huh, Y. Melt-spinning Basalt Fibers Based on Dielectric Heating and Steady-state Process Characteristics. *Fibers Polym.* **2013**, *14*, 1148–1156. [CrossRef]
2. Nikitina, F.M.; Kogano, V. Solubility of chrysotile asbestos and basalt fibres in relation to their fibrogenic and carcinogenic action. *Environ. Health Perspect.* **1994**, *102*, 205–206. [CrossRef]
3. Zhao, G.; Li, M.; Wang, S.; Peng, X.; Wang, L.; Li, X.; Zhao, Y.; Gao, Y.; Zhang, Y.; Zheng, J. Effect of interlaminar basalt fiber veil reinforcement on mode I fracture toughness of basalt fiber composites. *Polym. Compos.* **2024**, *45*, 4985–4993. [CrossRef]
4. Yang, C.; Liu, L.; Liu, Z.; Huang, Y.; Yu, S.; Fu, Y. Study on the mechanism of bond strength generation and debonding failure between basalt fiber and asphalt based on molecular dynamics. *Case Stud. Constr. Mater.* **2023**, *19*, e02493. [CrossRef]
5. Abiola, O.S.; Kupolati, W.K.; Sadiku, E.R.; Ndambuki, J.M. Utilisation of natural fibre as modifier in bituminous mixes: A review. *Constr. Build. Mater.* **2014**, *54*, 305–312. [CrossRef]
6. Sarayu, K.; Gopinath, S.; Ramachandra, M.A.; Iyer, N.R. Structural stability of basalt fibers with varying biochemical conditions-A invitro and invivo study. *J. Build. Eng.* **2016**, *7*, 38–45. [CrossRef]
7. Liu, Y.; Zhang, M.; Liu, H.; Tian, L.; Liu, J.; Fu, C.; Fu, X. Properties of Basalt Fiber Core Rods and Their Application in Composite Cross Arms of a Power Distribution Network. *Polymers* **2022**, *14*, 2443. [CrossRef]
8. Li, Y.; Wang, Q.; Xu, S.; Song, Q. Study of eco-friendly fabricated hydrophobic concrete containing basalt fiber with good durability. *J. Build. Eng.* **2023**, *65*, 105759. [CrossRef]
9. Fiore, V.; Scalici, T.; Bella, G.D.; Valenza, A. A review on basalt fibre and its composites. *Compos. B Eng.* **2015**, *74*, 74–94. [CrossRef]
10. Zhou, S.; Gao, J.; Wang, J.; Wang, S.; Liu, Y. Polydopamine-coupling of carbon nanotubes onto microscaled basalt fiber to enhancing mechanical, thermal and tribological properties of composite materials. *Mater. Res. Express* **2019**, *6*, 0850g6. [CrossRef]
11. Celauro, C.; Praticò, F.G. Asphalt mixtures modified with basalt fibres for surface courses. *Constr. Build. Mater.* **2018**, *170*, 245–253. [CrossRef]
12. Xie, T.; Wang, L. Optimize the design by evaluating the performance of asphalt mastic reinforced with different basalt fiber lengths and contents. *Constr. Build. Mater.* **2023**, *363*, 129698. [CrossRef]
13. Pei, Z.; Lou, K.; Kong, H.; Wu, B.; Wu, X.; Xiao, P.; Qi, Y. Effects of Fiber Diameter on Crack Resistance of Asphalt Mixtures Reinforced by Basalt Fibers Based on Digital Image Correlation Technology. *Materials* **2021**, *14*, 7426. [CrossRef]
14. Lou, K.; Xiao, P.; Wu, B.; Kang, A.; Wu, X.; Shen, Q. Effects of fiber length and content on the performance of ultra-thin wearing course modified by basalt fibers. *Constr. Build. Mater.* **2021**, *313*, 125439. [CrossRef]
15. Sun, X.; Qin, X.; Chen, Q.; Ma, Q. Investigation of enhancing effect and mechanism of basalt fiber on toughness of asphalt material. *Pet. Sci. Technol.* **2018**, *36*, 1710–1717. [CrossRef]
16. Zhang, J.; Huang, W.; Zhang, Y.; Lv, Q.; Yan, C. Evaluating four typical fibers used for OGFC mixture modification regarding drainage, raveling, rutting and fatigue resistance. *Constr. Build. Mater.* **2020**, *253*, 119131. [CrossRef]
17. Wang, W.; Cheng, Y.; Tan, G. Design Optimization of SBS-Modified Asphalt Mixture Reinforced with Eco-Friendly Basalt Fiber Based on Response Surface Methodology. *Materials* **2018**, *11*, 1311. [CrossRef]
18. Gao, L.; Hu, G.; Xu, N.; Fu, J.; Xiang, C.; Yang, C. Experimental Study on Unconfined Compressive Strength of Basalt Fiber Reinforced Clay Soil. *Adv. Mater. Sci. Eng.* **2015**, 561293. [CrossRef]
19. Lou, K.; Wu, X.; Xiao, P.; Kang, A.; Wu, Z.; Xia, Y. Comprehensive Study about Effect of Basalt Fiber, Gradation, Nominal Maximum Aggregate Size and Asphalt on the Anti-Cracking Ability of Asphalt Mixtures. *Appl. Sci.* **2021**, *11*, 2289. [CrossRef]
20. Li, Z.; Shen, A.; Wang, H.; Guo, Y.; Wu, H. Effect of basalt fiber on the low-temperature performance of an asphalt mixture in a heavily frozen area. *Constr. Build. Mater.* **2020**, *253*, 119080. [CrossRef]
21. Cetin, A.; Evirgen, B.; Karslioglu, A.; Tuncan, A. The Effect of Basalt Fiber on the Performance of Stone Mastic Asphalt. *Period. Polytech-Civ.* **2021**, *65*, 299–308. [CrossRef]
22. Thanh, D.V.; Feng, C.P. Study on Marshall and Rutting test of SMA at abnormally high temperature. *Constr. Build. Mater.* **2013**, *47*, 1337–1341. [CrossRef]
23. Wu, B.; Pei, Z.; Luo, C.; Xia, J.; Chen, C.; Kang, A. Effect of different basalt fibers on the rheological behavior of asphalt mastic. *Constr. Build. Mater.* **2022**, *318*, 125718. [CrossRef]
24. Shi, C.; Wang, J.; Sun, S.; Lv, D.; Xu, L.; Zhang, S. Research on properties of basalt fiber-reinforced asphalt mastic. *Front. Mater.* **2023**, *10*, 1277634. [CrossRef]
25. Kou, C.; Chen, Z.; Kang, A.; Zhang, M.; Wang, R. Rheological behaviors of asphalt binders reinforced by various fibers. *Constr. Build. Mater.* **2022**, *323*, 126626. [CrossRef]
26. Gu, Q.; Kang, A.; Li, B.; Xiao, P.; Ding, H. Effect of fiber characteristic parameters on the high and low temperature rheological properties of basalt fiber modified asphalt mortar. *Case Stud. Constr. Mater.* **2022**, *17*, e01247. [CrossRef]

27. *JTG E20-2011;* Standard Test Methods of Bitumen and Bituminous Mixtures for Highway Engineering. Occupation Standard of the People's Republic of China: Beijing, China, 2011.
28. *JTG E42-2005;* Test Methods of Aggregate for Highway Engineering. Ministry of Communications of the People's Republic of China: Beijing, China, 2011.
29. Liu, K.; Li, T.; Wu, C.; Jiang, K.; Shi, X. Bamboo fiber has engineering properties and performance suitable as reinforcement for asphalt mixture. *Constr. Build. Mater.* **2021**, *290*, 123240. [CrossRef]
30. Lou, K.; Xiao, P.; Ong, G.P.; Li, B.; Kang, A.; Wu, Z. Micromechanical behavior of single fiber-asphalt mastic interface: Experimental studies by self-designed innovative pullout test. *Constr. Build. Mater.* **2024**, *414*, 134873. [CrossRef]
31. Wang, S.; Kang, A.; Xiao, P.; Li, B.; Fu, W. Investigating the Effects of Chopped Basalt Fiber on the Performance of Porous Asphalt Mixture. *Adv. Mater. Sci. Eng.* **2019**, 2323761. [CrossRef]
32. *AASHTO T315;* Standard Method of Test for Determining the Rheological Properties of Asphalt Binder Using a Dynamic Shear Rheometer (DSR). American Association of State Highway and Transportation Officials: Washington, DC, USA, 2020.
33. *AASHTO T350;* Standard Method of Test for Multiple Stress Creep Recovery (MSCR) Test of Asphalt Binder Using a Dynamic Shear Rheometer (DSR). American Association of State Highway and Transportation Officials: Washington, DC, USA, 2020.
34. Peng, Y.; Xia, S.; Xu, Y.; Lu, X.; Li, Y. Discrete-Element Modeling of Influence of Void Characteristics on Uniaxial Penetration Strength of Asphalt Mixtures. *J. Mater. Civ. Eng.* **2021**, *33*, 04020399. [CrossRef]
35. *JTG D50-2017;* Specifications for Design of Highway Asphalt Pavement. Ministry of Communications of the People's Republic of China: Beijing, China, 2017.
36. Ma, Z.; Liu, L.; Sun, L. Investigation of top-down cracking performance of in-situ asphalt mixtures based on accelerated pavement testing and laboratory tests. *Constr. Build. Mater.* **2018**, *179*, 277–284. [CrossRef]
37. *ASTM D8225-19;* Standard Test Method for Determination of Cracking Tolerance Index of Asphalt Mixture Using the Indirect Tensile Cracking Test at Intermediate Temperature. ASTM International: West Conshohocken, PA, USA, 2019.
38. Chen, H.; Zhang, Y.; Bahia, H.U. The role of binders in mixture cracking resistance measured by ideal-CT test. *Int. J. Fatigue* **2021**, *142*, 105947. [CrossRef]
39. Zhang, K.; Li, W.; Han, F. Performance deterioration mechanism and improvement techniques of asphalt mixture in salty and humid environment. *Constr. Build. Mater.* **2019**, *208*, 749–757. [CrossRef]
40. Gao, L.; Wang, Z.; Liu, Y.; Zheng, J.; Li, H. Influence of Binder Property and Mortar Thickness on High-Temperature Performance of Cold Recycled Mixtures with Asphalt Emulsion. *Materials* **2019**, *12*, 2718. [CrossRef] [PubMed]
41. Li, S.; Shi, X.; Si, C.; Bao, B.; Hu, M. Correlation between the Rheological Properties of Asphalt Mortar and the High-Temperature Performance of Asphalt Mixture. *Coatings* **2023**, *13*, 1058. [CrossRef]
42. Pirmohammad, S.; Amani, B.; Shokorlou, Y.M. The effect of basalt fibres on fracture toughness of asphalt mixture. *Fatigue. Fract. Eng. Mater. Struct.* **2020**, *43*, 1446–1460. [CrossRef]
43. Chen, H.; Xu, Q.; Chen, S.; Zhang, Z. Evaluation and design of fiber-reinforced asphalt mixtures. *Mater. Des.* **2009**, *30*, 2595–2603. [CrossRef]
44. Yang, K.; He, Z.; Li, D.; Xu, H.; Kong, L. Experimental Study on Basalt Fiber Crack Resistance of Asphalt Concrete Based on Acoustic Emission. *Materials* **2021**, *14*, 4096. [CrossRef]
45. Wang, W.; Wang, L.; Xiong, H.; Luo, R. A review and perspective for research on moisture damage in asphalt pavement induced by dynamic pore water pressure. *Constr. Build. Mater.* **2019**, *204*, 631–642. [CrossRef]

materials

MDPI

Article

Preliminary Study on Multi-Scale Modeling of Asphalt Materials: Evaluation of Material Behavior through an RVE-Based Approach

Ahmed Ibrahim Hassanin Mohamed *, Oliver Giraldo-Londoño, Baolin Deng, Zhen Chen, Punyaslok Rath and William G. Buttlar

Department of Civil and Environmental Engineering, University of Missouri, Columbia, MO 65211, USA; ogiraldo@missouri.edu (O.G.-L.); dengb@missouri.edu (B.D.); chenzh@missouri.edu (Z.C.); rathp@missouri.edu (P.R.); buttlarw@missouri.edu (W.G.B.)

* Correspondence: aihnng@missouri.edu; Tel.: +1-573-5306666

Abstract: This study employs a microstructure-based finite element modeling approach to understand the mechanical behavior of asphalt mixtures across different length scales. Specifically, this work aims to develop a multi-scale modeling approach employing representative volume elements (RVEs) of optimal size; this is a key issue in asphalt modeling for high-fidelity fracture modeling of heterogeneous asphalt mixtures. To determine the optimal RVE size, a convergence analysis of homogenized elastic properties is conducted using two types of RVEs, one made with polydisperse spherical inclusions, and another made with polydisperse truncated cylindrical inclusions, each aligned with the American Association of State Highway and Transportation Official's maximum density gradation curve for a 12.5 mm Nominal Maximum Aggregate Size (NMAS). The minimum RVE lengths for this NMAS were found to be in the range of 32–34 mm. After the optimal RVE size for each inclusion shape is obtained, computational models of heterogeneous Indirect Tensile Asphalt Cracking Test samples are then generated. These models include the components of viscoelastic mastic, linear elastic aggregates, and cohesive zone modeling to simulate the rate-dependent failure evolution from micro- to macro-cracking. Examination of load-displacement responses at multiple loading rates shows that both heterogeneous models replicate experimentally measured data satisfactorily. Through micro- and macro-level analyses, this study enhances our understanding of the composition-performance relationships in asphalt pavement materials. The procedure proposed in this study allows us to identify the optimal RVE sizes that preserve computational efficiency without significantly compromising their ability to capture the asphalt material behavior under specific operational conditions.

Keywords: representative volume element; heterogeneous model; asphalt pavement; fracture; finite element modeling; RVE-based approach

Citation: Ibrahim Hassanin Mohamed, A.; Giraldo-Londoño, O.; Deng, B.; Chen, Z.; Rath, P.; Buttlar, W.G. Preliminary Study on Multi-Scale Modeling of Asphalt Materials: Evaluation of Material Behavior through an RVE-Based Approach. *Materials* **2024**, *17*, 5041. https://doi.org/10.3390/ma17205041

Academic Editor: Marco Corradi

Received: 28 August 2024
Revised: 27 September 2024
Accepted: 9 October 2024
Published: 15 October 2024

1. Introduction

Asphalt is a complex composite material, the performance of which depends strongly on the morphology of its microstructure. Considering heterogenous material properties at the micro level allows for the creation of multi-scale modeling approaches to better evaluate the composite behavior and overall pavement response as compared to traditional modeling with fully homogenized material properties. By accounting for real microstructural characteristics, such as constituent phases, sizes, shapes, orientations, and local properties, multi-scale modeling has the potential to provide more accurate predictions of mixture performance properties that emerge from these microstructural factors. Moreover, it is particularly useful for modeling pavements, which requires the simultaneous investigation of meter-scale domains along with mm-scale responses i.e., micro-fracture, coalesced cracks, dislocations, etc.

Multi-scale modeling concepts have been applied to asphalt materials and pavements [1–3]. These approaches can be divided into two types. In the first type, the constitutive behavior of an infinitesimal material point at the macro-scale is described by the homogenized response of a representative sample of the heterogeneous material structure at the micro-scale [4–6]. An important assumption of this approach is the scale separation, which is obtained if the structural dimensions are significantly larger than the dimensions of the material inhomogeneities [7–9]. To achieve this, a representative volume element (RVE) is defined, such that it statistically represents the entire composite. This RVE should contain sufficient inclusions to ensure that the overall moduli are effectively independent of the surface values of traction and displacement [10–12]. Thus, the RVE should be large enough when compared to the characteristic dimensions of the microstructure (e.g., particle sizes) but small enough when compared to the entire body. In the second type, the assumption of scale separation is not required. This can be addressed using multi-scale modeling, which iteratively solves the micro- and macro-problems, informing one another through a feedback loop [13,14].

Multi-scale modeling of granular composite materials, such as asphalt pavements, offers an avenue to gain a much-needed fundamental understanding of their behavior at different length scales. This understanding is pivotal for optimizing pavement design, assessing performance, and defining maintenance strategies, as shown in a recent review by Wang W. and Wang L. [15]. The scope of the review was focused on developing a multi-scale modeling framework involving the use of RVEs for micro-scale characterization to represent various asphalt pavement constituents.

The concept of RVE was developed in the field of materials science in the mid-20th century, with important early contributions by Hershey [16] on the elasticity of anisotropic cubic crystals. While Hershey did not address asphalt materials specifically, the principal approach of RVE is becoming popular in various material fields, including those dealing with asphalt pavement, to study distribution of stresses and other mechanical properties (observed macroscopically) in microstructural behaviors [16,17]. Building upon these works, more recent studies conducted micromechanical analyses on RVEs and derived Young's modulus, revealing the potential of utilizing crystal properties to estimate asphalt behavior [18–20].

Many studies have investigated the microstructural behavior of asphalt materials to determine the appropriate size of the RVE. These studies used various experimental and numerical techniques to study asphalt materials. For instance, by using geometrical analysis and FE simulations, a study by Kim et al. [21] verified that a size of 50 mm is effective as an RVE for dense-graded Superpave mixtures, but much larger scales are found to be required in the case of stone matrix asphalt. A further study by Kim et al. [22] employed a purely statistical-numerical approach based on volume fractions (not grain size) for multigrain features like aggregate area fraction, gradation, orientation, and spatial distribution, and identified RVE sizes of approximately 60 mm. Additionally, Kanit et al. [23] and Pelissou et al. [24] addressed the representative size of the RVE and the uncertainty in estimating effective properties. They also suggested a statistical theory to calculate the minimum RVE size and they found that an RVE of 20 mm $\times$ 20 mm $\times$ 20 mm was able to represent asphalt mixture behavior.

With values of 50 mm for Superpave and 60 mm RVE sizes for stone matrix asphalt mixtures, several studies use experimental testing in combination with numerical approaches, such as FEM or digital image correlation (DIC), to estimate the suitable RVE size [25–28]. Further research by Marasteanu et al., Liu et al., Szydłowski et al., Fadil et al., and Li et al. used a number of modeling approaches, such as computational microstructural modeling and fracture simulations, to refine these estimates [29–34].

Despite many improvements in this field, multi-scale modeling of asphalt pavement and its integration with experimental observations still has major research gaps. For instance, there is no standard RVE size estimation technique that could be used for all the different types of asphalt mixtures. The current literature has provided varying recommendations, suggesting a lack of consensus and the need for more comprehensive studies to establish uniform guidelines. Furthermore, multi-scale modeling provided an attractive solution in simulating asphalt behavior, but efforts that integrate these models with real field data are limited. Additionally, it imposes considerable challenges for dealing with computational complexity. Research into more efficient computational algorithms and the use of advanced computational resources is essential to overcome these problems. Finally, multi-scale modeling has not yet been adequately employed in advancing sustainable pavement design. Understanding how these models can facilitate a greater incorporation of recycled materials and other sustainable practices during pavement design could offer significant environmental benefits.

To address the growing challenges in asphalt pavement design, this study focuses on developing a multi-scale modeling approach to predict damage evolution in heterogeneous asphaltic mixtures. The main objective of this study is to further develop the understanding of how micro-structural features such as particle distribution and binder characteristics control mechanical behavior under realistic loading conditions, as shown in Figure 1.

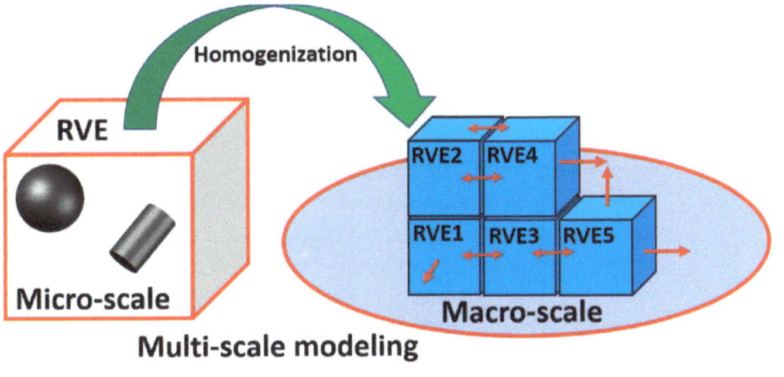

Figure 1. Scheme of multi-scale modeling through RVE based approach.

Building on the literature findings, this study proposes a multi-scale modelling framework for asphalt pavements using RVEs and particle size distributions based on the American Association of State Highway and Transportation Officials (AASHTO) gradation chart. The proposed framework uses RVEs of optimal sizes to create a full-scale IDEAL-CT model that considers viscoelastic mastic behavior and fracture evolution via cohesive zone modeling, for evaluating various asphalt pavement behaviors and mechanical properties under varying loading rates.

One key contribution of this study is the introduction of an approach to determine the RVE size for heterogeneous asphalt mixtures by considering different inclusion shapes (spherical and cylindrical). Our results indicate that the optimal RVE size depends on the inclusion shape, which has not been extensively explored in other studies [4,35–42]. This highlights the increase in sensitivity of microstructural variability, such as inclusion shape, in the modeling of asphalt mixture behavior. A second contribution of this study is the creation of an RVE-based modeling framework capable of reproducing the mechanical behavior of heterogeneous asphalt mixtures under various loading rates. The predictive ability of the proposed framework is demonstrated by comparisons with experimental data obtained from full-scale IDEAL-CT specimens. This represents a critical advance in multiscale modeling of heterogeneous asphalt mixtures as it validates the use of RVEs for evaluating their rate-dependent fracture behavior, a challenge hardly addressed using traditional homogenized models. This work is, however, limited to the scope of determining the optimal RVE sizes based on fixed values for the Young's modulus of the aggregates and mastic phases. Consequently, the applicability of these optimal RVE sizes may be constrained for different types of aggregates and mastic, or at different temperatures. The latter is an important issue as the apparent elastic properties of the mastic are highly temperature dependent. Further research will aim to address this limitation by exploring the impact of variations in material properties and environmental conditions on the optimal RVE sizes.

2. Background

Asphalt pavement consists of aggregate particles coated with a bituminous binder mastic. With such a complex microstructure, homogenization techniques based on RVEs have become an interesting and valuable modeling approach to determine and evaluate the effective properties of asphalt mixtures. In recent years, most of the research studies were focused on the viscoelastic properties of asphalt mixtures, fracture mechanics, and damage evolution models. Some studies took into consideration the effects of binder content, aggregate gradation, and temperature. However, most of the proposed models in past studies lacked detailed simulations at the microstructural level, and such modeling will be in the focus of the present study. Analytically, early models included the rule of mixtures and its variations, including the modified rule of mixtures, along with Halpin Tsai equations for the foundation of estimating elastic properties of composite materials [43–46]. However, such classical laminate theories experience accuracy limitations, especially when attempting to predict out-of-plane properties. Although the classical theories do provide useful numerical results, they ignore constituent interactions.

Models like the Chamis model, elasticity approach model, self-consistent model, and Mori–Tanaka model are noted for their better prediction capability [39,47–49]. The Chamis model has been established as one of the best micromechanical models for determining all five independent elastic properties. While using a similar formulation of the rule of mixtures to estimate Young's modulus E_{11}, which represents the Young's modulus in the 1-direction along the primary material axis and indicating the material's stiffness in that direction, and Poisson's ratio ν_{12}, which describes the ratio of the transverse strain to axial strain when stress is applied in the 1-direction, this model uses different methods for predicting other elastic moduli such as E_{22}, which is the Young's modulus in the 2-direction transverse to the primary axis, representing the stiffness perpendicular to the 1-direction, and G_{12} and G_{23}, which represnt shear modulus in the 1–2 plane and 2–3 plane, respectively, representing the material's resistance to shear deformation between the 1, 2, and 3 directions. The Mori–Tanaka model is a self-consistent method that assumes that each inclusion is embedded in an equivalent particle with the average strain of the composite. This method yields closed-form analytical solutions for computing the five independent elastic moduli of a composite with spherical- or ellipsoidal-shaped inclusions. However, the use of numerical procedures is needed for the composite elastic constants with non-spherical inclusions.

With advancements in computational power, the finite element method (FEM) has become one of the most powerful numerical modeling tools used to model RVEs. FEM facilitates the study of damage mechanisms in composites and complex microstructures [50]. However, generating accurate RVE models is difficult because correctly representing the fine microstructural details of heterogeneous materials requires large databases and computational times. This involves modeling the inter-phase and phase-constituent interactions within the composite (i.e., statistical representativeness, multi-scale material response, etc.). In addition, the computational cost of high-resolution simulations and the challenge to validate models using exactly observed experimental data further complicates this process. While numerous algorithms have been proposed to generate particulate RVE models, these are limited primarily to simple inclusion shapes and low volume fractions [51–53]. Computational tools, such as ANSYS Material Designer [54], could be a possible solution to these problems and could facilitate the generation of RVEs tailored for heterogeneous asphalt mixtures. In this study, ANSYS Material Designer is utilized to create a variety of RVEs for homogenized elastic property calculation and for identifying those optimal RVEs that can accurately model heterogeneous asphalt mixtures.

3. RVE Modeling and Selection

The development of accurate multi-scale models for asphalt pavement modeling requires an appropriate selection of RVE size while considering adequate aggregate (inclusion) shapes and distribution. In the remainder of this section, we discuss our approach to determining the optimal size of the RVE while considering various types of aggregate shapes with size distributions adhering to AASHTO gradation charts. As will be discussed in detail later, the RVE geometry, as well as the effective elastic properties of the asphalt mixture, will be obtained using ANSYS Material Designer.

3.1. RVE Modeling in ANSYS

Two crucial components for accurately predicting the mechanical behavior of asphalt materials within a multi-scale modeling framework are the determination of effective mechanical properties and the selection of a suitable RVE [28,55–57].

ANSYS (2022 R2) commercial finite element software was used in this study for modeling RVEs and for simulating fracture behavior in asphalt mixtures. To find the effective mechanical properties of the RVEs, this study employs the ANSYS Material Designer tool. This tool is employed because it simplifies the creation of many types of RVE models and automates the calculation of effective mechanical properties based on aggregate size, distribution, and the material properties of each phase in the RVE. These effective properties could be directly used in macro-scale simulations, or the actual RVEs could be used in heterogeneous modeling. It not only offers several predefined RVE templates, but also allows modeling custom-made RVEs to suit specific needs in an adaptive and efficient way.

While ANSYS is typically used for the macro-mechanical analysis of structures, our work further extends its application by incorporating microstructural features into the RVEs and cohesive zone models (CZM) for modeling of laboratory-scale fracture tests (see Section 4). This provides an approach to not only capture the microscale interactions of aggregates and mastic but also the macroscale performance of asphalt mixture. The use of ANSYS also allowed us to model the viscoelastic behavior of the mastic phase, the most important feature of asphalt mixtures, without requiring additional software development to handle viscoelasticity. Furthermore, certain features of ANSYS, such as the implementation of the exponential cohesive zone model, allowed us to run more realistic fracture simulations. This cohesive law is one of the most widely used laws in modeling fractures in asphalt and other quasi-brittle materials such as concrete, and proved effective in capturing the crack propagation behavior in our asphalt mixture samples [54,58–60].

As discussed later, the numerical modeling of typical fracture tests, such as the IDEAL-CT (Indirect Tensile Asphalt Cracking Test), which is widely used within the asphalt pavement community, is performed based on RVEs of optimal sizes obtained in this section. Although ANSYS provides tools for RVE-based modeling, its limitations in full-scale pavement modeling must be acknowledged. For instance, a fully heterogeneous model of an entire pavement section would require the explicit modeling of thousands or even millions of aggregate particles, which is computationally prohibitive. Instead, this study focuses on determining optimal RVE sizes and their applications in laboratory-scale fracture tests. Although out of the scope of the present study, results from such tests could guide the development of equivalent homogeneous fracture models for future full-scale applications, eliminating the need for explicit aggregate modeling.

ANSYS Material Designer determines the effective material properties of the RVE via homogenization. To do so, the software performs six finite element analyses: tensile tests in the X, Y, and Z directions and shear tests in the XY, YZ, and ZX planes. Figure 2 depicts two representative load cases used to obtain the effective mechanical properties in ANSYS Material Designer. For each test case, a macroscopic unit strain is applied, and the corresponding reaction forces on the boundary faces of the RVE are computed. For each of these analyses, the software imposes periodic boundary conditions on all faces of the RVE. Periodic boundary conditions are typically employed in homogenization approaches to simulate infinite media and mitigate boundary effects [61–63]. The material stiffness matrix is then constructed using these reaction forces, from which engineering constants can be derived. The material stiffness matrix D for an orthotropic material is expressed by the below matrix, in which σ_x, σ_y, and σ_z are the normal stresses in the x, y, and z axes, respectively, and σ_{xy}, σ_{yz}, and σ_{zx} the shear stresses in the respective planes. The terms ε_x, ε_y, and ε_z are the normal strains, while γ_{xy}, γ_{yz}, and γ_{zx} are the respective shear strains. This equation is obtained from the principles of elasticity theory, applying to orthotropic materials, and it describes how the stress components relate to the applied strain components under various loadings [64,65].

$$\begin{bmatrix} \sigma_x \\ \sigma_y \\ \sigma_z \\ \sigma_{xy} \\ \sigma_{yz} \\ \sigma_{zx} \end{bmatrix} = \begin{bmatrix} D_{11} & D_{12} & D_{13} & 0 & 0 & 0 \\ D_{21} & D_{22} & D_{23} & 0 & 0 & 0 \\ D_{31} & D_{32} & D_{33} & 0 & 0 & 0 \\ 0 & 0 & 0 & D_{44} & 0 & 0 \\ 0 & 0 & 0 & 0 & D_{55} & 0 \\ 0 & 0 & 0 & 0 & 0 & D_{66} \end{bmatrix} \begin{bmatrix} \varepsilon_x \\ \varepsilon_y \\ \varepsilon_z \\ \gamma_{xy} \\ \gamma_{yz} \\ \gamma_{zx} \end{bmatrix}$$

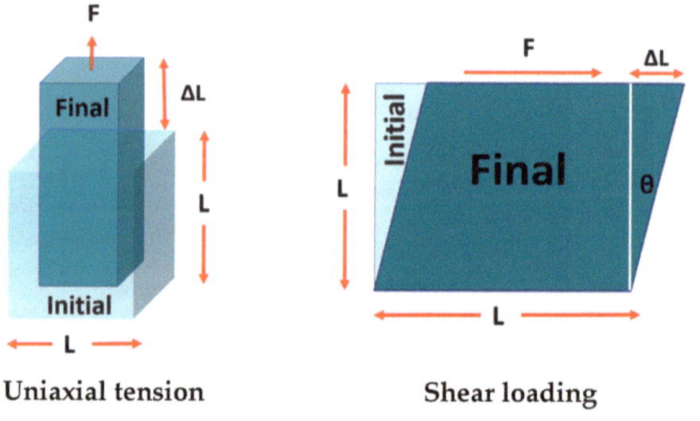

Figure 2. Mechanical characterization of periodic unit cell: load in extension-compression (**left**) and loading in shear (**right**).

Furthermore, the model setup includes full details of the displacement fields at the RVE boundaries to ensure that rigid body motions are restricted, and the periodicity requirements are met. These specifications are critical as they prevent non-physical deformations and maintain the integrity of the homogenization process, thereby providing a more reliable and accurate prediction of the material properties in practical engineering applications [54,55,59,60,66].

3.2. RVE Types

As shown in Figure 3, this study considers two types of RVEs. The first considers polydisperse spherical inclusions (Figure 3a), whereas the second considers polydisperse truncated cylindrical inclusions (Figure 3b). The first RVE was considered to capture the effect of different aggregate sizes within the matrix, whereas the second was considered to account for specific material features such as the angularity of aggregate particles. The evaluation of these two types of RVE provided valuable insights concerning the influence of aggregate shape on effective elastic mechanical properties on asphalt pavement materials.

ANSYS Material Designer requires the user to specify the elastic properties of particles and matrix to obtain the effective mechanical properties of the composite. While this work focuses primarily on the mechanical behavior under the specific loading conditions mentioned, future research will incorporate temperature-dependent properties to simulate asphalt performance in varying thermal conditions. For instance, properties such as thermal expansion coefficient, heat conductivity, and temperature-dependent elastic moduli for the different material phases could be included to account for thermal effects in the RVE model. A model incorporating these thermal properties would allow for analyzing the influence of temperature variations on the optimal RVE size needed for modeling heterogeneous asphalt mixtures. For the purposes of this work, the material properties for both particles (aggregates) and matrix (mastic) are adopted from the literature and are shown in Table 1.

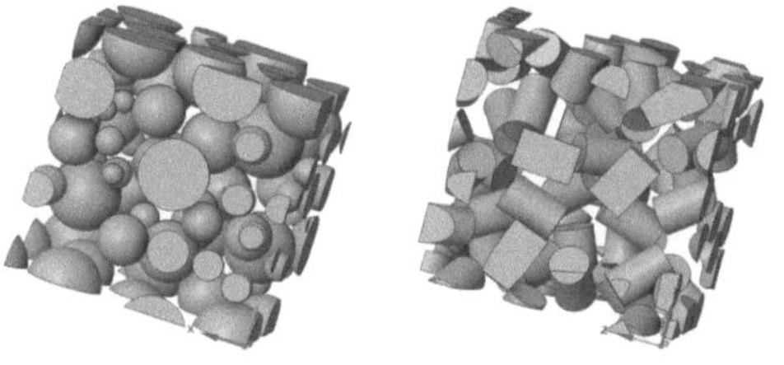

(**a**) (**b**)

Figure 3. Types of inclusions considered in this study: (**a**) Polydisperse spherical particles and (**b**) truncated cylindrical particles.

Table 1. Elastic material properties of the homogenized RVE components.

Material	Density (kg/m^3)	Young's Modulus (GPa)	Poisson's Ratio
Aggregate	2300	40.50	0.2
Mastic	1050	2.025	0.3

Material properties for the aggregates were selected based on typical values of Young's modulus (40.5 GPa), and Poisson's ratio (0.2). For limestone aggregates commonly used in asphalt mixtures, the modulus might vary between 33 and 48 GPa depending on the source material and its composition. In previous studies, the aggregate modulus for asphalt concrete mixtures has been assumed to be approximately 40 GPa. This value agrees well with our chosen value of 40.5 GPa, selected to maintain consistency with the standard industrial practices during the modeling of the elastic behavior of the aggregates in asphalt [67–70].

This work does not aim to perform a sensitivity analysis of the variation on the influence of Young's modulus and Poisson's ratio of aggregates on the optimum size of RVEs. Instead, the effort is focused on using typical values of elastic properties for the different material phases (i.e., aggregates and mastic) and investigating the effect of aggregate shape on mechanical behavior when both aggregate shapes meet the target gradation curves commonly used for asphalt pavements design. The selected elastic properties values have, therefore, been chosen under realistic conditions so that the microstructural interactions of the asphalt mixture can be effectively modeled and analyzed.

The predefined parameters for generating the RVEs containing each of the two inclusion types included volume fraction, initial size, and diameter of particles [55,56]. Next, within the cubical modeled volume, Material Designer randomly places non-overlapping particles to create the RVEs. This process of randomly inserting particles continues until either the volume fraction (i.e., filling level), percolation limit (jamming), or any other primary requirement is satisfied [71,72]. It is important to note that, in all our models, the inclusions were fully contained within the boundaries of the cube. Also, both the matrix and the inclusions were meshed using 10-node tetrahedral elements [58,59,73].

3.3. Particle Gradation

The size distribution of aggregates plays a crucial role in the performance and engineering properties of asphalt mixtures. Therefore, an appropriate RVE model should accurately reflect the gradations of particles found in real asphalt mixtures. In this context, for the two types of RVEs considered in this study, the percentage of particles present for each sieve size is specified, following a NMAS of 12.5 mm (see Figure 4).

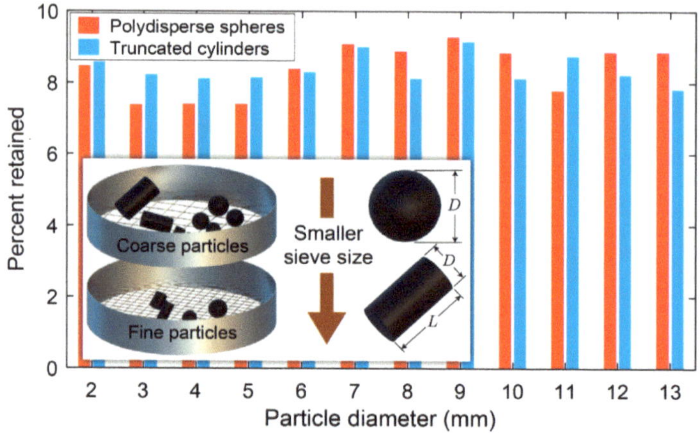

Figure 4. Presentation of each inclusion size in the two selected RVEs.

The particle distribution shown in Figure 4 is used to generate gradation curves for each RVE model. These curves were created by first calculating the percentage of particles within the RVE that fall into the 1 mm size range. Particles smaller than 2.36 mm were grouped and treated as homogenized mastic for simplicity, while discrete particles in the range of 2.36–12.5 mm were considered individually [67,74]. Then, by multiplying the percent passing value for each increment by its size, summing these values throughout the range, and dividing by the overall percent passing for the range, the weighted average particle size within each size range was determined.

As depicted in Figure 5, the gradation curve obtained for each RVE is then compared against the 0.45 power maximum density gradation curve per the 12.5 mm size chart specified in AASHTO M323 [75,76]. This 0.45 power curve represents the optimal gradation for dense-graded asphalt mixes. This gradation curve was plotted based on the 0.45 power law, which is generally applied for asphalt mixture designs to achieve optimum packing of aggregates. The law calculates the percentage of aggregate passing through each sieve size so that a well-graded asphalt mixture may be achieved. The X-axis of the plot shows the particle sizes in mm, while the y-axis shows the percentage passing through the respective sieves. The plot has included both the RVE gradation curve and the target 0.45 power gradation curve for comparison. The gradation curve showed a good agreement with the 0.45 power target curve line across all particle sizes for polydisperse spherical inclusion RVEs. Significantly, the truncated cylindrical inclusion RVE was also well-aligned with the 0.45 power gradation curve and had weighted averages matching specified ranges. As the gradation curves of both spherical and cylindrical inclusion RVE agreed well with the target gradation curve, it can be concluded that these two groups of models are appropriate for the representation of real asphalt mixture morphologies. Air voids were not explicitly considered in this study but will be addressed in follow-up research. Air voids are known to reduce the stiffness and strength of asphalt paving materials, and their impact will be an important factor to be considered in subsequent studies [77–79].

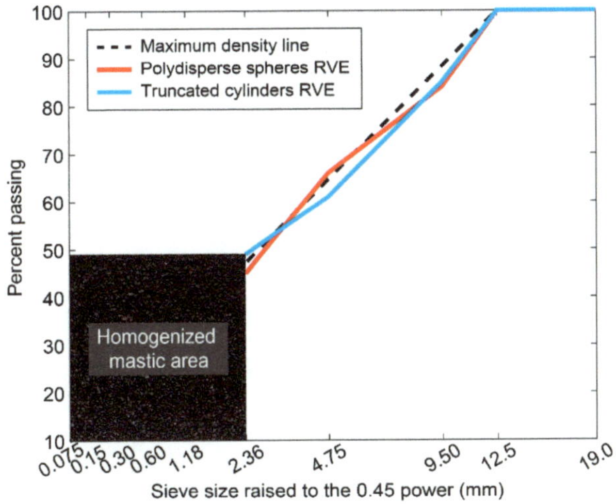

Figure 5. Gradation of RVE particle distribution based on 0.45 power gradation chart.

3.4. Optimal RVE Size Selection

To obtain effective elastic properties of heterogeneous asphalt materials using a homogenization approach, the RVE size should be sufficiently large relative to the aggregate size yet small relative to the macroscopic dimensions of the pavement structure. We generated a series of RVE models with side lengths ranging from 24 to 36 mm, in increments of 2 mm, aligning with the NMAS used in our gradation chart. Moreover, for each RVE size, we considered four volume fractions (0.4, 0.5, 0.6, and 0.7).

The optimal RVE size for each type was determined iteratively by increasing their size until the computed effective elastic properties converged, while choosing appropriate aggregate volume fractions that ensure the converged properties align with expected values from the literature. Effective properties for asphalt, typically within 5–7 GPa for E, 1.5–2 GPa for G, and 0.30–0.35 for ν at a standard reference temperature of 25 °C [80–83], guided our selection. This method allowed us to identify the minimum RVE size necessary to achieve statistically homogeneous and stable material property estimates that are close to the expected values discussed above.

Figure 6 shows the convergence analysis of the homogenized elastic properties for spherical and cylindrical RVEs, where the particle distributions are based on the 0.45 power gradation chart. Based on these results and the selection criteria discussed above, the recommended optimal RVE configurations are a 0.5 volume fraction with 32 mm size for the polydisperse spherical RVE and 0.6 volume fraction with 34 mm size for the truncated cylindrical RVE.

For the polydisperse spherical RVE with a 0.5 volume fraction and a 32 mm size, the converged effective properties are $E = 5.90$ GPa (Young's modulus), $G = 1.82$ GPa (shear modulus), and $\nu = 0.35$ (Poisson's ratio), all of which are within the ranges discussed above. For the truncated cylindrical RVE with 0.6 volume fraction and a 34 mm size, the converged properties are $E = 6.11$ GPa, $G = 1.93$ GPa, and $\nu = 0.30$, which are also within the expected ranges. The convergence analysis of homogenized elastic properties shows that the RVE size predictions chosen, 32 mm for spherical inclusions and 34 mm for cylindrical inclusions, can ensure the obtainment of reliable mechanical performances, laying in the expected range given in the literature [25–28]. The good accuracy of RVE-based models is also ensured by the satisfactory match between the load-displacement curves obtained from output of numerical simulations with those obtained from experimental test results [67] and shown in the following section.

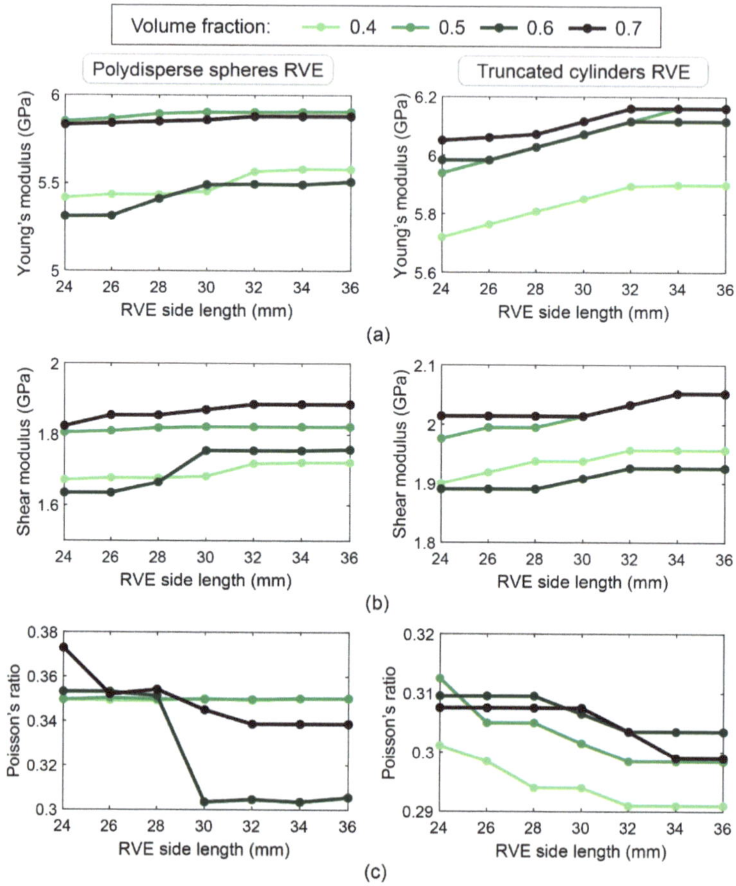

Figure 6. Engineering constants for the two types of RVEs vs. RVE side length for various aggregate volume fractions: (**a**) Young's modulus, (**b**) Shear modulus, and (**c**) Poisson's ratio.

4. Full-Scale Models

The optimal RVEs discussed in the previous section were used to create two high-fidelity heterogenous models for an IDEAL-CT. The IDEAL-CT is conducted according to the industry standards, specifically ASTM D8225 [84]. This test evaluates the cracking potential of asphalt mixtures under Mode-I tensile stresses, which simulate the tensile conditions experienced in the field.

The first model considers the optimal RVE with polydisperse spherical inclusions, while the second one considers the optimal RVE with truncated cylindrical inclusions. Unlike the earlier RVE models, which assume linear elastic material properties, these full-scale models incorporate viscoelastic properties for the mastic phase and fracture behavior defined through a cohesive zone model.

Appropriate viscoelastic and cohesive zone properties are defined for the mastic and interfaces, respectively, based on a previous study [67]. Additionally, a mesh convergence analysis is conducted to determine the optimal mesh size. The results from the full-scale models are then compared against experimental results reported in [67]. This validation step involves comparing the load-displacement responses obtained from the models with those obtained experimentally for a range of loading rates. In the remainder of this section, details of the full-scale models and the validated results are provided.

4.1. Model Geometry

The IDEAL-CT test sample consists of a cylindrical specimen with a diameter of 150 mm and a thickness of 62 mm. Both full-scale models are created using ANSYS workbench. Each model is created by replicating their respective optimal RVE in the x, y, and z directions to fill the volume of the test sample (refer to Figure 7a–c). Given that the optimal RVE size is different when considering polydisperse spherical inclusions or truncated cylindrical inclusions, the number of RVEs occupying each model domain is different for each RVE type. Once the RVEs are replicated, all elements outside the initial cylindrical domain are removed from the model. Finally, for computational efficiency, only half of the IDEAL-CT domains were modeled (refer to Figure 7d).

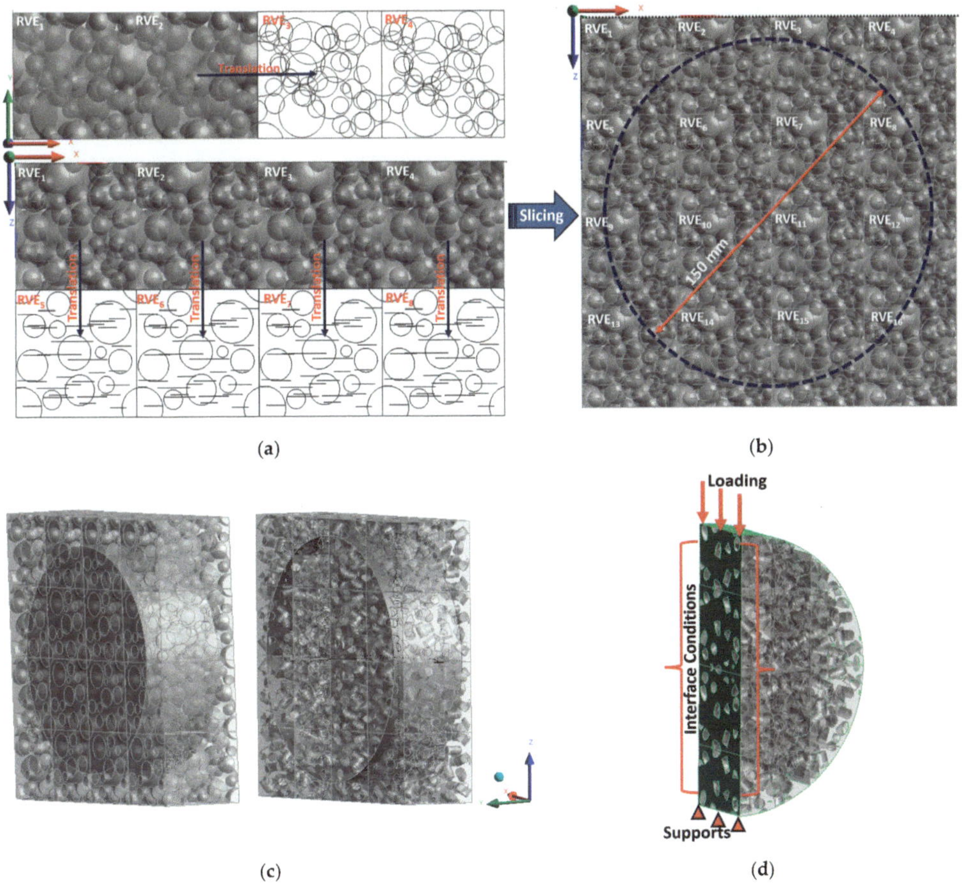

Figure 7. Creation of IDEAL CT full-scale model: (**a**) replicating unit RVE to extend model domain beyond 150 mm × 150 mm × 62 mm, (**b**) slicing cylindrical specimen from extended model domain, (**c**) generated full scale models, and (**d**) applying boundary conditions.

Once the domains are generated, cohesive elements are inserted at the mid-plane of the specimens (see Figure 7d) to simulate crack propagation throughout the entire specimen. This was then implemented in the full-scale model validation (see Figure 8) to depict the level of damage on the symmetry plane.

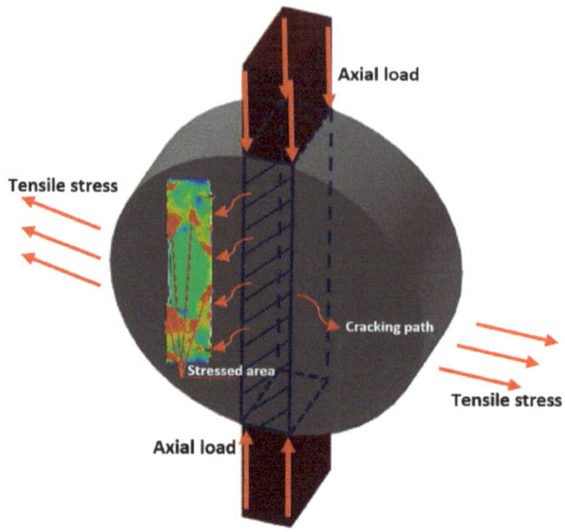

Figure 8. Schematic illustration of a full-scale model depicting the level of damage on the symmetry plane.

4.2. Material Modeling

For the full-scale models, the aggregates are modeled as elastic inclusions with a Young's modulus of 40,500 MPa and Poisson's ratio equal to 0.2 [67]. The mastic phase, which is comprised of the asphalt binder, filler content, and fine aggregates, is characterized as a homogeneous viscoelastic material with an 11-term Prony series representation of the relaxation modulus listed in Table 2. Exponential CZMs were used to model the interfaces between aggregate and mastic to simulate fracture propagation at the symmetry plane of the IDEAL-CT specimen.

Table 2. Prony series terms for binder relaxation modulus [67].

Term i	Relaxation Modulus E_i (MPa)	Relaxation Time τ_i (s)
∞	57.36	
1	23.28	1.098×10^6
2	34.56	2.307×10^4
3	64.32	1.056×10^3
4	162.5	1.440×10^2
5	246.0	13
6	6082	1.05
7	4488	0.15
8	5947	0.016
9	6744	0.00029
0	5700	0.000026

The values of relaxation modulus and relaxation time used in this study were obtained from data reported by Vasconcelos De Souza et al. [67] in their work on heterogeneous viscoelastic asphaltic mixtures. In their study, to obtain the Prony series terms of the binder's relaxation modulus, frequency sweep tests were carried out in shear assuming a constant Poisson's ratio of 0.30. The tests were carried out for a large range of relaxation times over the binder time-dependent viscoelastic behavior. This was used in the modeling of the viscoelastic behavior of asphalt mastic within our numerical simulations so that the relaxation properties exhibited by the binder are correctly captured for analysis.

The cohesive zone approach introduces interface elements that describe the progressive separation of fracturing surfaces through a cohesive traction-separation law. The exponential cohesive zone model (CZM) is advantageous for modeling fracture in asphalt materials exhibiting a gradual softening and damage accumulation ahead of the crack tip [41,67,85]. The exponential CZM defined in ANSYS considers a potential function of the form [86]:

$$\phi(\delta_n, \delta_t) = e\sigma_{n_{\max}}\overline{\delta}_n\left[1 - (1 + \Delta_n)e^{-\Delta_n}e^{-\Delta_t^2}\right], \tag{1}$$

where $\overline{\delta}_n$ is the characteristic normal separation corresponding to the maximum normal traction, $\sigma_{n_{\max}}$, and $\Delta_n = \delta_n/\overline{\delta}_n$ and $\Delta_t = \delta_t/\overline{\delta}_t$ are, respectively, the normalized normal and tangential crack opening displacements, which are expressed in terms of the normal separation, δ_n, and tangential separation, δ_t. The normal and tangential cohesive tractions are obtained from the potential function in Equation (1). Specifically, the normal traction is given by Equation (2)

$$T_n = e\sigma_{n_{\max}}\Delta_n e^{-\Delta_n}e^{-\Delta_t^2}, \tag{2}$$

where $\sigma_{n_{\max}}$ is the maximum normal traction, and the shear traction is given by Equation (3)

$$T_t = 2e\sigma_{t_{max}}\frac{\overline{\delta}_n}{\overline{\delta}_t}\Delta_t(1 + \Delta_n)e^{-\Delta_n}e^{-\Delta_t^2}, \tag{3}$$

where $\sigma_{t_{\max}}$ is the maximum tangential traction. The fracture energy in mode I is expressed in Equation (4)

$$\phi_n = e\sigma_{n_{\max}}\overline{\delta}_n, \tag{4}$$

and the fracture energy in mode II is expressed as presented in Equation (5)

$$\phi_t = \sqrt{e/2}\,\sigma_{t_{\max}}\overline{\delta}_t, \tag{5}$$

For both full-scale models, we considered $\sigma_{n_{\max}} = \sigma_{t_{\max}} = 1.32$ MPa and $\phi_n = \phi_t = 960.7\ \mathrm{J/m^2}$. These values were selected based on reasonable tensile and normal stress values found in the literature for asphalt mixtures subjected to tensile loading. This fracture energy value reflects the energy required to propagate a crack per unit area in the modeled asphalt mixture [87–89].

As outlined in our current study, the next phase of research will involve the application of homogenization techniques to translate the mesoscale behaviors observed in our Representative Volume Element (RVE) models into effective macroscale thermomechanical properties. This transition is crucial for accurately simulating the behavior of asphalt pavements under real-world conditions, where both thermal and mechanical loads play critical roles.

4.3. Mesh Convergence Analysis

To ensure the accuracy of our numerical solutions, we conducted a mesh convergence analysis for both full-scale models. As shown in Figure 9, we assessed convergence of the numerical results by monitoring the predicted peak load and maximum displacement as a function of the average element size. This analysis was conducted for both full-scale models by monitoring the predicted peak load and maximum displacement as a function of the element size. For each response parameter, the value from the coarsest mesh (3.5 mm element size) was set as the baseline, and we gradually refined the mesh to assess the variations in the response parameters. Coarse, medium, and fine convergence criteria of 10–15% and 2–3% normalized variation for peak load and maximum displacement from the baseline, respectively, were established based on recommendations in the literature [90–93]. Initial coarse meshes showed a more than 12% difference in peak load between the 3.5 mm and 2.5 mm elements, reducing below 8% when refined to 1.5 mm elements. Further refinement to 1.25 mm, 1 mm, and 0.5 mm resulted in less than 3% variation, indicating convergence as per the criteria. The optimized mesh size of 1.25 mm was then implemented in the full-scale model validation.

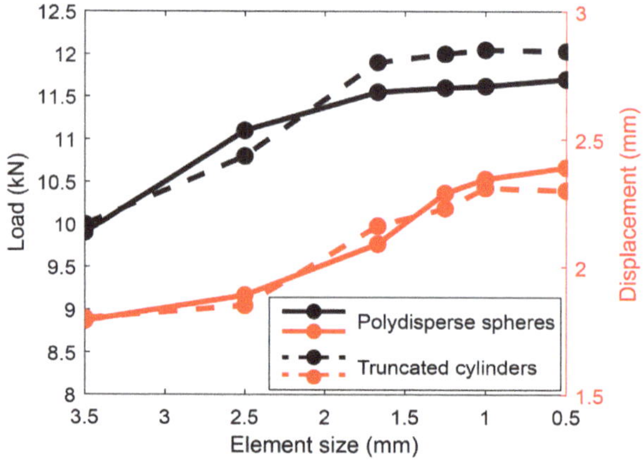

Figure 9. Element size in relation to peak load and maximum displacement.

4.4. Validation and Experimental Considerations

This section deals with the validation of the full-scale models presented previously and provides numerical results. These numerical results were validated by comparing them with experimental results provided in Vasconcelos de Souza et al. [67] and Soares et al. [68]. Proprietary testing was carried out in laboratory-controlled conditions on asphalt mixtures with a nominal maximum aggregate size of 12.5 mm at various displacement rates, 0.8 mm/s, 1.6 mm/s, and 3.2 mm/s, recording the corresponding load-displacement curves. These experimental data provided a direct comparison to the numerical simulation results and ensured that the rate-dependent fracture behavior of the asphalt mixture was captured by the model. Figure 10 presents stress and deformation contours extracted from the finite element models. These results provide the areas of high stress and can be used to gain an understanding of how cracks initiate and propagate. As shown by the results in Figure 10a–c, the load-displacement responses predicted by both full-scale models (i.e., the one using polydisperse spherical inclusions and the one using truncated cylindrical inclusion) are in reasonable agreement with the experimental results. In Figure 10d, the contour plots illustrate the total deformation (in mm) at the peak load for each of the displacement rates. The contour plots clearly depict the areas of large deformation, particularly around the midplane of the specimens, which become more pronounced as the displacement rate decreases.

In addition to qualitatively comparing the load-displacement curves at different loading rates, we provide a quantitative assessment of the predictive capabilities of each full-scale model. To do so, we computed the root mean squared error (RMSE) for each model and for each loading rate. The computed RMSE values for each model are shown in Figure 11. As shown in the figure, for the spherical inclusions model, the RMSE values generally increased as the loading rate increased, demonstrating the reduced accuracy at higher loading rates. This trend is also observed for the truncated cylinder model, though the RMSEs were consistently lower in these models, indicating an improved fit to the experimental results.

The results discussed above indicate that the full-scale models, built using the optimal RVEs obtained in Section 3, are adequate to predict the behavior of the IDEAL-CT tests under various displacement rates. However, it is important to note that the current models with the assumed RVE sizes may be inadequate to model the behavior of IDEAL-CT tests at different temperatures. This is because the optimal RVE size may be different for different temperatures since the Young's modulus of the mastic is temperature dependent. This observation points to the need for further studies investigating the optimal RVE size for heterogeneous asphalt mixtures at different temperatures, which can enable the creation of reliable full-scale models for heterogeneous asphalt modeling. Besides the possibility of temperature-dependent RVE sizes, heterogeneous fracture modeling at various temperatures may require more sophisticated cohesive zone models that can capture rate- and temperature-dependent fracture behavior. Our previous work has investigated rate-dependent cohesive-zone models [94,95] and can be used as a basis for future modeling efforts in this area.

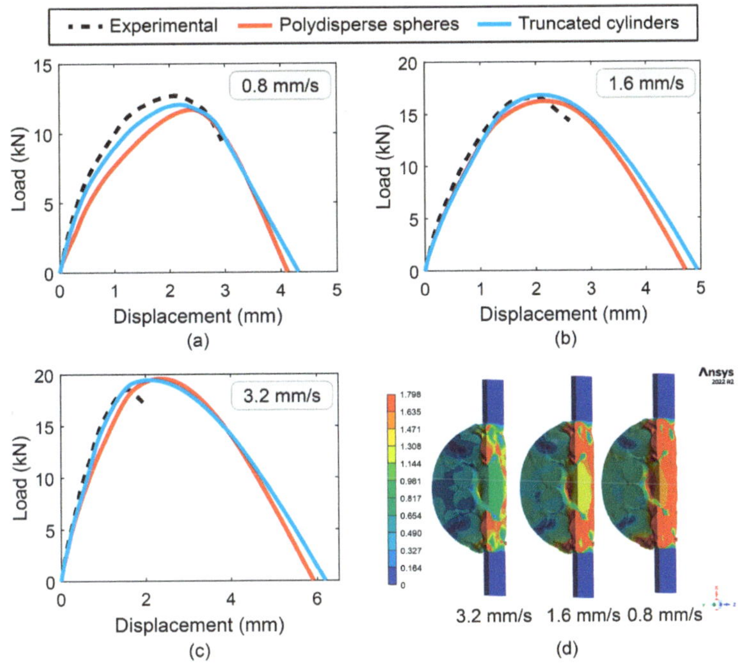

Figure 10. Numerical simulation vs. experiment for load-displacement relation at different loading rates: (**a**) 0.8 mm/s, (**b**) 1.6 mm/s, (**c**) 3.2 mm/s, and (**d**) total deformation (in mm) at the peak load for all loading rates.

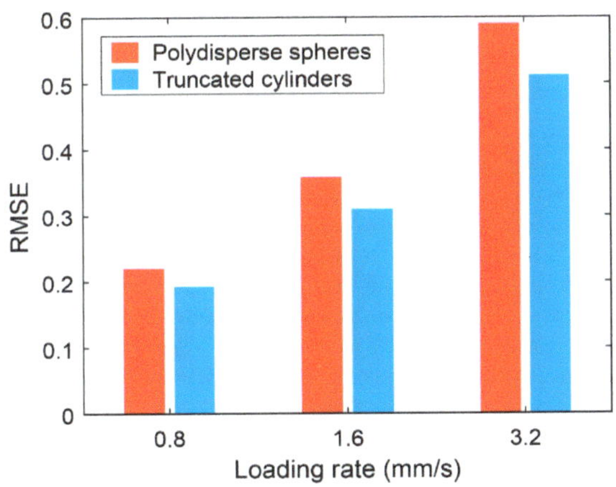

Figure 11. Root mean squared error comparison between the two proposed models vs. the applied loading rates.

5. Conclusions

The main objective of this study was to develop a multi-scale modeling framework for characterizing microstructure-performance relationships in asphalt pavement materials. An RVE-based approach was adopted by combining micromechanical homogenization and full-scale modeling. At the micro-scale, optimal RVE sizes were identified through a size-convergence analysis. Convergence analysis was assessed by considering different inclusion shapes and volume fractions. The first type of RVE considered polydisperse spherical inclusions whereas the second type considered truncated cylindrical inclusions. The inclusion size distribution for each RVE type was selected to agree with target gradation curves from AASHTO. The results from the size convergence analysis showed that the minimum RVE size for spherical inclusions was 32 mm, whereas that for truncated cylindrical inclusions, the size was 34 mm. The converged RVE sizes were obtained by analyzing the convergence of the effective elastic properties of the asphalt composite.

Upon determining the optimal sizes of the two types of RVEs, we created two full-scale, heterogeneous models for an IDEAL-CT test. The first model considered polydisperse spherical inclusions whereas the second considered truncated cylindrical inclusions. The heterogeneous models were created by replicating the optimal RVEs in the three special dimensions and performing a Boolean operator whereas elements outside the IDEAL-CT specimen geometry were removed. Unlike the RVE size convergence study, the full-scale models considered material nonlinearities and fracture. The models consisted of aggregates (either spherical or cylindrical), mastic, and interfaces. The aggregates were modeled as linear elastic materials whereas the mastic was modeled by using a viscoelastic material model. To simulate fracture evolution, we inserted cohesive elements at the interface between aggregates and mastic and at the mid-plane of the specimen geometry, where cracks are expected to propagate. The numerical results from the full-scale models were compared against experimental results obtained from the literature, providing a means of validating our multi-scale modeling approach.

While the current model focused on the mechanical behavior of asphalt mixtures without considering variations in chemical composition, future work will explore how changes in binder type, aggregate content, and additives may influence the viscoelastic properties. Similarly, variations in aggregate gradation can impact the load distribution and damage evolution within the material.

The key findings from the study are listed as follows:

1. From the sensitivity analyses, it was concluded that polydisperse spherical and truncated cylindrical RVEs are valid representations of asphalt mixture microstructure at minimum sizes of 32 mm and 34 mm, which fall within the typical range of 20 mm to 60 mm found in the literature.
2. Full-scale, heterogeneous IDEAL-CT finite element models were developed that accurately represented specimen geometry, material phasing, and test characteristics for relevant loading rate effects.
3. Simulations of experimental monotonic fracture tests confirmed the effectiveness of truncating cylindrical inclusion modeling, based on the proposed RVE-based multi-scale approach, to adequately capture the overall behavior, as compared with polydisperse spherical inclusion modeling.
4. The cylindrical inclusion model was found to be slightly more computationally efficient over the spherical inclusion model, which was based on the computed RMSE in terms of predicted vs. modeled load-deflection curves.
5. The proposed multi-scale modeling framework could link microstructure-performance and allow for virtual characterization through full-scale analyses, establishing a foundation for prediction and evaluation of full-scale composite system designs.
6. Potential future work could involve including temperature and moisture effects along with considering the addition of recycled materials as a third phase for better understanding of the effect of varying environmental conditions on durability and fracture behavior of asphalt mixtures.

Author Contributions: Conceptualization, O.G.-L., B.D., Z.C. and W.G.B.; Methodology, A.I.H.M., Z.C., P.R. and W.G.B.; Software, A.I.H.M.; Validation, A.I.H.M.; Formal analysis, W.G.B.; Resources, B.D.; Data curation, O.G.-L., P.R. and W.G.B.; Writing—original draft, A.I.H.M.; Writing—review & editing, A.I.H.M., O.G.-L., Z.C., P.R. and W.G.B.; Visualization, A.I.H.M.; Supervision, O.G.-L., B.D., Z.C. and W.G.B. All authors have read and agreed to the published version of the manuscript.

Funding: The authors are grateful for the financial support provided by the Department of Civil and Environmental Engineering at the University of Missouri. O.G-L., W.G.B., and P.R. are grateful for the endowments provided by the James W. and Joan M. O'Neill Faculty Scholar in Engineering and the Glen Barton Faculty Chair in Flexible Pavement Systems at the University of Missouri.

Institutional Review Board Statement: Not applicable.

Informed Consent Statement: Not applicable.

Data Availability Statement: The original contributions presented in the study are included in the article, further inquiries can be directed to the corresponding author.

Conflicts of Interest: The authors declare no conflicts of interest. The findings and opinions expressed in this study solely represent those of the authors and do not necessarily reflect the views of sponsors or sponsoring agencies.

References

1. Cucalon, L.G.; Rahmani, E.; Little, D.N.; Allen, D.H. A multiscale model for predicting the viscoelastic properties of asphalt concrete. *Mech. Time Depend. Mater.* **2016**, *20*, 325–342. [CrossRef]
2. Souza, F.V.; Allen, D.H. Multiscale modeling of impact on heterogeneous viscoelastic solids containing evolving microcracks. *Int. J. Numer. Methods Eng.* **2009**, *82*, 464–504. [CrossRef]
3. Teixeira, J.E.S.L.; Kim, Y.-R.; Souza, F.V.; Allen, D.H.; Little, D.N. Multiscale model for asphalt mixtures subjected to cracking and viscoelastic deformation. *Transp. Res. Rec. J. Transp. Res. Board* **2014**, *2447*, 136–145. [CrossRef]
4. Nie, F.; Chow, C.L.; Lau, D. A Review on Multiscale Modeling of Asphalt: Development and Applications. *Multiscale Sci. Eng.* **2022**, *4*, 10–27. [CrossRef]
5. Sun, Y.; Zhang, Z.; Gong, H.; Zhou, C.; Chen, J.; Huang, B. 3D Multiscale Modeling of Asphalt Pavement Responses under Coupled Temperature–Stress Fields. *J. Eng. Mech.* **2022**, *148*, 04022010. [CrossRef]
6. Tan, Z.; Yang, B.; Leng, Z.; Jelagin, D.; Cao, P.; Li, R.; Zou, F. Multiscale characterization and modeling of aggregate contact effects on asphalt concrete's tension–compression asymmetry. *Mater. Des.* **2023**, *232*, 112092. [CrossRef]
7. Nemat-Nasser, S.; Lori, M.; Datta, S.K. Micromechanics: Overall Properties of Heterogeneous Materials. *J. Appl. Mech.* **1996**, *63*, 561. [CrossRef]
8. Zaoui, A. Continuum Micromechanics: Survey. *J. Eng. Mech.* **2002**, *128*, 808–816. [CrossRef]
9. Bhattacharya, P.; Li, Q.; Lacroix, D.; Kadirkamanathan, V.; Viceconti, M. A systematic approach to the scale separation problem in the development of multiscale models. *PLoS ONE* **2021**, *16*, e0251297. [CrossRef]
10. Hill, R. Elastic properties of reinforced solids: Some theoretical principles. *J. Mech. Phys. Solids* **1963**, *11*, 357–372. [CrossRef]
11. Benseddiq, N.; Belayachi, N.; Abdelaziz, M.N. Multiscale approach to the behaviour and damage of the heterogeneous elastic–viscoplastic materials. *Theor. Appl. Fract. Mech.* **2006**, *46*, 15–25. [CrossRef]
12. Key, C.T.; Lua, J. Constituent based analysis of composite materials subjected to fire conditions. *Compos. Part A Appl. Sci. Manuf.* **2006**, *37*, 1005–1014. [CrossRef]
13. Geers, M.; Kouznetsova, V.; Brekelmans, W. Multi-scale computational homogenization: Trends and challenges. *J. Comput. Appl. Math.* **2010**, *234*, 2175–2182. [CrossRef]
14. Kanit, T.; Forest, S.; Galliet, I.; Mounoury, V.; Jeulin, D. Determination of the size of the representative volume element for random composites: Statistical and numerical approach. *Int. J. Solids Struct.* **2003**, *40*, 3647–3679. [CrossRef]
15. Wang, W.; Wang, L. Review on Design, Characterization, and Prediction of Performance for Asphalt Materials and Asphalt Pavement Using Multi-Scale Numerical Simulation. *Materials* **2024**, *17*, 778. [CrossRef]
16. Hershey, A.V. The Elasticity of an Isotropic Aggregate of Anisotropic Cubic Crystals. *J. Appl. Mech.* **1954**, *21*, 236–240. [CrossRef]
17. Qayyum, F.; Chaudhry, A.A.; Guk, S.; Schmidtchen, M.; Kawalla, R.; Prahl, U. Effect of 3D representative volume element (RVE) thickness on stress and strain partitioning in crystal plasticity simulations of multi-phase materials. *Crystals* **2020**, *10*, 944. [CrossRef]
18. Huang, G.; Zhang, J.; Hui, B.; Zhang, H.; Guan, Y.; Guo, F.; Li, Y.; He, Y.; Wang, D. Analysis of Modulus Properties of High-Modulus Asphalt Mixture and Its New Evaluation Index of Rutting Resistance. *Sustainability* **2023**, *15*, 7574. [CrossRef]
19. Ferretti, P.; Santi, G.M.; Leon-Cardenas, C.; Fusari, E.; Donnici, G.; Frizziero, L. Representative volume element (RVE) analysis for mechanical characterization of fused deposition modeled components. *Polymers* **2021**, *13*, 3555. [CrossRef]
20. Wei, X.; Sun, Y.; Gong, H.; Li, Y.; Chen, J. 3D Mesomechanical Simulation–Based Approach to Determining Representative Volume Element of Asphalt Concrete. *J. Transp. Eng. Part B Pavements* **2024**, *150*, 04024001. [CrossRef]

21. Kim, Y.-R.; Lutif, J.E.S.; Allen, D.H. Determining representative volume elements of asphalt concrete mixtures without damage. *Transp. Res. Rec. J. Transp. Res. Board* **2009**, *2127*, 52–59. [CrossRef]
22. Kim, Y.; Lee, J.; Lutif, J.E.S. Geometrical evaluation and experimental verification to determine representative volume elements of heterogeneous asphalt mixtures. *J. Test. Eval.* **2010**, *38*, 660–666. [CrossRef]
23. El Moumen, A.; Kanit, T.; Imad, A. Numerical evaluation of the representative volume element for random composites. *Eur. J. Mech.-A/Solids* **2021**, *86*, 104181. [CrossRef]
24. Pelissou, C.; Baccou, J.; Monerie, Y.; Perales, F. Determination of the size of the representative volume element for random quasi-brittle composites. *Int. J. Solids Struct.* **2009**, *46*, 2842–2855. [CrossRef]
25. Romero, P.; Masad, E. Relationship between the Representative Volume Element and Mechanical Properties of Asphalt Concrete. *J. Mater. Civ. Eng.* **2001**, *13*, 77–84. [CrossRef]
26. Coleri, E.; Harvey, J.T. Analysis of Representative Volume Element for Asphalt Concrete Laboratory Shear Testing. *J. Mater. Civ. Eng.* **2011**, *23*, 1642–1653. [CrossRef]
27. Casillas, S.E.; Morales, A.M.; Braham, A.F. Evaluating aggregate size effects on dynamic modulus using gray scale analysis. In *Asphalt Paving Technology: Association of Asphalt Paving Technologists-Proceedings of the Technical Sessions*; DEStech Publications, Inc.: Lancaster, PA, USA, 2019.
28. Despringre, N.; Chemisky, Y.; Bonnay, K.; Meraghni, F. Micromechanical modeling of damage and load transfer in particulate composites with partially debonded interface. *Compos. Struct.* **2016**, *155*, 77–88. [CrossRef]
29. Marasteanu, M.; Falchetto, A.C.; Velasquez, R.; Le, J.-L. On the representative volume element of asphalt concrete at low temperature. *Mech. Time Depend. Mater.* **2016**, *20*, 343–366. [CrossRef]
30. Liu, P.; Hu, J.; Wang, D.; Oeser, M.; Alber, S.; Ressel, W.; Falla, G.C. Modelling and evaluation of aggregate morphology on asphalt compression behavior. *Constr. Build. Mater.* **2017**, *133*, 196–208. [CrossRef]
31. Szydłowski, C.; Górski, J.; Stienss, M.; Smakosz, Ł. Numerical simulation of asphalt mixtures fracture using continuum models. *AIP Conf. Proc.* **2018**, *1922*, 130012. [CrossRef]
32. Fadil, H.; Jelagin, D.; Larsson, P.-L.; Partl, M.N. Measurement of the viscoelastic properties of asphalt mortar and its components with indentation tests. *Road Mater. Pavement Des.* **2019**, *20*, S797–S811. [CrossRef]
33. Li, H.; Luo, X.; Zhang, Y. Predicting fatigue crack growth rate of cement-based and asphalt paving materials based on indirect tensile cyclic loading tests. *Eng. Fract. Mech.* **2024**, *303*, 110146. [CrossRef]
34. Li, H.; Luo, X.; Gu, Z.; Chen, Q.; Zhang, Y. Predicting crack growth of paving materials under indirect tensile fatigue loads. *Int. J. Fatigue* **2023**, *175*, 107818. [CrossRef]
35. Ghuzlan, K.A.; Al-Khateeb, G.G.; Sharo, A.A.; Shwaikeh, M.E. Viscoelastic-based finite element rutting predictive models for asphalt pavements. *Emergent Mater.* **2023**, *6*, 337–354. [CrossRef]
36. Masad, E.; Scarpas, A.; Rajagopal, K.R.; Kassem, E.; Koneru, S.; Kasbergen, C. Finite element modelling of field compaction of hot mix asphalt. Part II: Applications. *Int. J. Pavement Eng.* **2013**, *17*, 24–38. [CrossRef]
37. Kim, M.; Tutumluer, E.; Kwon, J. Nonlinear Pavement Foundation Modeling for Three-Dimensional Finite-Element Analysis of Flexible Pavements. *Int. J. Géoméch.* **2009**, *9*, 195–208. [CrossRef]
38. Eghbalpoor, R.; Baghani, M.; Shahsavari, H. An implicit finite element framework considering damage and healing effects with application to cyclic moving load on asphalt pavement. *Appl. Math. Model.* **2019**, *70*, 139–151. [CrossRef]
39. Luo, Y. An Accuracy Comparison of Micromechanics Models of Particulate Composites against Microstructure-Free Finite Element Modeling. *Materials* **2022**, *15*, 4021. [CrossRef]
40. Luo, Y. Microstructure-Free Finite Element Modeling for Elasticity Characterization and Design of Fine-Particulate Composites. *J. Compos. Sci.* **2022**, *6*, 35. [CrossRef]
41. Chen, J.; Ouyang, X.; Sun, X. Numerical Investigation of Asphalt Concrete Fracture Based on Heterogeneous Structure and Cohesive Zone Model. *Appl. Sci.* **2022**, *12*, 11150. [CrossRef]
42. Leng, Z.; Tan, Z.; Cao, P.; Zhang, Y. An efficient model for predicting the dynamic performance of fine aggregate matrix. *Comput. Civ. Infrastruct. Eng.* **2021**, *36*, 1467–1479. [CrossRef]
43. Affdl, J.C.H.; Kardos, J.L. The Halpin-Tsai equations: A review. *Polym. Eng. Sci.* **1976**, *16*, 344–352. [CrossRef]
44. Saeb, S.; Steinmann, P.; Javili, A. Aspects of computational homogenization at finite deformations: A unifying review from Reuss' to Voigt's bound. *Appl. Mech. Rev.* **2016**, *68*, 050801. [CrossRef]
45. Reuss, A. Berechnung der Fließgrenze von Mischkristallen auf Grund der Plastizitätsbedingung für Einkristalle. *Zamm-J. Appl. Math. Mech./Z. Für Angew. Math. Und Mech.* **1929**, *9*, 49–58. [CrossRef]
46. Ghafaar, M.A.; Mazen, A.A.; El-Mahallawy, N.A. Application of the rule of mixtures and Halpin-Tsai equations to woven fabric reinforced epoxy composites. *JES J. Eng. Sci.* **2006**, *34*, 227–236. [CrossRef]
47. Hashin, Z.; Rosen, B.W. The elastic moduli of fiber-reinforced materials. *J. Appl. Mech.* **1964**, *31*, 223–232. [CrossRef]
48. Michel, J.; Moulinec, H.; Suquet, P. Effective properties of composite materials with periodic microstructure: A computational approach. *Comput. Methods Appl. Mech. Eng.* **1999**, *172*, 109–143. [CrossRef]
49. Younes, R.; Hallal, A.; Fardoun, F.; Hajj, F. Comparative Review Study on Elastic Properties Modeling for Unidirectional Composite Materials. In *Composites and Their Properties*; IntechOpen: London, UK, 2012. [CrossRef]
50. Otero, F.; Oller, S.; Martinez, X.; Salomón, O. Numerical homogenization for composite materials analysis. Comparison with other micro mechanical formulations. *Compos. Struct.* **2015**, *122*, 405–416. [CrossRef]

51. Sangani, A.S.; Mo, G. Elastic interactions in particulate composites with perfect as well as imperfect interfaces. *J. Mech. Phys. Solids* **1997**, *45*, 2001–2031. [CrossRef]
52. Segurado, J.; Llorca, J. A numerical approximation to the elastic properties of sphere-reinforced composites. *J. Mech. Phys. Solids* **2002**, *50*, 2107–2121. [CrossRef]
53. Tanemura, M. Random packing and random tessellation in relation to the dimension of space. *J. Microsc.* **1988**, *151*, 247–255. [CrossRef]
54. Ansys Inc. *Corporate Philanthropy Report 38*; Wiley: Hoboken, NJ, USA, 2023; Volume 38. [CrossRef]
55. Srivastava, V.K.; Gabbert, U.; Berger, H. Representative volume element analysis for the evaluation of effective material properties of fiber and particle loaded composites with different shaped inclusions. In *Mechanics of Time-Dependent Materials and Processes in Conventional and Multifunctional Materials*; Conference Proceedings of the Society for Experimental Mechanics Series; Springer: New York, NY, USA, 2011. [CrossRef]
56. Berger, H.; Kari, S.; Gabbert, U.; Rodriguez-Ramos, R.; Bravo-Castillero, J.; Guinovart-Diaz, R. Calculation of effective coefficients for piezoelectric fiber composites based on a general numerical homogenization technique. *Compos. Struct.* **2005**, *71*, 397–400. [CrossRef]
57. Berger, H.; Kari, S.; Gabbert, U.; Rodriguez-Ramos, R.; Bravo-Castillero, J.; Guinovart-Diaz, R.; Sabina, F.J.; Maugin, G.A. Unit cell models of piezoelectric fiber composites for numerical and analytical calculation of effective properties. *Smart Mater. Struct.* **2006**, *15*, 451–458. [CrossRef]
58. Mali, R.; Pise, U. Modeling elastic properties of biocomposites using various analytical models and ansys material designer. *Mater. Today Proc.* **2023**, *72*, 1372–1378. [CrossRef]
59. Bin Rayhan, S.; Rahman, M. Modeling elastic properties of unidirectional composite materials using ansys material designer. *Procedia Struct. Integr.* **2020**, *28*, 1892–1900. [CrossRef]
60. ANSYS Inc. *Material Designer User's Guide*; ANSYS: Canonsburg, PA, USA, 2023.
61. Ye, Y.; Cai, Y.Z. Laboratory investigate of the effect of fine aggregates on asphalt mixture materials. *Adv. Mater. Res.* **2011**, *225*, 577–580. [CrossRef]
62. Zhu, Z.R.; Ye, Y. Performance investigate of fine aggregate size and content on asphalt mastic. *Adv. Mater. Res.* **2012**, *424*, 7–10. [CrossRef]
63. Ye, Y.; Chen, H.K.; Cai, Y.Z. The effect of coarse aggregate gradation degradation on the mechanical behavior of asphalt mixture. *Adv. Mater. Res.* **2013**, *811*, 223–227. [CrossRef]
64. Maceri, A. *Theory of Elasticity*; Springer Nature: Dordrecht, The Netherlands, 2010. [CrossRef]
65. Mercier, J.P.; Zambelli, G.; Kurz, W. Elastic behaviour of solids. In *Introduction to Materials Science*; Elsevier: Amsterdam, The Netherlands, 2002; pp. 121–150. [CrossRef]
66. ANSYS Inc. *ANSYS Workbench User's Guide*; ANSYS: Canonsburg, PA, USA, 2023; Volume 15317, pp. 724–746.
67. de Souza, F.V.; Soares, J.B.; Allen, D.H.; Evangelista, F. Model for predicting damage evolution in heterogeneous viscoelastic asphaltic mixtures. *Transp. Res. Rec. J. Transp. Res. Board* **2004**, *1891*, 131–139. [CrossRef]
68. Soares, J.B.; Allen, D.H. Crack modeling of asphaltic mixtures considering heterogeneity of the material. In Proceedings of the TRB 2003 Annual Meeting, Washington, DC, USA, 12–16 January 2003.
69. Allen, D.; Jones, R.; Boyd, J. Micromechanical analysis of a continuous fiber metal matrix composite including the effects of matrix viscoplasticity and evolving damage. *J. Mech. Phys. Solids* **1994**, *42*, 505–529. [CrossRef]
70. Soares, J.B.; de Freitas, F.A.C.; Allen, D.H. Considering Material Heterogeneity in Crack Modeling of Asphaltic Mixtures. *Transp. Res. Rec. J. Transp. Res. Board* **2003**, *1832*, 113–120. [CrossRef]
71. Schneider, M.; Josien, M.; Otto, F. Representative volume elements for matrix-inclusion composites—A computational study on the effects of an improper treatment of particles intersecting the boundary and the benefits of periodizing the ensemble. *J. Mech. Phys. Solids* **2022**, *158*, 104652. [CrossRef]
72. Meakin, P.; Jullien, R. Random sequential adsorption of spheres of different sizes. *Phys. A Stat. Mech. Its Appl.* **1992**, *187*, 475–488. [CrossRef]
73. Hershey, A.V.; Cooper, D.W.; Gitman, I.M.; Askes, H.; Sluys, L.J.; Valls, O.L.; Pelissou, C.; Baccou, J.; Monerie, Y.; Perales, F.; et al. Numerical evaluation of the representative volume element for random composites. *Constr. Build Mater* **2021**, *20*, 75–86. [CrossRef]
74. Kim, H.; Buttlar, W.G. Discrete fracture modeling of asphalt concrete. *Int. J. Solids Struct.* **2009**, *46*, 2593–2604. [CrossRef]
75. *AASHTO MP19-10*; Standard Specification for Performance-Graded Asphalt Binder Using Multiple Stress Creep Recovery (Mscr) Test. AASHTO: Washington, DC, USA, 2010.
76. *AASHTO M 332-14*; Standard Specification for Performance-Graded Asphalt Binder Using Multiple Stress Creep Recovery (MSCR) Test. AASHTO: Washington, DC, USA, 2018.
77. Kim, M.; Buttlar, W.G. Stiffening Mechanisms of Asphalt–Aggregate Mixtures. *Transp. Res. Rec. J. Transp. Res. Board* **2010**, *2181*, 98–108. [CrossRef]
78. Kim, M.; Buttlar, W.G. Differential Scheme Effective Medium Theory for Hot-Mix Asphalt $|E^*|$ Prediction. *J. Mater. Civ. Eng.* **2011**, *23*, 69–78. [CrossRef]
79. Kim, M. *Development of Differential Scheme Micromechanics Modeling Framework for Predictions of Hot-Mix Asphalt (HMA) Complex Modulus and Experimental Validations*; University of Illinois at Urbana-Champaign: Champaign, IL, USA, 2009.

80. Maher, A.; Bennert, T. *Evaluation of Poisson's Ratio for Use in the Mechanistic Empirical Pavement Design Guide*; Rutgerds: New Brunswick, NJ, USA, 2008.
81. Brown, S.F. Determination of young's modulus for bituminous materials in pavement design. In Proceedings of the 52nd Annual Meeting of the Highway Research Board, Washington, DC, USA, 22–26 January 1973; pp. 38–49.
82. Wang, X.; Feng, J.; Wang, H. Stress Analysis of Asphalt Concrete Deck Pavement on Steel Bridge based on Burgers Model and interlayer contact. In Proceedings of the 2017 6th International Conference on Energy and Environmental Protection (ICEEP 2017), Shanghai, China, 23–24 April 2017. [CrossRef]
83. Cai, X.; Chen, L.; Zhang, R.; Huang, W.; Wu, K.; Ye, X. Estimation of shear modulus of asphalt mixture based on the shear strength of the aggregate interface. *Constr. Build. Mater.* **2020**, *248*, 118695. [CrossRef]
84. *ASTM D8225-19*; Determination of Cracking Tolerance Index of Asphalt Mixture Using the Indirect Tensile Cracking Test at Intermediate Temperature. ASTM: West Conshohocken, PA, USA, 2019.
85. Park, K.; Paulino, G.H. Cohesive zone models: A critical review of traction-separation relationships across fracture surfaces. *Appl. Mech. Rev.* **2011**, *64*, 060802. [CrossRef]
86. Cerrone, A.; Wawrzynek, P.; Nonn, A.; Paulino, G.H.; Ingraffea, A. Implementation and verification of the Park–Paulino–Roesler cohesive zone model in 3D. *Eng. Fract. Mech.* **2014**, *120*, 26–42. [CrossRef]
87. Shi, X.; Mirsayar, M.; Mukhopadhyay, A.; Zollinger, D. Characterization of two-parameter fracture properties of portland cement concrete containing reclaimed asphalt pavement aggregates by semicircular bending specimens. *Cem. Concr. Compos.* **2019**, *95*, 56–69. [CrossRef]
88. Tekalur, S.A.; Shukla, A.; Sadd, M.; Lee, K.W. Mechanical characterization of a bituminous mix under quasi-static and high-strain rate loading. *Constr. Build. Mater.* **2008**, *23*, 1795–1802. [CrossRef]
89. Amirdehi, H.R.F.; Ghoreishi, S.M.N.; Aliha, M.R.M.; Hesami, S.; Haghighatpour, P.J.; Choupani, N. On comparison of fracture energy measured for six hot mix asphalt composite mixtures using force–load line displacement and force–crack mouth opening displacement curves at intermediate temperatures and different loading rates. *Fatigue Fract. Eng. Mater. Struct.* **2024**, *47*, 3858–3873. [CrossRef]
90. Zienkiewicz, O.C.; Zhu, J.Z. The superconvergent patch recovery and *a posteriori* error estimates. Part 1: The recovery technique. *Int. J. Numer. Methods Eng.* **1992**, *33*, 1331–1364. [CrossRef]
91. Wiberg, N.; Li, X.D. Superconvergent patch recovery of finite-element solution and a posteriori L_2 norm error estimate. *Commun. Numer. Methods Eng.* **1994**, *10*, 313–320. [CrossRef]
92. Patil, H.; Jeyakarthikeyan, P.V. Mesh convergence study and estimation of discretization error of hub in clutch disc with integration of ANSYS. *IOP Conf. Ser. Mater. Sci. Eng.* **2018**, *402*, 012065. [CrossRef]
93. Alfano, G.; Crisfield, M.A. Finite element interface models for the delamination analysis of laminated composites: Mechanical and computational issues. *Int. J. Numer. Methods Eng.* **2001**, *50*, 1701–1736. [CrossRef]
94. Giraldo-Londoño, O.; Spring, D.W.; Paulino, G.H.; Buttlar, W.G. An efficient mixed-mode rate-dependent cohesive fracture model using sigmoidal functions. *Eng. Fract. Mech.* **2018**, *192*, 307–327. [CrossRef]
95. Giraldo-Londoño, O.; Paulino, G.H.; Buttlar, W.G. Fractional calculus derivation of a rate-dependent PPR-based cohesive fracture model: Theory, implementation, and numerical results. *Int. J. Fract.* **2019**, *216*, 1–29. [CrossRef]

materials

MDPI

Article

Experimental Research on the Anti-Reflection Crack Performance of Basalt Fiber Modified Rubber Asphalt Stress-Absorbing Layer

Cheng Shen [1], Zhengguang Wu [1,2], Peng Xiao [1,2,*], Aihong Kang [1,2] and Yangbo Wang [1]

1 College of Civil Science and Engineering, Yangzhou University, Yangzhou 225127, China; shenchengyzu@163.com (C.S.); zgwu@yzu.edu.cn (Z.W.); kahyzu@163.com (A.K.); wybyzu@163.com (Y.W.)
2 Research Center for Basalt Fiber Composite Construction Materials, Yangzhou 225127, China
* Correspondence: pengxiao@yzu.edu.cn

Abstract: Reflection cracks are one of the most common problems in semi-rigid base pavement. Setting a stress absorption layer can effectively delay the occurrence of reflection cracks, but further improvement is still needed in its interlayer bonding performance and anti-reflection crack performance. Considering the excellent crack resistance of basalt fibers and the good elastic recovery ability of rubber asphalt, it is considered worthwhile to incorporate them into traditional stress absorption layers to improve performance. To simulate the actual pavement layer effect, composite specimens consisting of a cement-stabilized macadam base + basalt fiber rubber asphalt stress-absorbing layer + AC-20 asphalt mixture surface layer were prepared to evaluate their performance through interlayer direct shear tests, interlayer tensile tests, three-point bending tests, and overlay tests (OTs). To determine the optimal fiber blending combination, four fiber lengths (3 cm, 6 cm, 9 cm, 12 cm) and four fiber proportions (120 g/m^2, 140 g/m^2, 160 g/m^2, 180 g/m^2) were selected respectively. The specific effects of basalt fibers with different lengths and dosages were analyzed. The results show that compared with the absence of fibers, the improvement of interlayer bonding performance of rubber asphalt with basalt fibers is not significant, and it has certain limitations; however, the improvement of anti-reflective crack performance is significant, with an increase of up to 305.5%. This indicates that the network structure formed by basalt fibers and rubber asphalt stress absorption layer can effectively absorb and disperse external loads, causing an excellent crack resistance effect. Meanwhile, the results indicate that the main factor affecting its interlayer bonding strength and anti-reflective crack performance is the fiber content. Based on the comprehensive analysis of the performance and economy of the stress absorption layer of basalt fiber rubber asphalt, the optimal fiber parameter combination recommended is as fiber length 9 cm and fiber content 160 g/m^2. These results can provide a reference for the design and performance evaluation of basalt fiber rubber asphalt stress absorption layer, and have certain application value.

Keywords: basalt fibers; rubber asphalt stress-absorbing layer; interlayer bonding performance; anti-reflection crack performance

Citation: Shen, C.; Wu, Z.; Xiao, P.; Kang, A.; Wang, Y. Experimental Research on the Anti-Reflection Crack Performance of Basalt Fiber Modified Rubber Asphalt Stress-Absorbing Layer. *Materials* **2024**, *17*, 2013. https://doi.org/10.3390/ma17092013

Academic Editor: Simon Hesp

Received: 19 March 2024
Revised: 21 April 2024
Accepted: 23 April 2024
Published: 25 April 2024

1. Introduction

Most Chinese high-grade highways use semi-rigid base asphalt pavement [1,2], which is extensively used because of its high overall stiffness, high compressive strength, and good load diffusion capacity, but at the same time, its lack of crack resistance cannot be ignored [3]. Generally, the semi-rigid base is prone to dry shrinkage and temperature shrinkage cracks, which easily transfer upward and expand to the asphalt pavement to form reflective cracks [3–5], which greatly reduce the pavement performance and durability. On the one hand, it damages the integrity and continuity of the pavement structure, reduces the overall smoothness of the pavement, and affects the service quality of the pavement; on the other hand, rainwater can continuously seep into the cracks and accumulate in the base layer. Due to the particularity of semi-rigid foundation material, rainwater cannot seep down and stay between the surface layer and the base layer. Under repeated vehicle load, the dynamic water pressure generated by it causes the base to be continuously scoured, weakening the adhesion between the surface layer and the base layer, thus reducing the stability of the pavement.

In order to prevent the early fracturing of asphalt pavement due to cracks in semi-rigid bases, scientists at home and abroad have put forward various anti-reflection crack measures [6–10], and paving modified asphalt fiber stress absorption layer [11,12] has been an effective method. Firstly, because the stress-absorbing-layer [13–15] itself had a certain anti-reflective crack effect. Secondly, by adding modified asphalt, the bonding performance between layers can be further improved, and the small deformation energy generated by the base layer can be elastically restored. The addition of fibers [16–18] can produce a significant crack resistance effect. The specific reason was because the stress-absorbing layer [19,20] had high stress-absorbing performance, which can prevent the reflection cracking of the semi-rigid base from cracking upwards. It also possessed excellent waterproof performance, preventing rainwater from infiltrating along the crack and damaging the base and good bonding performance, increasing the bonding performance of base and surface course and improving the integrity and bearing capacity. According to the anti-reflection crack mechanism of the stress-absorbing layer, it is proposed that the stress-absorbing layer must have the functional requirements of high elasticity, low temperature flexibility, water damage resistance, and interlayer bonding [21,22]. The fibers are added to the asphalt layer of the stress-absorbing layer. When the force is transferred from the asphalt to the fibers, the deformation of the fibers needs to consume some energy. Once the asphalt is stretched and broken, the fibers will connect to the crack surface. Because the fibers has a strong tensile capacity, it can more effectively prevent or slow down the propagation of cracks [22–24].

Basalt fiber is a kind of inorganic high-performance fiber material gradually rising in popularity in recent years. Because of its excellent mechanical properties, stable material performance, and wide distribution of raw materials, it has received more and more attention in asphalt pavement construction [25–28]. At this stage, scholars at home and abroad have conducted a lot of research on the performance of freshly mixed asphalt mixture at all levels and found that adding basalt fibers into the asphalt mixture can comprehensively improve the road performance of asphalt concrete pavement, such as fatigue cracking resistance, rutting resistance, low-temperature cracking resistance, and water damage resistance [29–31], but the research on using basalt fibers in the stress absorption layer has just started.

The main purpose of this study is to develop a type of basalt fiber-modified stress-absorbing layer and subsequently investigate its interlaminar shear resistance, reflective crack resistance, etc. by interlaminar direct shear test, interlaminar pull-out test, three-point bending test, and overlay test (OT). The influence of different fiber parameters on the performance of the stress-absorbing layer is comprehensively compared, and the best basalt fiber parameter combination is selected. It is of great significance to the development of stress-absorbing layer technology and pavement overlay.

2. Raw Material Performance Test and Mix Proportion Composition Design

2.1. Raw Material Performance Test

The raw materials used in this paper include coarse and fine aggregates, mineral powder, rubber asphalt, basalt fibers, etc. According to the performance requirements of relevant test raw materials, they are tested according to the test methods specified in the test procedures. The test results of all raw materials used in the test meet the requirements of the current specifications. See Tables 1–5 for the specific performance test results.

Table 1. Aggregate test results.

Types		Apparent Relative Density	Relative Density of Gross Volume
Limestone	1# (19–13.2 mm)	2.724	2.683
	2# (13.2–2.36 mm)	2.715	2.695
Basalt	3# (4.75–0.15 mm)	2.975	2.888
	4# (0.15–0.075 mm)	2.970	2.863

Table 2. Technical indicators of coarse aggregate quality.

Test Index	Test Result	Specification Requirements	Test Method
Crushing value/%	14	≤26	T0316
Los Angeles abrasion value/%	16	≤28	T0317
Apparent relative density/g/m^2	2.883	≥2.6	T0304
Water absorption/%	0.7	≤2	T0305
Adhesion to asphalt/grade	5	≥4	T0616
Water washing method < 0.075 particle content/%	0.4	≤1	T0310

Table 3. Technical indicators of fine aggregate quality.

Test Index	Test Result	Specification Requirements	Test Method
Apparent relative density/g/m^2	2.762	≥2.5	T0330
Firmness (>0.3)/%	9	≤12	T0340
Sand equivalent/%	66	≥60	T0334

Table 4. Performance indicators and test results of rubber asphalt.

Test Index		Test Result	Specification Requirements	Test Method
Penetration (25 °C)/0.1 mm		44	30~60	T0604
Penetration index (PI)		0.8	≥0	T0604
Softening point/°C		68	≮60	T0606
Ductility (5 cm/min, 5 °C)/cm		9	≮5	T0605
Elastic recovery (25 °C)/%		86	≮60	T0662
Segregation (Softening point difference)/°C		1.8	≯3	T0661
Residue after RTFOT	Quality change/%	−0.06	±1.0	T0610
	Penetration ratio/%	76	≮60	T0604
	15 °C residual ductility/cm	28	≮10	T0605

Table 5. Performance indexes and test results of basalt fibers.

Test Index	Specification Requirements	Test Result
Elongation at break/%	≤3.1	2.71
Breaking strength/MPa	≥1200	2218
Oil absorption rate/%	≥50	52
Heat resistance, breaking strength retention/%	≥85	93
Alkali resistance, breaking strength retention/%	≥75	89

2.2. Mix Proportion Design and Composite Specimen Forming

Considering the particularity of the position of the stress absorption layer, which is located between the surface layer and the base layer, the base layer is usually a cement-stabilized crushed stone base layer, and the surface layer is usually the AC-20 mixture of the lower layer. Therefore, the overall mix design is divided into three parts: the mix design of the cement-stabilized crushed stone base layer, the AC-20 asphalt mixture, and the design of the basalt fiber rubber asphalt stress absorption layer.

2.2.1. Mix Proportion Design of Cement-Stabilized Macadam Base

The aggregate grading type of cement-stabilized macadam mixture is skeleton dense grading (C-B-3) recommended by the current specification technical rules for construction of highway pavement base (JTG/T F20-2015). The composite gradation of cement-stabilized crushed stone mixture is: 1 #:2 #:3 #:4 # =10:29:33:28, and the cement content is 5%. The composite grading curve of cement-stabilized crushed stone base is shown in Figure 1.

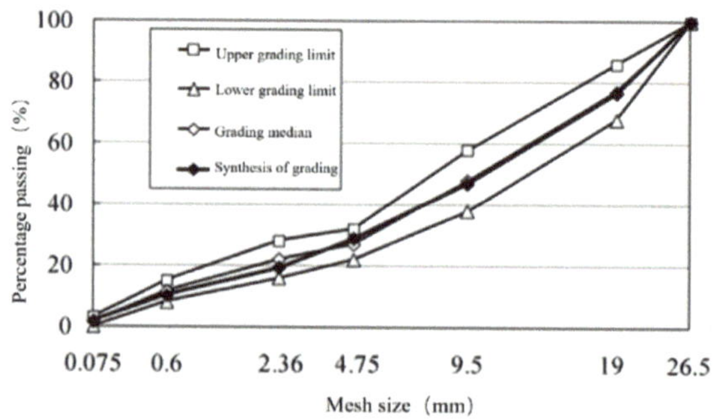

Figure 1. Grading curve of cement-stabilized macadam base.

Calculate the unconfined compressive strength of the specimen according to the formula. The strength of the cement-stabilized crushed stone base mixture meets the 7-day unconfined compressive strength standard of 4.0~6.0 MPa for cement-stabilized materials used in super heavy traffic road base of Class II and below highways, with a guarantee rate of 95%, which meets the specification requirements.

2.2.2. Mix Proportion Design of AC-20 Asphalt Mixture Surface Course

An AC-20 asphalt mixture surface layer is equipped with 4.75 mm as the key sieve, the coarse mixture with the key sieve passing rate less than 45%, and the initially proposed asphalt aggregate ratio is 4.4%. The composite grading of the mixture is 1 #:2 #:3 #:4 #:mineral powder = 23:32:13:29:3. The composite grading curve of AC-20 asphalt mixture surface layer and the upper, lower, and median curves of the grading range are shown in Figure 2.

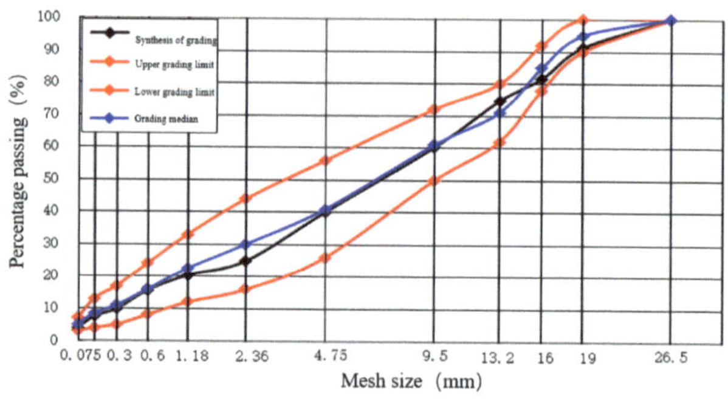

Figure 2. AC-20 asphalt mixture surface gradation curve.

At the same time, the standard Marshall test method is used to determine the best asphalt aggregate ratio of the AC-20 asphalt mixture. According to the asphalt consumption corresponding to the maximum value of Marshall stability, gross bulk density, and void fraction, the average value of asphalt consumption corresponding to the median value of asphalt saturation oac1 is 4.4%. Therefore, considering the factors of temperature and rutting, the optimum asphalt dosage (OAC) is selected as 4.4%.

2.2.3. Preparation of Basalt Fiber Rubber Asphalt Stress-Absorbing Layer Composite Board

Considering the comprehensive factors such as interlayer performance test and anti-reflection crack performance test, and because the spread gravel is a single-layer gravel stress absorption layer, the gravel particle size range is 4.75–9.5 mm, and the spread amount of gravel is 14 kg/m^2.The selection range of rubber asphalt dosage is 2.2–2.4 kg/m^2, and an average value of 2.3 kg/m^2 is selected in this paper.

According to the specific requirements of the later test on the test piece, 300 mm is uniformly used for the test × 300 mm × 100 mm large rut plate forming. According to the requirements of layered pouring and molding, due to the different height of each layer, the height shall be raised separately. Firstly, mix water, cement, and graded crushed stone according to the measured ratio, and the mixing time is two minutes. After the mixture is fully mixed and uniform, put the mixture into a mold, using the static pressure molding method to form a cement-stabilized crushed stone layer with a thickness of 25 mm. After 24 h, remove the mold and conduct surface polishing treatment and send it to a standard curing room for standard curing for 28 days.

Secondly, the basalt fiber rubber asphalt stress-absorbing layer studied is paved onto the cement-stabilized gravel layer under standard curing. After screening, the crushed stone with particle size ranging from 4.75 mm to 9.5 mm is added, with 70 # base asphalt at 0.3% of asphalt to stone ratio for premixing, so as to effectively enhance its adhesion with rubber asphalt. At the same time, the total amount of each fiber is divided into three equal parts. After the first layer of rubber asphalt is distributed, the basalt fibers is distributed in two-dimensional random direction. Generally, the preparation method of rubber asphalt, basalt fibers, rubber asphalt, and ready-mixed crushed stone from the bottom to the top is followed and finally rolled, as shown in Figure 3. In order to ensure that the climbing height of asphalt is 2/3 of the aggregate height and the overall coverage rate in the stress-absorbing layer reaches 96%, three iterations of positive and negative rolling are adopted for rolling formation.

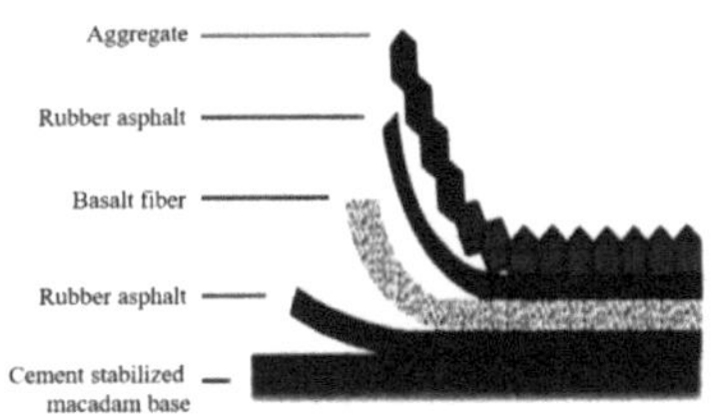

Figure 3. Paving of basalt fiber rubber asphalt stress absorption layer.

Finally, add 60 mm AC-20 asphalt mixture surface layer above the stress absorption layer, use QCX-4 produced by Jingu Shenjian Company in Beijing, China portable combined rut sample forming machine for rolling and molding, and after cooling for 24 h, conduct demolding treatment. Finally, the whole basalt fiber rubber asphalt stress absorption layer composite plate is formed, as shown in Figure 4.

Figure 4. Basalt fiber rubber asphalt stress absorption layer specimen.

3. Test Methods

The interlayer performance of the stress-absorbing layer ensures good coupling between the asphalt layer and the base, improves the service life of the asphalt pavement, and reduces the possibility of interlayer sliding and fracture. It can effectively reduce the interlaminar tensile stress and strain caused by the change of load and moving temperature, and also reduce the possibility of interlaminar slip, so that the overall structural strength of asphalt pavement is enhanced.

The anti-reflective crack performance of the stress-absorbing layer can effectively solve the problem that the semi-rigid base is prone to dry shrinkage cracks and low-temperature shrinkage cracks, and the cracks expand to the asphalt pavement to form reflective cracks under the coupling effect of traffic load and temperature load. The anti-reflection crack ability of the stress absorption layer can effectively inhibit and delay the upward transmission of this kind of reflection crack, and also has the effect of stress absorption.

In this paper, the interlaminar properties of stress absorption are tested by interlaminar direct shear test and interlaminar pull-out test. The anti-reflective crack performance of the stress-absorbing layer is tested by three-point bending test and OT. The overall test process is shown in Figure 5.

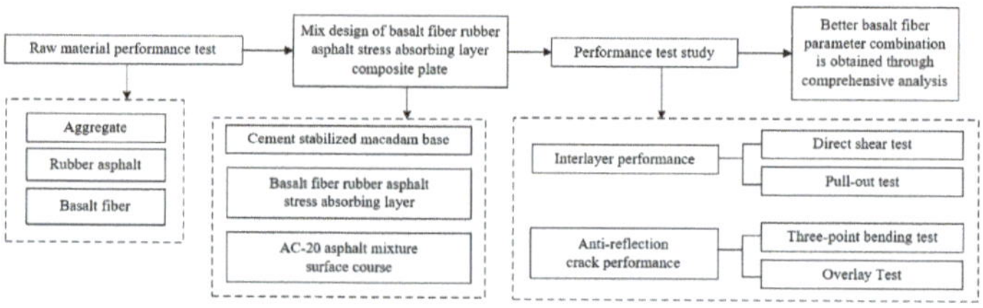

Figure 5. Overall test flow chart.

3.1. *Raw Material Performance Test*

The maximum shear stress that a bonding material can bear between layers is called the interlaminar shear strength, which is usually used to evaluate its interlaminar shear resistance. In this paper, the direct shear test is used to test the shear resistance between stress absorption layers [32]. The direct shear test maintains the position of the specimen by fixing the fixture, and directly loads the load onto the failure surface of the specimen, so as to test its maximum failure load. Through formula calculation, the corresponding interlaminar shear strength is obtained.

The specimen is obtained by drilling the core of basalt fiber rubber asphalt stress absorption layer composite plate. The specific size is a cylinder 100 mm in diameter and 100 mm high. During the test, one end of the test piece is fixed with two semicircular clamps, and the other half of the circular clamp is placed above the test piece through the iron bar to undertake the shear force transmitted by the UTM machine, as shown in Figures 6 and 7.

Figure 6. Fixing method of interlaminar shear test mold.

Figure 7. Loading mode of interlaminar shear test.

In this test, 25 °C was selected as the test environment to study the interlayer bonding performance. To ensure uniform heating of the test piece, the test piece shall be placed in an oven at the specified temperature for more than 4 h before the interlaminar shear test. Combined with the actual factors, the fast loading mode is adopted, and the loading rate is set to 50 mm/min.

Conduct the interlaminar direct shear test under set conditions until failure. According to the load peak value obtained by the software, the interlaminar shear strength of the stress absorption layer is obtained through Formula (1), which is used as an index to evaluate the interlaminar shear strength of the stress absorption layer.

$$T = \frac{F/1000}{S} \tag{1}$$

In the formula: T—interlaminar shear strength, MPa; F—peak load, KN; S—shear area; for 100 mm diameter test pieces, the contact area between layers is 0.00785 m^2.

3.2. Interlayer Pull-Out Test

The semi-rigid base course easily produces load-type reflection cracks and temperature-type reflection cracks due to its own material characteristics. If the stress absorption layer has good bonding performance, it can reduce the tensile stress and shear stress caused by temperature shrinkage and driving load. The adhesion performance of interlayer materials can be evaluated by the pull-out strength measured by the pull-out tester of pavement interlayer adhesion [33,34].

Drill the core of the prepared composite plate test piece. The thickness of the drill core is the thickness of the asphalt mixture, which is 6 cm. One end of the test piece is bonded with a steel plate under the oil cylinder with epoxy resin. After the epoxy resin forms the bonding strength, the test piece is insulated. The XH-15 T produced by Beijing Shengshi Weiye Technology Co., Ltd. in Beijing, China subgrade and pavement bonding performance tester is used to carry out the pull-out test on the composite test piece, as shown in Figure 8.

Figure 8. XH-15 T subgrade and pavement adhesion performance tester.

During the test, one oil circuit is connected to a pressure sensor and the other is connected with the jack to drive the pull rod to apply tension to the test piece. As the tension gradually increases, when the test piece is stripped, because the pressure on the sensor is equal to the pressure on the jack, the maximum oil pressure at the moment of stripping of the test block can be read out through the digital display circuit. The schematic diagram of pull-out test is shown in Figure 9.

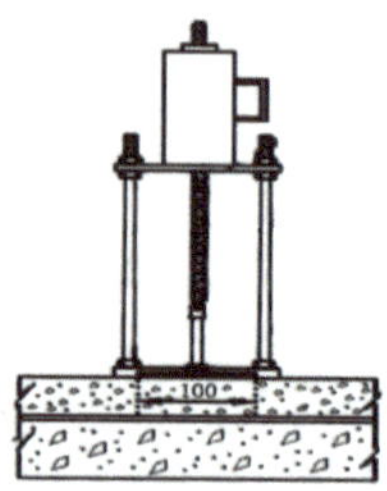

Figure 9. Schematic diagram of drawing test (unit: mm).

In this test, 25 °C is selected to represent the normal temperature environment. In order to ensure the uniform heating of the test piece, the test piece is kept warm for more than 4 h. During the test, tensile force is applied to the bonding end of the test piece until the test piece is damaged, and the maximum tensile force when it is damaged is measured.

Under the set conditions, apply tension to the bonding end of the specimen until the specimen is destroyed, measure the maximum tension when it is destroyed, calculate the pull-out bonding strength according to Formula (2), and select the pull-out strength as the index to evaluate the bonding performance between layers of basalt fiber stress absorption.

$$\sigma = \frac{F}{S} \tag{2}$$

In the formula: σ—interlayer bonding strength, MPa; F—peak load, KN; S—shear area; for 100 mm diameter test pieces, the contact area between layers is 0.00785 m^2.

3.3. *Three-Point Bending Test*

The three-point bending test consists of placing the test piece on two fixed supports with a certain distance and applying a downward load to the test piece along the midpoint direction of the two fixed supports [28]. The specific size is 300 mm long, 100 mm wide, and 100 mm high. A 5 mm crack is pre-cut in the water-stabilized gravel span to ensure that the test piece starts to expand from the tip of the pre-cut crack.

The test piece has three contact points with the support and the compression device, as shown in Figure 10. When two equal cracks are formed, the specimen will break at the midpoint; that is, it will form a three-point bend, as shown in Figure 11, enabling us to measure its disturbance, bending force, and other indicators.

Figure 10. Three-point bending test loading die.

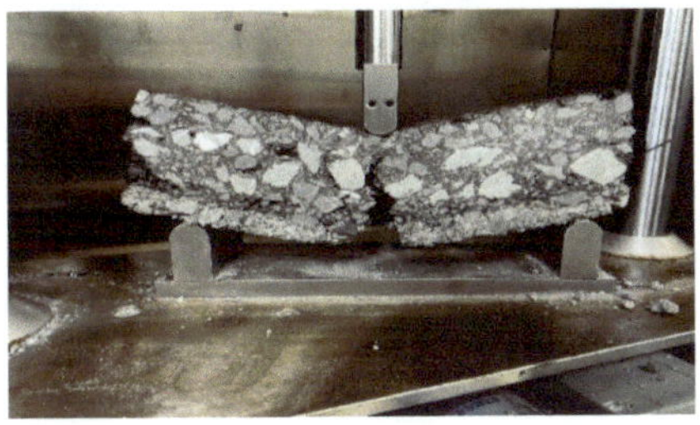

Figure 11. Failure of test piece.

The external force acting on the three-point bending test piece consists of three parts: the work carried out by the load applied to the test piece, the work carried out by the weight m_1 of the beam between the supports, and the work carried out by the weight m_2 of the loading part. Before loading, the force F_1 on the beam is strong, which can be calculated by Formula (3).

$$F_1 = \frac{1}{2}m_1 g + m_2 g \tag{3}$$

In the formula: g—gravitational acceleration.

The measured load deflection curve of the test piece is O'ab. When considering the self-weight of the test piece and the weight of the loading part, the load deflection curve is O'ac, as shown in Figure 12.

Figure 12. Load deflection curve of three-point bending.

It can be seen from the figure that the total energy absorbed by the fracture surface of the test piece is A; see Formula (4).

$$A = A_0 + A_1 + A_2 \tag{4}$$

In the formula, A_0 is the area under curve O'ab; see Formula (5) for A_1 and A_2.

$$A_1 = A_2 = F_1\delta_0 = \left[\frac{1}{2}m_1 g + m_2 g\right]\delta_0 \tag{5}$$

The fracture energy G_F specified in the test standard can be calculated by Formula (6).

$$G_F = \frac{A_0 + mg\delta_0}{A_{lig}} \tag{6}$$

In the formula, $m = m_1 + 2m_2$, A_{lig} is the stress area.

A three-point bending die with a mid-span length of 250 mm was used in the test. A temperature of 15 °C is selected for the test to represent the conventional environment. In order to ensure that the test piece is heated evenly, the test piece is placed in an oven at the specified temperature for more than 4 h. Fast loading mode is adopted, and the loading rate is set to 50 mm/min.

According to the requirements of the three-point bending test specification, when a concentrated load is applied to the midspan section of the test piece at the specified rate, when the test piece breaks in the midspan, it is considered that the test piece is damaged, as shown in Figure 11. The data acquisition system records the load–disturbance curve in real time, and the data system provides the maximum bending load, maximum deformation disturbance, and other data results.

3.4. Overlay Test

The three-point overlay test (OT) simulates the tension of asphalt pavement under the test, tests the real impact of reflection cracks on the pavement surface, and can readily evaluate the anti-reflection crack ability of the stress absorption layer [35–37]. As shown in Figure 13, the whole test mold is composed of a fixed plate, movable plate, bonding plate, and fixed rivet.

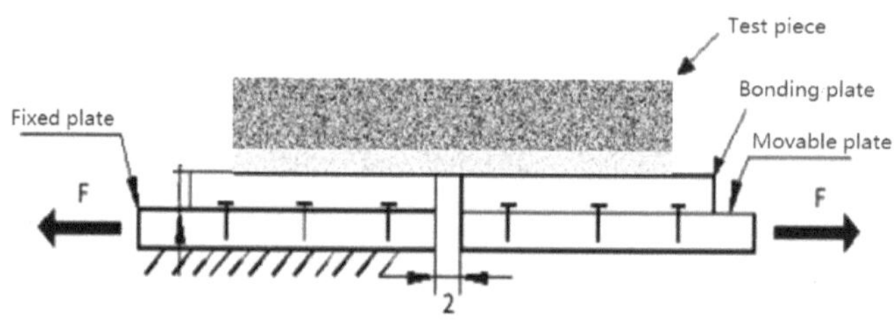

Figure 13. Schematic diagram of overlay test mold.

The specific size of the test piece is a length of 150 mm, width of 76 mm, and height of 38 ± 0.5 mm. The prepared OT specimen is bonded to two movable plates with epoxy resin, and a 10 kg weight is placed on each specimen to ensure its firmness under epoxy resin bonding, as shown in Figure 14. After standing and pressurizing for more than eight hours at room temperature, when the bonding strength is sufficient, take the test piece out of the mold.

The test mold is compared to the OT loading mold as shown in Figure 15. There is a gap of 2–3 mm in the middle of the mold. The lower part of the mold is fixed, and the upper part is stretched upward, simulating the effect of a lower reflection crack on the stress absorption layer.

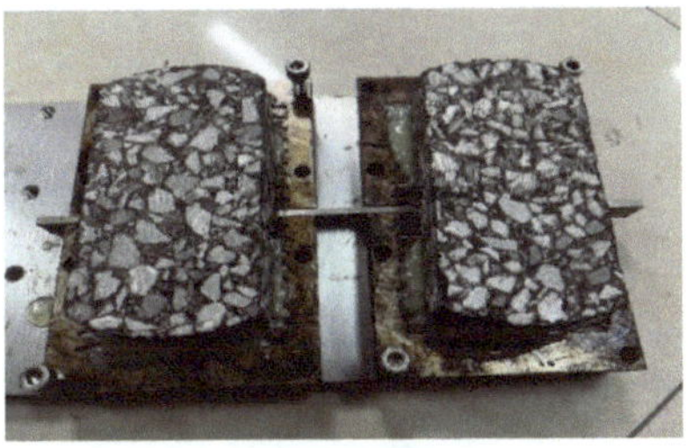

Figure 14. Epoxy resin fixed test piece.

Figure 15. OT loading mold and method.

In the test, 25 °C is selected to represent the normal temperature environment. In order to ensure that the test piece is heated evenly, the test piece is placed in an oven at the specified temperature for more than 4 h. The test load loading waveform adopts triangular wave, and the load cycle is 10 s. The data acquisition system records the cycle–displacement–load curve in real time, and the data system provides data results such as the number of test cycles, load loss rate, the maximum load value of the first cycle, etc.

According to the requirements of OT specification [20], taking the length of crack displacement as the control index, the target displacement value of the crack is 0.635 mm, and the closing movement of the reflection crack is simulated through the above parameters. When the maximum load of the specimen in a certain cycle decreases by more than 93% compared with the maximum load of the initial cycle, it is considered that the specimen is damaged and the test will stop automatically. At the same time, the test operation cycle is set as 1000 iterations in the specification. When the actual operation cycle is reached, the system will also recognize that the test has stopped automatically.

In OT, there are two different stages of cracking, namely, the crack initiation stage and the crack propagation stage.

The fracture energy at the crack initiation stage is calculated by Formula (7).

$$G_A = \frac{1}{tb}\int_{W_1}^{W_2} f(w)dw \tag{7}$$

The fracture energy at the crack growth stage is calculated by Formula (8).

$$G_B = \frac{1}{tb}\int_{W_2}^{W_3} f(w)dw \tag{8}$$

Therefore, the calculation formula of total fracture energy is shown in Formula (9).

$$G_f = G_A + G_B = \frac{1}{tb}\int_{W_1}^{W_3} f(w)dw \tag{9}$$

In the formula: $F(w)$—load–displacement curve; t, b—thickness and width of test specimen; W_1, W_2, W_3—different displacements corresponding to load–displacement curve.

4. Test Results and Analysis

4.1. Interlaminar Direct Shear Test

The effect of different fiber contents on interlaminar shear strength is shown in Figure 16.

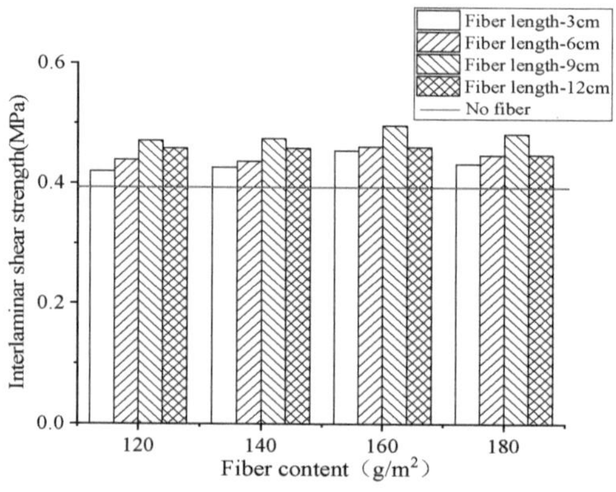

Figure 16. Effect of fiber content on interlaminar shear strength.

As can be seen from Figure 16, the interlaminar direct shear strength without adding fibers was 0.396 MPa. Compared with the control group without fibers, when the fiber content was 160 g/m^2, the interlaminar shear strength of the basalt fiber rubber asphalt stress-absorbing layer with a length of 3 cm, 6 cm, 9 cm, and 12 cm was increased by 14.6%, 16.4%, 25.3%, and 16.1%, respectively. Compared with other dosage groups, its improvement effect was better.

It can be seen from Figure 16 that when the fiber content was 160 g/m^2, the interlaminar shear strength of the basalt fiber rubber asphalt stress-absorbing layer with the length of 3 cm, 6 cm, 9 cm, and 12 cm increased by 14.6%, 16.4%, 25.3%, and 16.1%, respectively, compared with other content, which was better than other content.

The interlaminar shear strength of rubber asphalt stress-absorbing layer with different basalt fiber content increased at first and then decreased with the increase of content. This was because the increase of the content of the mixture led to the gradual reduction of the interlayer porosity, which made the upper and lower bond of the asphalt layer firmer and increased the interlayer shear strength. However, when the fiber content was increased to 180 g/m^2, in the case of uniformly distributed fibers, the fibers were surplus, and there was a lot of overlap between the fibers, which made the asphalt layer at the overlap develop delamination, thus reducing the interlaminar shear strength. The three situations of sparse, uniform, and excessive fiber distribution are shown in Figure 17.

In general, when the fiber length was 9 cm and the fiber content was 160 g/m^2, the interlaminar shear resistance of the rubber asphalt stress absorption layer was the strongest. This was because the fiber length was compatible with the gravel particle size, and the optimal fiber length was closely related to the maximum nominal particle size. When the particle size of crushed stone was 4.75~9.5 mm, with the increase of fiber length, the number of crushed stones that can be completely connected by the fibers gradually increases, making the shear resistance gradually improve. When the fiber length was 9 cm, the number of crushed stones that could be connected was the largest, and the compatibility with crushed stones within the particle size range was the highest. When the stress-absorbing layer was sheared, the maximum shear stress could be borne.

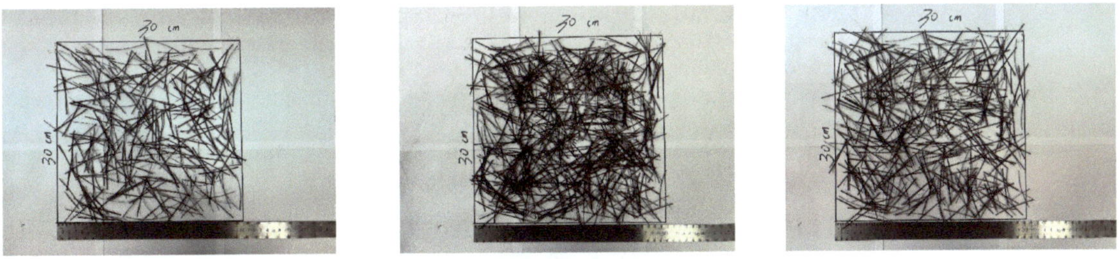

(a) (b) (c)

Figure 17. (**a**) Insufficient fibers leads to sparsity. (**b**) Excessive fibers leads to overlap. (**c**) Uniform distribution of fibers.

We used a scanning electron microscope (as shown in Figure 18) to take photos of the micro-morphology of the fibers, and at the same time, we took photos of the micro-morphology observed under the contact conditions between the fibers and asphalt binder. We analyzed the surface micro-morphology characteristics, and the micro-morphology image is shown in Figure 19.

Figure 18. Scanning electron microscope.

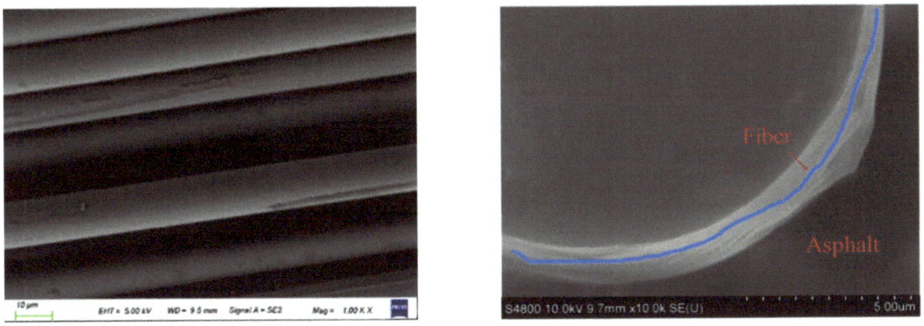

(a) (b)

Figure 19. Microscopic morphology of basalt fibers: (**a**) Basalt fibers. (**b**) Basalt fiber–rubber asphalt.

At the micro scale, scanning electron microscopy experiments can help reveal the true state of the fibers. From Figure 19, it can be seen that the short-cut basalt fibers do not undergo bending and winding and can form a good bridging effect in the mixture. The flocculent basalt fibers were cotton-like fibers formed by bending basalt fibers, which helped to form a three-dimensional network structure, allowing asphalt to adsorb to the fiber surface, making the distribution of asphalt slurry between aggregates more stable, thereby increasing the thickness of the asphalt film and improving the bonding strength between asphalt and aggregates. At the same time, the above viewpoint can be verified through microscopic images of fiber asphalt slurry. Both types of basalt fibers and asphalt binders had clear distinguishing areas, which effectively increased the contact surface between fibers and asphalt, generated anchoring effects during the interaction between fibers and asphalt, and further improved the adhesion between fibers and asphalt.

4.2. Interlayer Pull-Out Test

The effect of different fiber contents on the bonding performance between stress-absorbing layers is shown in Figure 20.

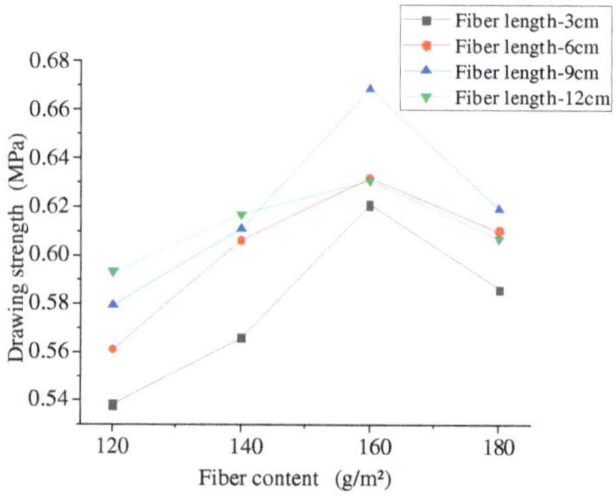

Figure 20. Effect of fiber content on drawing strength.

The experimental data showed that the interlaminar tensile strength without fiber incorporation was 0.312 MPa. Compared with the blank group without fiber incorporation, with fiber incorporation, fiber combinations with different parameters improved interlaminar tensile strength with a significant increase in the range of 72% to 114% and with a significant improvement in tensile strength.

It can be seen from Figure 20 that the influence of different fiber lengths on the pull-out strength of rubber asphalt stress-absorbing layer had certain similarity with the change of the content. The overall curve trend was gradually increasing at first, and when the fiber content was 160 g/m^2, it reached the peak value at the same length, and then gradually decreased.

When the fiber content was 120 g/m^2 and 140 g/m^2, the reinforcement effect of basalt fibers on drawing strength was linear with the increase of fiber length, and increased with the increase of fiber length. When the fiber content was 160 g/m^2 and 180 g/m^2, the reinforcement effect of basalt fiber increased first and then decreased with the increase of fiber length, and decreased after the peak value at the fiber length of 9 cm. Different from the results of direct shear test, when the fiber content was 120 g/m^2 and 140 g/m^2, the interlaminar tensile strength gradually increased. This was because the fiber content was related to the total contact area of asphalt. When the fiber content was small, the greater the fiber dispersion was under the same spreading area, which encouraged more fibers to make contact with rubber asphalt, absorbed the light components in asphalt, increased the thickness of asphalt film, and thus enhanced the adhesion between layers.

On the whole, compared with the blank group without fibers, when the fiber length was 9 cm and the fiber dosage was 160 g/m^2, the basalt fibers significantly improved the interlayer bonding performance of the stress absorption layer, because the basalt fibers itself had better asphalt adsorption capacity, and the asphalt could evenly wrap around the fiber surface and produce a physical reaction with it, forming a structural asphalt layer with a stronger bonding force and increasing the interlayer bonding performance.

4.3. Three-Point Bending Test

4.3.1. Test Results

The test parameters that can be directly read on the test instrument selected in this test were maximum bending load *F* and maximum deformation disturbance *L*. According to the relevant research regarding the three-point bending test, the bending fracture energy W_0 was selected as the evaluation index. The bending fracture energy W_0 was the ratio of the area surrounded by the load disturbance curve of the specimen to the stress area of the specimen. See Figure 21 for different groups of load deflection curves.

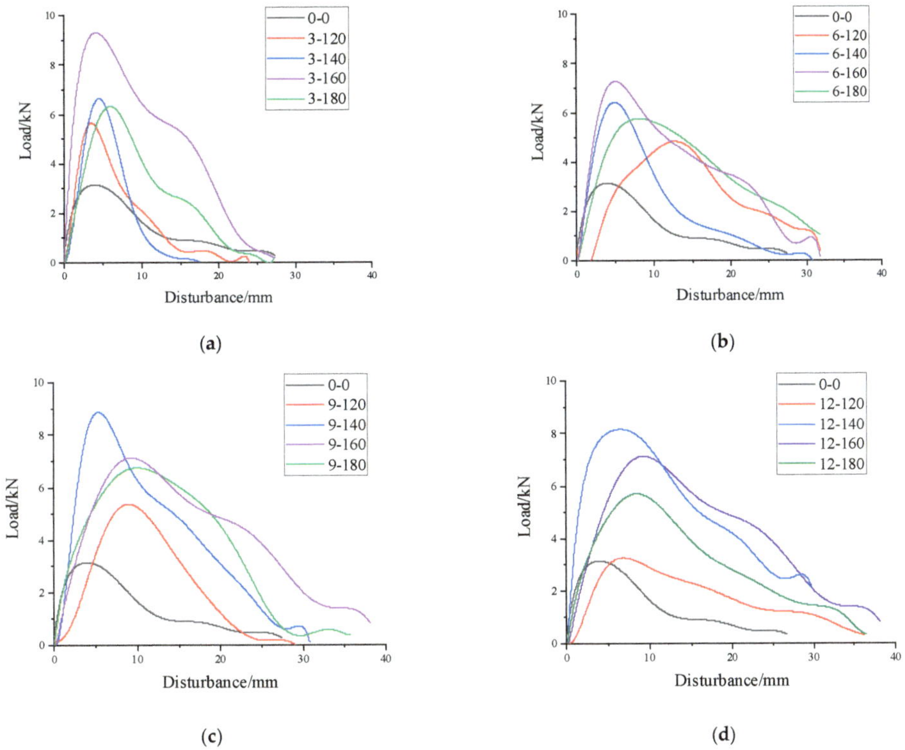

Figure 21. Load-deflection curve: (**a**) 3 cm group; (**b**) 6 cm group; (**c**) 9 cm group; (**d**) 12 cm group.

4.3.2. Effect of Basalt Fiber on Bending Fracture Energy

The influence of different fiber contents on the bending fracture energy of the stress-absorbing layer is shown in Figure 22.

As can be seen from Figure 22, the three-point bending fracture energy without fiber incorporation was 5036.7 $N{\cdot}m^{-1}$. Compared with the blank group without fiber incorporation, with fiber incorporation, fiber combinations with different parameters had a significant improvement in fracture energy, and the increase was significant.

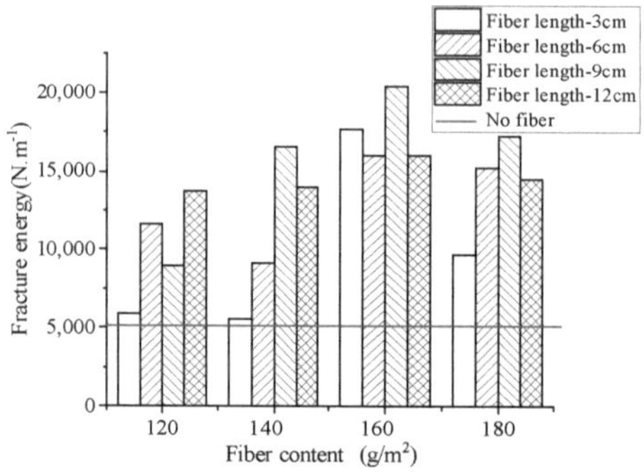

Figure 22. Effect of fiber content on bending fracture energy.

When the fiber content was 160 g/m^2, the bending fracture energy of the stress-absorbing layer was significantly improved by 249.3%, 218.9%, 305.5%, and 217.6% for rubber asphalt stress-absorbing layers mixed with basalt fibers of 3 cm, 6 cm, 9 cm, and 12 cm length, respectively. In general, when the fiber length was 9 cm and the fiber content was 160 g/m^2, basalt fibers had the best effect on improving the anti-cracking performance of the stress absorption layer.

With the increase of fiber length, the distribution area of fibers decreased at the same dosage, resulting in fewer fibers at the crack to prevent cracking. According to the observations of the test, when the test piece started to break, the fibers of each length can produce a crack resistance effect at the crack, but with the continuous expansion of the crack, the basalt fibers with the lengths of 3 cm and 6 cm were gradually separated, while the basalt fibers with the length of 9 cm and 12 cm could still be connected at the crack until the test piece was completely broken.

With the increase of fiber content, the bending fracture energy of the rubber asphalt stress-absorbing layer first increased and then decreased. This was because the increase of the fiber content made the fiber distribution density increase, and there were more fibers at the crack to play the role of "reinforcement and crack resistance" [38], but the excessive content reduced the overall orientation of the fiber asphalt layer, reducing the tensile performance of the fiber asphalt layer at the crack.

4.4. Overlay Test

4.4.1. Test Results

The test parameters that can be directly read on the test instrument selected in this test were the number of test cycles (N), the load loss rate (R), and the maximum load (F) in the first cycle. At the same time, according to the relevant research of OT, the total fracture energy (G) was selected to represent the total fracture energy during cracking, which was used as an index to evaluate the anti-reflection crack performance of the stress absorption layer. The OT load cycle curve under different mix combinations is shown in Figure 23.

It can be seen from Figure 24 that when the number of test cycles of 17 groups of stress-absorbing layers reached 1000, the load loss rate of the stress-absorbing layer without fibers was the least, and the R value was 60.05%. With the increase of fiber length and fiber content, the R value mostly shows a trend of rising first and then declining. When the fiber length was 9 cm and the fiber content was 160 g/m^2, the load loss rate was the largest, and the R value was 85.22%, which indicates that in the whole failure cycle of reflection crack, the fiber length was 9 cm. When the fiber content was 160 g/m^2, the stress-absorbing layer needed the most energy to destroy, and more force and energy were needed to form reflective cracks.

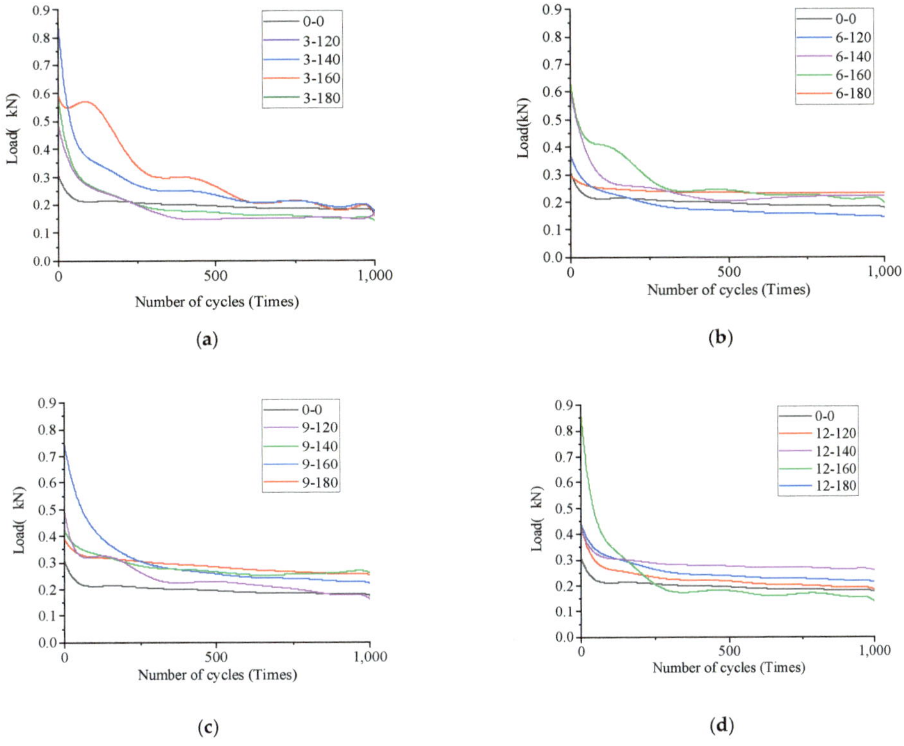

Figure 23. Load-cycle iteration curve: (**a**) 3 cm group; (**b**) 6 cm group; (**c**) 9 cm group; (**d**) 12 cm group.

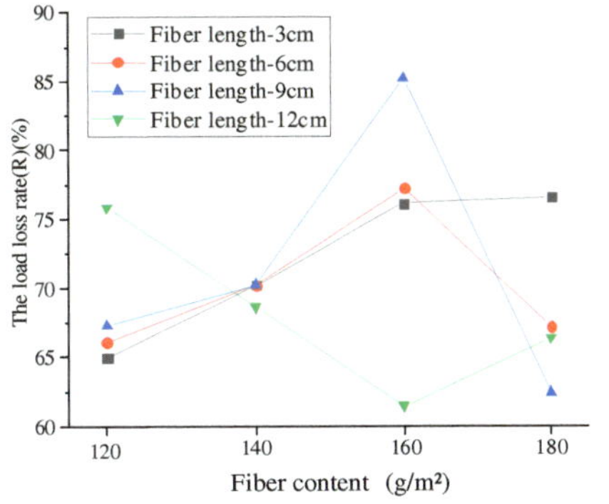

Figure 24. The load loss rate (*R*).

However, the R value of the 12 cm fiber shows a trend of first decreasing and then increasing, because at the same dosage, the excessive fiber length leads to excessive concentration of fatigue cracks and development along the fiber length. The shorter the length of the fiber, the more it can disperse and form a strong network structure during distribution. However, the 12 cm fibers could only form a smaller fiber bundle-like structure at the ends of the fibers, while the middle part of the fiber is still a ball and not effectively separated. However, when the dosage increases to a certain amount, this undivided fiber forms a larger fiber skeleton, similar to the "reinforcement" effect, which can play a certain role in delaying load loss, but the effect is not significant and the improvement is not significant.

4.4.2. Effect of Basalt Fiber on Anti-Reflection Crack Performance

The influence of different fiber contents on the anti-reflection crack performance of rubber asphalt stress absorption layer is shown in Figure 25.

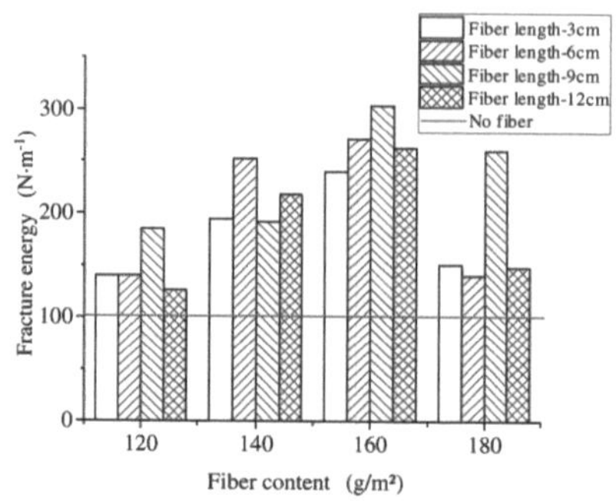

Figure 25. Effect of fiber content on total breaking energy.

It can be seen from Figure 25 that the addition of basalt fibers can significantly improve the anti-reflection crack performance of the stress absorption layer. The total fracture energy of the stress-absorbing layer without fibers was 101.05 N·m^{-1}. Under the condition of fixed fiber length, the total breaking energy first increased and then decreased with the increase of fiber content. When the fiber length was 9 cm and the content was 120 g/m^2, 140 g/m^2, 160 g/m^2 and 180 g/m^2, the total fracture energy of the rubber asphalt stress-absorbing layer increased by 82.3%, 90.31%, 200.26% and 157.3%, respectively.

In general, when the fiber length was 9 cm and the fiber content was 160 g/m^2, basalt fibers presented the best effect on the anti-reflection crack performance of the stress absorption layer. This was because the increase of fiber content made the stress dissipation rate of rubber asphalt stress absorption layer gradually increased. The faster the stress dissipation rate was, the less likely the stress absorption layer was to crack, but too much fiber content was not conducive to deformation recovery.

On the other hand, when the fiber content was fixed, with the increase of fiber length, the total fracture energy of the rubber asphalt stress absorption layer showed a trend of rising first and then falling. However, when the fiber length was too long to cause fatigue cracks in the specimen, the cracks were excessively concentrated and developed along the fiber length, thus reducing the anti-reflection crack performance. At the same time, this rule was also consistent with the change rule of test R value (load loss rate measured by the system).

4.4.3. OT Test Curve Fitting

According to Figure 24, it can be found that the maximum load cycle curve of 17 basalt fiber stress absorption layers with different length and content under conventional conditions conformed to the change law of power function, so the power function shown in Formula (10) was selected to fit the test curve [39].

$$y = ax^{-b} \tag{10}$$

In the formula, a, b were parameters. Among them, parameter "a" represents the decline rate of the curve. The larger its value, the faster the decline rate of the curve. At the same time, it also indicates that the faster the load attenuation rate of the stress absorption layer was, the faster the damage of the stress absorption layer was, and the worse the anti-reflection crack performance was. We fit the OT curve and obtained the fitting results shown in Table 6.

Table 6. Fitting results of OT curve.

Length (cm)–Content (g/m²)	Fitting Formula	R^2	*a*	*b*
0–0	$y = 1.8246x^{-0.6364}$	0.9783	1.8246	0.6364
3–120	$y = 1.364x^{-0.4644}$	0.9657	1.364	0.4644
3–140	$y = 1.266x^{-0.4732}$	0.9816	1.266	0.4732
3–160	$y = 0.8547x^{-0.3558}$	0.976	0.8547	0.3558
3–180	$y = 1.343x^{-0.419}$	0.9872	1.343	0.419
6–120	$y = 1.2139x^{-0.4738}$	0.9693	1.2139	0.4738
6–140	$y = 1.4131x^{-0.4121}$	0.9693	1.4131	0.4121
6–160	$y = 0.9539x^{-0.3162}$	0.9859	0.9539	0.3162
6–180	$y = 1.3x^{-0.3312}$	0.9739	1.3	0.3312
9–120	$y = 1.3307x^{-0.3503}$	0.9869	1.3307	0.3503
9–140	$y = 1.157x^{-0.287}$	0.9554	1.157	0.287
9–160	$y = 0.6807x^{-0.1319}$	0.9865	0.6807	0.131
9–180	$y = 1.4184x^{-0.175}$	0.9782	1.4184	0.175
12–120	$y = 1.5192x^{-0.431}$	0.9782	1.5192	0.431
12–140	$y = 1.762x^{-0.412}$	0.9861	1.762	0.412
12–160	$y = 1.389x^{-0.373}$	0.9713	1.389	0.373
12–180	$y = 1.837x^{-0.427}$	0.9725	1.837	0.427

Through the curve fitting results, it was found that the variation law of load cycle curve of basalt fiber stress absorption layer was very consistent with the variation law of power function. It can be seen from Table 6 that the value of the parameter "*a*" of the stress absorption layer without fibers was the largest, reaching 1.824, while the *a* value was the smallest when the fiber dosage was 160 g/m² and the fiber length was 9 cm, reaching 0.6807, which meant that the stress absorption layer without basalt fibers had the fastest destruction speed, the stress absorption layer had the slowest destruction speed when the fiber dosage was 160 g/m² and the fiber length was 9 cm, and the anti-reflection crack performance was the strongest, which was basically consistent with the influence law of the above fiber parameters on the fracture energy.

4.5. Comprehensive Engineering Analysis

4.5.1. Engineering Performance Analysis

In this paper, the performance of basalt fiber rubber asphalt stress-absorbing layer was evaluated and analyzed by interlaminar direct shear test, interlaminar pull-out test, three-point bending test and OT. Overall, fiber content played the major affecting role on the performance of rubber asphalt stress absorption layer rather than the fiber length.

The fiber parameter combination with the best improvement effect of fiber length of 9 cm and fiber content of 160 g/m² was selected to analyze the test results of basalt fiber rubber asphalt stress-absorbing layer under four tests. The improvement ranges of different properties under the same content are shown in Table 7.

Table 7. Performance improvement range of basalt fiber rubber asphalt stress absorption layer.

Performance Index	Fiber Length	Fiber Content	Average Value of Test Data	Increase Range Compared with That without Fiber
Interlaminar shear strength	9 cm	160 g/m²	0.496	25.2%
Interlayer bonding strength			0.669	114.4%
Bending fracture energy			20,422.454	305.5%
Total breaking energy			303.405	200.2%

It can be seen from Table 7 that compared with the blank group without fibers, the properties of the rubber asphalt stress-absorbing layer mixed with basalt fibers had been improved to a certain extent, and the increase of interlaminar shear strength was small, only 25.2%. The increase of bending fracture energy and total fracture energy was obvious, both being greater than 200%, and the improvement effect of bending fracture energy was the best, up to 305.5%. This showed that the basalt fiber combination with fiber length of 9 cm and fiber content of 160 g/m^2 can significantly enhance the cracking resistance of the rubber asphalt stress absorption layer. This was because the basalt fiber acts as a "bridge" in the cracking process of the specimen, thus effectively preventing or delaying the speed of crack initiation and propagation in the rubber asphalt stress absorption layer.

As the number of cycles increased, the stress drop in the blank sample suddenly rose, and then the pavement became cracked (after about 70 cyclic loading times) [14,20]. In comparison, the rate at which the stress dropped in the test sample with a CSAL was significantly slower and the load decreased smoothly to 7% of the initial value after 1000 cycles. The fatigue life of the sample with the RASAL resisting the formation of reflective cracks was more than 10 times that of the blank sample at this lower temperature. In the same situation, the stress absorption layer of basalt fiber rubber asphalt has less load loss after 1000 cycles of action, which means it has stronger stress absorption effect. If reflection cracks are not to cause damage, a stronger ability is needed.

4.5.2. Engineering Economic Analysis

Taking the asphalt pavement maintenance project of the Lima Expressway in 2022 as an example, a rubber asphalt stress absorption layer was used in both the overlay maintenance plan and the service area square pavement and ramp disposal planed in this major maintenance project. The construction plan design diagram is shown in Table 8. According to the bill of quantities for the 2022 Lima Expressway Asphalt Pavement Overhaul Project, a total of approximately 32,000 m^2 rubber asphalt stress-absorbing layers were used in this project, with a total contract amount of over CNY 1.4 million. The average unit price of the rubber asphalt stress-absorbing layer during this process was calculated to be approximately CNY 43.5 /m^2.

Table 8. Design Table for Asphalt Pavement Structure of Lima Expressway in 2022.

Project	**Pavement Layer**	**Pavement Plan**	**Gate Treatment Plan**
		Double Layer Cover Scheme	**Milling and Laying Double Layers + Cover Scheme**
Pavement Structure Design Drawing	1	Overlay 4 cm SMA-13 (PG76-22 + 3‰ Anti stripping agent)	Overlay 4 cm modified asphalt SMA-13 (PG76-22 + 3‰ Anti stripping agent)
	2	Pave 6 cm modified asphalt high modulus material	Milling and laying back 10 cm modified asphalt Sup-20 (+3‰ basalt fiber)
	3	Rubber asphalt stress absorption layer	Rubber asphalt stress absorption layer
	4	Original surface layer	Original surface layer
Applicable road section		K284 + 398~K286 + 762	Service area runs through lanes and gates

From the comparison of raw material composition and dosage, it can be seen that compared with the ordinary rubber asphalt stress absorption layer, the basalt fiber rubber stress absorption layer only added a layer of basalt fiber layer in the middle of the rubber asphalt layer, with a fiber length of 9 cm and a dosage of 160 g/m^2, while maintaining the same type and dosage of other raw materials. According to the current market price, the selling price of basalt fiber was about CNY 20,000/ton, so the cost of selecting basalt fiber rubber asphalt stress absorption layer was about CNY 47/m^2. Compared to the unit price of CNY 43.5/m^2 for ordinary rubber asphalt stress absorption layer, the cost of basalt fiber rubber asphalt stress absorption layer only increased by CNY 3/m^2. Based on the performance analysis above, it can be seen that the crack resistance and reflection crack resistance of the stress-absorbing layer had been improved by 305.5% and 200.2%, respectively, with significant economic benefits. Therefore, the basalt fiber rubber asphalt stress absorption layer was a low-cost and efficient reflection crack prevention and control measure.

According to relevant engineering research data, compared to pavement structures without a stress absorption layer, laying a rubber asphalt stress absorption layer on the basis of a semi-rigid base will greatly improve the service life of the road structure. By relying on its high-fatigue performance, the overall fatigue extension life of the structure can be increased by 36%. At the same time, the production of rubber powder has become the leading direction for the reuse of waste tires. Vigorously promoting the application of rubber powder and rubber asphalt is not only conducive to promoting the recycling and reuse of resources, achieving the goal of saving natural resources, but also conducive to reducing environmental pollution and improving human living environment.

Therefore, on the one hand, it reflects the construction requirements of Chinese "long life roads", effectively reducing the frequency of road upgrades and maintenance; on the other hand, it also fundamentally reduces the consumption of natural resources and energy, indirectly playing a role in energy conservation, low-carbon and environmental protection, in line with the current strategic goal of peak carbon and carbon neutrality.

5. Conclusions

In this paper, a new type of basalt fiber modified rubber asphalt stress-absorbing layer was prepared. It was applied into the composite plate in the form of a cement-stabilized macadam base + basalt fiber rubber asphalt stress-absorbing layer + AC-20 asphalt mixture surface layer. The interlaminar shear resistance, interlaminar bonding performance and anti-reflection crack performance were investigated, and the influence of basalt fiber length and content on the performance of the absorbing layer was analyzed. The main conclusions can be drawn as follows:

- From the perspective of interlayer bonding performance, compared with the neat samples without fiber addition, the improvement in interlayer bonding strength of basalt fiber modified samples significantly achieves up to 114%, while the improvement of interlayer direct shear strength is not very significant, with a maximum reinforcement amplitude of only 20%. The reinforcement of the addition of basalt fiber mainly reflect in the improvement of the cohesion of asphalt binder rather than the interface strength.
- Based on the three-point bending test, the fracture energy of basalt fiber modified samples increases by 305.5% compared with the neat samples. Basalt fiber can enhance the anti-cracking performance some extent under one-time failure loads. Furthermore, optimal fiber content should be determined since excessive fiber will cause negative effect on the anti-cracking performance.
- Based on the overlay test, the maximum increase in fracture energy of basalt fiber-modified samples can reach up to 200.3%, compared with the neat samples. Basalt fibers possess a significant impact on the anti-reflection cracking performance under cyclic loads. Similarly, optimal fiber content should be determined since the fracture energy first increases and then decreases with an increasing fiber content.
- According to range analysis, the main factor affecting the anti-reflection cracking performance was the fiber content rather than the fiber length.
- Based on comprehensive performance and cost-effectiveness analysis, it was recommended that the basalt fiber parameter combination with the length of 9 cm and fiber content of 160 g/m^2 can be used to prepare the new type of basalt fiber modified rubber asphalt stress absorption layer.

Author Contributions: Conceptualization, P.X. and A.K. Data curation, P.X. and Z.W. Formal analysis, C.S. and Z.W. Funding acquisition, P.X. and A.K. Investigation, C.S., Y.W. and Z.W. Methodology, C.S., P.X. and Z.W. Project administration, P.X., A.K. and C.S. Supervision, P.X. Visualization, Y.W., C.S. and Z.W. Writing—original draft, C.S. Writing—review and editing, P.X. and Z.W. All authors have read and agreed to the published version of the manuscript.

Funding: This study was funded by the National Natural Science Foundation of China [grant number 52178439] and the Postgraduate Research & Practice Innovation Program of Jiangsu Province (Yangzhou University) [grant number SJCX22_1774].

Institutional Review Board Statement: Not applicable.

Informed Consent Statement: Not applicable.

Data Availability Statement: Data are contained within the article.

Conflicts of Interest: The authors declare no conflict of interest.

References

1. Lv, S.; Yuan, J.; Peng, X.; Zhang, N.; Liu, H.; Luo, X. A structural design for semi-rigid base asphalt pavement based on modulus optimization. *Constr. Build. Mater.* **2021**, *302*, 124216. [CrossRef]
2. Lv, S.; Guo, Y.; Xia, C.; Wang, X.; You, L.; Cabrera, M.B.; Li, J. Unified approach to characterize the strength of cement stabilized macadam subjected to different loading modes. *Constr. Build. Mater.* **2020**, *265*, 120143. [CrossRef]
3. Wang, X.; Zhong, Y. Reflective crack in semi-rigid base asphalt pavement under temperature-traffic coupled dynamics using XFEM. *Constr. Build. Mater.* **2019**, *214*, 280–289. [CrossRef]
4. Gonzalez-Torre, I.; Calzada-Perez, M.A.; Vega-Zamanillo, A.; Castro-Fresno, D. Evaluation of reflective cracking in pavements using a new procedure that combine loads with different frequencies. *Constr. Build. Mater.* **2015**, *75*, 368–374. [CrossRef]
5. Shan, J.; Zhang, Y.; Wu, S.; Wu, Q.; Jiao, Y.; Du, Y. Cracking behavior of asphalt pavement with a graded gravel layer based on computational granular mechanics. *Constr. Build. Mater.* **2022**, *345*, 128199. [CrossRef]
6. Zhang, G.; Zhang, H.; Sun, W. Effect of coating optimization on performances of glass fiber geogrid for semi-rigid base asphalt pavement. *Geotext. Geomembr.* **2022**, *50*, 337–346. [CrossRef]
7. Wang, X.; Zhong, Y. Influence of tack coat on reflective cracking propagation in semi-rigid base asphalt pavement. *Eng. Fract. Mech.* **2019**, *213*, 172–181. [CrossRef]
8. Das, A.; Bhuyan, M.R.; Khattak, M.J.; Zhang, Q. Mitigating reflective cracking in composite pavements through the use of a ductile concrete interlayer. *Constr. Build. Mater.* **2020**, *259*, 120383. [CrossRef]
9. Gonzalez-Torre, I.; Calzada-Perez, M.A.; Vega-Zamanillo, A.; Castro-Fresno, D. Experimental study of the behaviour of different geosynthetics as anti-reflection cracking systems using a combined-load fatigue test. *Geotext. Geomembr.* **2015**, *43*, 345–350. [CrossRef]
10. Zamora-Barraza, D.; Calzada-Pérez, M.A.; Castro-Fresno, D.; Vega-Zamanillo, A. Evaluation of anti-reflection cracking systems using geosynthetics in the interlayer zone. *Geotext. Geomembr.* **2011**, *29*, 130–136. [CrossRef]
11. Zhu, Z.; Xiao, P.; Kang, A.; Wu, Z.; Kou, C.; Ren, Z. Research on the road performance of self-adhesive basalt fiber geotextiles based on the background of long-life pavements. *Constr. Build. Mater.* **2023**, *392*, 131776. [CrossRef]
12. Hou, F.; Li, T.; Li, X.; Li, Y.; Guo, M. Research on the Anti-Reflective Cracking Performance of a Full-Depth Asphalt Pavement. *Sustainability* **2021**, *3*, 9499. [CrossRef]
13. Bao, C.; Liu, Q.; Xia, Y.; Cui, Y.; Cao, Z.; Qian, Y.; Liu, M.; Mu, C.; Wang, H. Influence of Crack Spacing/Layer Thickness Value on Reflection Crack Propagation Mechanism Under Low Temperatures. *Front. Earth Sci.* **2021**, *9*, 810964. [CrossRef]
14. Sun, Y.; Yan, T.; Wu, C.; Sun, X.; Wang, J.; Yuan, X. Analysis of the Fatigue Crack Propagation Process of the Stress-Absorption Layer of Composite Pavement Based on Reliability. *Appl. Sci.* **2018**, *8*, 2093. [CrossRef]
15. Tsai, F.-L.; Lytton, R.L.; Lee, S. Prediction of reflection cracking in hot-mix asphalt overlays. *Transp. Res. Rec.* **2010**, *2155*, 43–54. [CrossRef]
16. Tanzadeh, R.; Tanzadeh, J.; Honarmand, M.; Tahami, S.A. Experimental study on the effect of basalt and glass fibers on behavior of open-graded friction course asphalt modified with nano-silica. *Constr. Build. Mater.* **2019**, *212*, 467–475. [CrossRef]
17. Yang, K.; He, Z.; Li, D.; Xu, H.; Kong, L. Experimental Study on Basalt Fiber Crack Resistance of Asphalt Concrete Based on Acoustic Emission. *Materials* **2021**, *14*, 4096. [CrossRef] [PubMed]
18. Wu, B.; Meng, W.; Xia, J.; Xiao, P. Influence of Basalt Fibers on the Crack Resistance of Asphalt Mixtures and Mechanism Analysis. *Materials* **2021**, *15*, 744. [CrossRef]
19. Pan, R.; Li, Y. Effect of warm mix rubber modified asphalt mixture as stress absorbing layer on anti-crack performance in cold region. *Constr. Build. Mater.* **2020**, *251*, 118985. [CrossRef]
20. Zhang, H.; Gao, P.; Zhang, Z.; Pan, Y. Experimental study of the performance of a stress-absorbing waterproof layer for use in asphalt pavements on bridge decks. *Constr. Build. Mater.* **2020**, *254*, 119290. [CrossRef]
21. Han, L.; Han, L.; Zhang, S.; Zhang, S.; Zhang, Z.; Zhang, Z.; Gao, T.; Gao, T. Mechanical analysis of preventing reflection cracks based on stress absorbing layer. *Adv. Civ. Eng.* **2022**, *2022*, 8016215. [CrossRef]
22. Nie, D.; Wang, S.; Sun, P.; Huang, C. Study on anti-crack effect of semi-rigid base pavement with stress absorbing layer. *J. Eng. Appl. Sci.* **2023**, *70*, 45. [CrossRef]
23. Norambuena-Contreras, J.; Gonzalez-Torre, I.; Fernandez-Arnau, D.; Lopez-Riveros, C. Mechanical damage evaluation of geosynthetics fibres used as anti-reflection cracking systems in asphalt pavements. *Constr. Build. Mater.* **2016**, *109*, 47–54. [CrossRef]

24. Xie, P.; Wang, H. Finite element analysis of thermal-in duced reflective cracking in composite pavement with mitigation strategies. *Eng. Fract. Mech.* **2022**, *266*, 108396. [CrossRef]
25. Zhang, Y.; Gu, Q.; Kang, A.; Ding, X.; Ma, T. Characterization of mesoscale fracture damage of asphalt mixtures with basalt fiber by environmental scanning electron microscopy. *Constr. Build. Mater.* **2022**, *344*, 128188. [CrossRef]
26. Gu, Q.; Kang, A.; Li, B.; Xiao, P.; Ding, H. Effect of fiber characteristic parameters on the high and low temperature rheological properties of basalt fiber modified asphalt mortar. *Case Stud. Constr. Mater.* **2022**, *17*, e01247. [CrossRef]
27. Wu, B.; Pei, Z.; Xiao, P.; Lou, K.; Wu, X. Influence of fiber-asphalt interface property on crack resistance of asphalt mixture. *Case Stud. Constr. Mater.* **2022**, *17*, e01703. [CrossRef]
28. Zhang, C.; Shi, F.; Cao, P.; Liu, K. The fracture toughness analysis on the basalt fiber reinforced asphalt concrete with prenotched three-point bending beam test. *Case Stud. Constr. Mater.* **2022**, *16*, e01079. [CrossRef]
29. Xie, T.; Wang, L. Optimize the design by evaluating the performance of asphalt mastic reinforced with different basalt fiber lengths and contents. *Constr. Build. Mater.* **2023**, *363*, 129698. [CrossRef]
30. Lou, K.; Xiao, P.; Tang, Q.; Wu, Y.; Wu, Z.; Pan, X. Research on the micro-nano characteristic of basalt fiber and its impact on the performance of relevant asphalt mastic. *Constr. Build. Mater.* **2022**, *318*, 126048. [CrossRef]
31. Lou, K.; Xiao, P.; Kang, A.; Wu, Z.; Li, B.; Lu, P. Performance evaluation and adaptability optimization of hot mix asphalt reinforced by mixed lengths basalt fibers. *Constr. Build. Mater.* **2021**, *292*, 123373. [CrossRef]
32. Li, S.; Liu, X.; Liu, Z. Interlaminar shear fatigue and damage characteristics of asphalt layer for asphalt overlay on rigid pavement. *Constr. Build. Mater.* **2014**, *68*, 341–347. [CrossRef]
33. Sudarsanan, N.; Fonte, B.R.; Kim, Y.R. Application of time-temperature superposition principle to pull-off tensile strength of asphalt tack coats. *Constr. Build. Mater.* **2020**, *262*, 120798. [CrossRef]
34. Noorvand, H.; Mobasher, B.; Underwood, S.; Kaloush, K. Development of an analytical framework for evaluation of critical fiber length in asphalt concrete with a fiber pullout test. *Constr. Build. Mater.* **2022**, *360*, 129561. [CrossRef]
35. Xiao, X.; Li, J.; Cai, D.; Cheng, H.; Xiao, F. Characterizing thermal fatigue behaviors of asphalt concrete waterproofing layer in high-speed railway using customized overlay test. *Int. J. Fatigue* **2022**, *165*, 107176. [CrossRef]
36. Li, J.; Oh, J.; Naik, B.; Simate, G.S.; Walubita, L.F. Laboratory characterization of cracking-resistance potential of asphalt mixes using overlay tester. *Constr. Build. Mater.* **2014**, *70*, 130–140. [CrossRef]
37. Walubita, L.F.; Faruk, A.N.; Alvarez, A.E.; Scullion, T. The Overlay Tester (OT): Using the Fracture Energy Index concept to analyze the OT monotonic loading test data. *Constr. Build. Mater.* **2013**, *40*, 802–811. [CrossRef]
38. Gou, H.; Zhu, H.; Zhou, H.; Yang, Z. Reinforcement mechanism of orientally distributed steel fibers on ultra-high-performance concrete. *Constr. Build. Mater.* **2021**, *281*, 122646. [CrossRef]
39. Walubita, L.F.; Fuentes, L.; Lee, S.I.; Guerrero, O.; Mahmoud, E.; Naik, B.; Simate, G.S. Correlations and preliminary validation of the laboratory monotonic overlay test (OT) data to reflective cracking performance of in-service field highway sections. *Constr. Build. Mater.* **2021**, *267*, 121029. [CrossRef]

materials

MDPI

Article

Investigation on the Performances of Esterified Waste Cooking Oil Rejuvenator and Recycled Asphalt

Junhui Wang [1], Qunshan Ye [1,*], Lingyi Fan [1], Cheng Xie [2] and Haobin Liu [2]

1 School of Traffic and Transportation Engineering, Changsha University of Science and Technology, Changsha 410114, China; 22201060242@stu.csust.edu.cn (J.W.); 21101030029@stu.csust.edu.cn (L.F.)
2 Guangxi Beitou Transportation Maintenance Technology Group Co., Ltd., Nanning 530201, China; xiecheng1357@163.com (C.X.); liuhaobin9203@163.com (H.L.)
* Correspondence: yequnshan@csust.edu.cn; Tel.: +86-136-8732-6030

Abstract: Waste cooking oil (WCO) recycled asphalt is facing issues regarding insufficient thermal oxidation stability and aging resistance. In this research, glycerol esterification was adopted to pretreat WCO, and the consequences of this treatment on the aging resistance and thermal stability of WCO were analyzed. The impacts of varying levels of esterification of WCO on the high-temperature, low-temperature performances, fatigue properties, and aging resistance of recycled asphalt were investigated. Furthermore, the mechanisms of regeneration and the anti-aging of deeply esterified WCO recycled asphalt were revealed by Fourier transform infrared spectroscopy (FTIR) and gel permeation chromatography (GPC) tests. The results indicated that variations in the physical properties of WCO during the aging process were reduced, and its aging resistance was improved following glycerol esterification therapy. The initial thermal decomposition temperature was increased by approximately 115 °C, which resulted in the enhancement of thermal stability significantly. Recycled asphalt obtained from deeply esterified WCO exhibited superior high-temperature, low-temperature performances, and fatigue properties. Moreover, the thermal oxidation stability and aging resistance of recycled asphalt with deep-esterified WCO could be promoted by reducing the oxidation and volatilization of light components during the aging process, with the complex modulus ageing resistance index decreasing by 13.27% and the phase angle ageing resistance index increasing by 14.71%.

Keywords: recycled asphalt; waste cooking oil; glycerol esterification; thermo-oxidative stability; anti-aging properties

Citation: Wang, J.; Ye, Q.; Fan, L.; Xie, C.; Liu, H. Investigation on the Performances of Esterified Waste Cooking Oil Rejuvenator and Recycled Asphalt. *Materials* **2024**, *17*, 4725. https://doi.org/10.3390/ma17194725

Academic Editor: Giovanni Polacco

Received: 5 September 2024
Revised: 18 September 2024
Accepted: 24 September 2024
Published: 26 September 2024

1. Introduction

According to the statistical bulletin on the development of the transportation industry published by the Ministry of Transport of China in 2022, 99.9% of China's gross roadway miles, which amount to over 5.3 million kilometers, are currently undergoing maintenance. Annually, approximately 300 million tons of reclaimed asphalt pavement (RAP) materials are generated during medium and major repair projects [1–3]. However, it is reported that less than 30% of pavement materials are recycled, which is significantly lower than the over 80% recycling rates prevalent in developed countries, resulting in enormous resource waste and ecological harm [4,5]. Consequently, it is acknowledged that recycling RAP on an extensive scale employing recycling technologies is an essential method to accomplish green construction in the highway transportation sector [6,7], motivate synergistic effects in reducing pollution and carbon emissions, as well as enable a comprehensive green transformation of social and economic development [8–12].

Existing research typically focuses on rejuvenating aging asphalt in RAP with low molecular weight rejuvenators to achieve objectives, including minimizing resource consumption, saving construction costs, and diminishing carbon emissions [13,14]. Based on research into the rheological properties and recycling mechanisms of aged asphalt, it has been discovered that the addition of light oils can enhance the relative contents of saturated and aromatic fractions in aged asphalt, adjust the colloidal structure, as well as restore the crack resistance, water damage resistance, and durability of the aged asphalt, facilitating it to satisfy road performance requirements [15,16]. Asphalt rejuvenators are currently classified into two main categories: petroleum-based and bio-based. However, petroleum-based rejuvenators (including aromatic oil extracts and naphthenic oils) possess deprived high-temperature stability, resulting in insufficient anti-aging performance of the recycled asphalt, as well as high material costs and non-renewability, which contradict the concepts of sustainability [17]. Consequently, the pursuit of a low-cost, environmentally friendly, and significantly effective asphalt rejuvenator is considered crucial for achieving the rational and efficient recycling of asphalt [18]. It has been estimated that the amount of WCO formed globally each year is enormous, with China alone producing up to 20 million tons of WCO each year, whereinthe major components of WCO are comparable to the lightweight components lost during the aging process of asphalt [19–21]. Moreover, compared to petroleum-based rejuvenators, a lower dosage of WCO is adequate to convey aged asphalt's penetration, softening point, and viscosity back up to virgin asphalt levels [22–25]. The dual waste utilization of RAP and WCO has been successfully established through the development and promotion of WCO recycled asphalt, offering significant benefits in terms of the economy, environment, and resources.

Nonetheless, the intricate origins of WCO have resulted in problems, including uneven quality and erratic rejuvenator effectiveness [26,27]. WCO rejuvenators with high acid values include substantial levels of free unsaturated fatty acids, which are susceptible to volatilization and oxidation, hastening the aging process and performance degradation of recycled asphalt [28,29]. As of this moment, dehydration and simple impurity removal are the most prevalent methods to handle WCO [30], but the impact that fluctuations in WCO acid values have on the rejuvenator and how effectively it performs with recycled asphalt have not been given sufficient attention. Thus, in order to enhance the durability of recycled asphalt, it is imperative to investigate effective pretreatment techniques for WCO rejuvenators.

In this study, WCO has been pre-treated using glycerol esterification techniques, and the impacts of various esterification levels on the properties of rejuvenated asphalt have been examined, along with the performance outcomes of esterified WCO rejuvenators in terms of thermal oxidative stability. Additionally, the mechanism of action of esterified WCO rejuvenators has been explored through microscopic experiments, revealing how the performance of aged asphalt could be improved. This research not only establishes a relatively reliable method for optimizing the performance of WCO rejuvenator but also provides a theoretical basis for the widespread application of WCO recycled asphalt.

2. Materials and Methods

2.1. Materials

2.1.1. Base Asphalt

The base asphalt is 70# A-grade road petroleum asphalt from Hunan Baoli Asphalt Co., Ltd. (Changsha, China), and the specific performance indicators are shown in Table 1.

Table 1. Performance indexes of matrix asphalt.

Technical Indexes	Results	Test Methods
Penetration at 25 °C (0.1 mm)	65	ASTM D5 [31]
Softening point (°C)	50.0	ASTM D36 [32]
Ductility at 15 °C (cm)	>100	ASTM D113 [33]
Viscosity at 135 °C (Pa·s)	611.9	ASTM D4402 [34]
Flash point (°C)	290	ASTM D92 [35]

2.1.2. Waste Cooking Oil

The waste cooking oil (WCO) provided by a waste oil recovery company in a certain region of Shandong is dark brown, with an average molecular weight of 890. Before the experiment, contaminants were removed using a simple filtration method, eliminating the moisture and impurity content of the WCO to mitigate more than 1%. The results of its conventional physical attributes are displayed in Table 2.

Table 2. Conventional physical attributes of WCO.

Technical Indexes	Results	Test Methods
Density at 15 °C (g/cm^3)	0.91	ASTM D1298 [36]
Viscosity at 60 °C (mPa·s)	19	ASTM D445 [37]
Acid value (mgKOH/g)	65.28	ASTM D974 [38]
Iodine value (g/100 g)	131.13	ASTM D5558 [39]
Color	blackish brown	-

2.2. *Test Methods*

2.2.1. Pre-Treatment of WCO

In this study, WCO was pretreated using the glycerol esterification method, and the optimum test conditions were 240 °C, 60 min, 500 r/min, and an oil–alcohol molar ratio of 1.1:1. The glycerol esterification reaction test equipment is depicted in Figure 1. Nitrogen gas was released through a wash bottle holding the organic solvent after it passed from the vent through the liquid's surface during the test. In this period, water vapor was produced, which was then condensed through a condensation tube and gathered in a collection device.

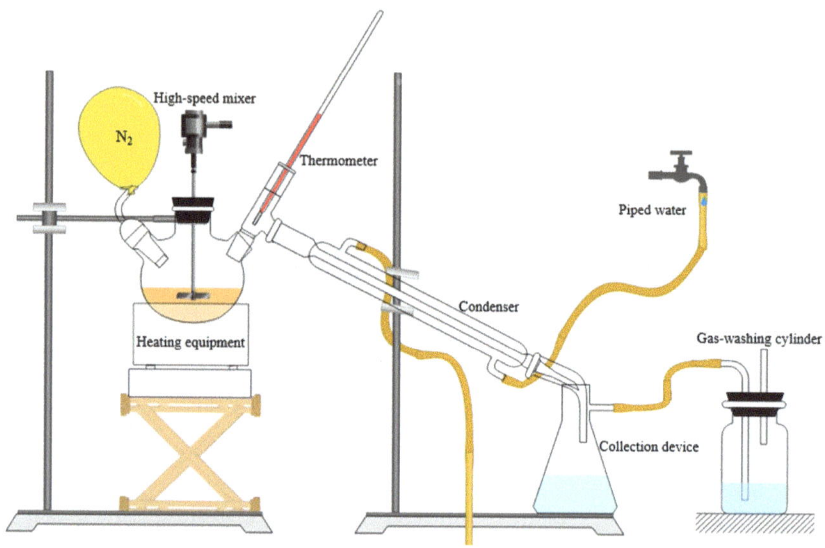

Figure 1. Esterification reaction test equipment.

2.2.2. Preparation of Aged Asphalt

Following ASTM D2872 [40], 35 g ± 0.5 g of virgin asphalt (VA) was aged for 85 min at 163 °C in a rotating thin-film oven (RTFOT) to generate short-term aged asphalt (SA). Subsequently, under ASTM D6521 [41], the collected short-term aged asphalt was placed in a pressure aging vessel (PAV) and aged for 20 h under conditions of 100 °C and 2.1 MPa pressure to obtain long-term aged asphalt (LA). The performance indicators of the virgin asphalt and the asphalt at different aging levels are shown in Table 3.

Table 3. Performance indicators of the virgin asphalt and the asphalt at different aging levels.

Type of Asphalt	Penetration (25 °C, 100 g, 5 s)/(0.1 mm)	Softening Point (Ring and Ball Method)/°C	Ductility (15 °C, 5 cm/min)/cm	Viscosity (mPa·s)
VA	57.5	50	>100	611.9
SA	49.4	55	7	826.4
LA	21.4	63	3.2	1130

2.2.3. Preparation of Recycled Asphalt

In order to investigate the impact of esterified WCO on the performance of recycled asphalt, three types of WCO were distinguished according to varying degrees of pretreatment: original WCO that was not treated (O-WCO), moderately esterified WCO (M-WCO) with a glycerol-to-fatty acid molar ratio of 0.5:1, and WCO that was deeply esterified (D-WCO), which satisfied the optimum pretreatment conditions. Referring to the optimal dosage range for restoring aged asphalt performance, these three rejuvenators with different degrees of esterification were added at a fixed dosage of 6% to 300g of long-term aged asphalt in molten form. Utilizing a high-speed shear mixer set to 2000 r/min for 30 min, the mixture was sheared to produce recycled asphalt, with the heating jacket maintained at a temperature of 135 °C ± 5 °C. Based on the esterification degree of the added rejuvenators, the recycled asphalts were named O-RA, M-RA and D-RA [42]. The acid value ranges of different degrees of esterified WCO rejuvenators and the corresponding names of the recycled asphalts are shown in Table 4.

Table 4. Names of rejuvenators and recycled asphalt with different acid value levels.

Name of Rejuvenator	Acid Value of WCO (mg KOH/g)	Name of Recycled Asphalt
O-WCO	60 ± 2	O-RA
M-WCO	30 ± 2	M-RA
D-WCO	4 ± 2	D-RA

2.2.4. Determination of the Acid Value of WCO

According to ASTM D974, the hot ethanol technique was utilized to determine the acid value of WCO under various reaction circumstances. To neutralize free fatty acids, a 0.1 mol/L sodium hydroxide solution was titrated in a moderately boiling ethanol–WCO solution. The volume of the standard solution needed to reach the titration's endpoint was measured to determine the acid value.

2.2.5. Viscosity Measurement of WCO

A Brookfield rotational viscometer from the USA was used to determine the 60 °C rotational viscosity of the same mass (50 g ± 0.5 g) of O-WCO and D-WCO placed in a film oven at 163 °C after 2, 4, 6, and 8 h of thermo-oxidative aging treatment, and then the differences in the viscosity changes of the different rejuvenators were analyzed.

2.2.6. Thermal Decomposition Test

The relationship between the thermal decomposition of WCO and temperature under a nitrogen atmosphere were obtained by thermogravimetry–derivative thermogravimetry (TG-DTG) simultaneous analyzer to determine the effect of glycerol esterification pretreatment on the thermo-oxidative stability of regeneration. We aimed for 800 °C as the target temperature, and we set the rate of temperature increase to 10 °C/min.

2.2.7. Rheological Properties Test of Asphalt

The high-temperature stability, low-temperature cracking resistance, fatigue performance, and resistance to thermal-oxidative aging of recycled asphalt made from WCO at varying degrees of esterification were all assessed in this study through an Anton Paar

MCR 302 rheometer via frequency scanning (temperature: 5, 10, 15, 28, 40, and 52 °C; frequency: 0.016 Hz~16 Hz), temperature scanning (temperature: 43 °C~82 °C; angular frequency: 10 rad/s; strain level: 6%), linear amplitude scans (temperature: 25 °C; strain level: 0.1%~30%), and multiple stress creep recovery tests (1. Stresses: 0.1 kPa, cyclicality: 20; 2. Stresses: 3.2 kPa, cyclicality: 10).

2.2.8. Microstructure Testing of Asphalt

The functional group properties of WCO, aged asphalt, and WCO-recycled asphalt were examined using Fourier transform infrared spectroscopy (Nicolet iS50) from Thermo Fisher Scientific Inc., Carlsbad, CA, USA. with the wavelength range of 500 cm^{-1} to 4000 cm^{-1}, which offers a qualitative analysis of the evolution of the microscopic structure during the recycling and aging processes of asphalt.

Gel permeation chromatography (PL-GPC50) was used to investigate the effects of the WCO rejuvenator on the components of aged asphalt, the aging mechanism of asphalt via molecular weight distribution, and the anti-aging performance of deeply esterified WCO-recycled asphalt. Tetrahydrofuran (THF) at a concentration of 2 mg/mL was chosen as the organic solvent. The elution procedure took place over 20 min, with a temperature control of 40 °C.

3. Results

3.1. Anti-Aging Properties of WCO Regenerations with Different Degrees of Esterification

The fluctuation patterns of the two WCOs' three physical properties (mass, viscosity, and acid value) against the thermo-oxidative aging period are shown in Figures 2–4, which show how the esterification treatment affects the WCO regenerator's performance.

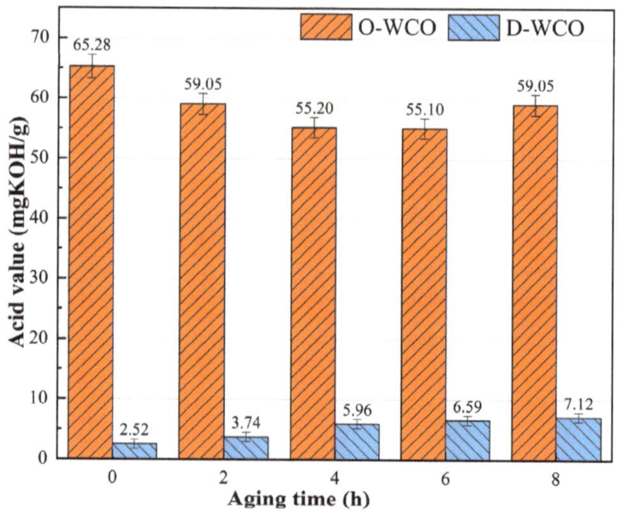

Figure 2. Variation of WCO acid value with varying esterification levels.

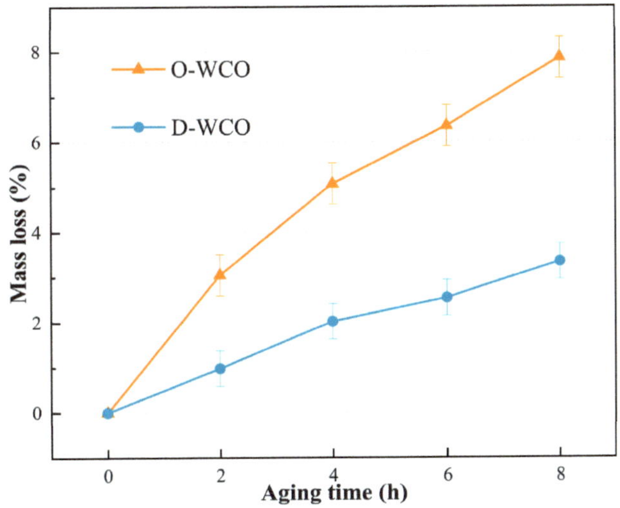

Figure 3. Variation of WCO mass loss rate with varying esterification levels.

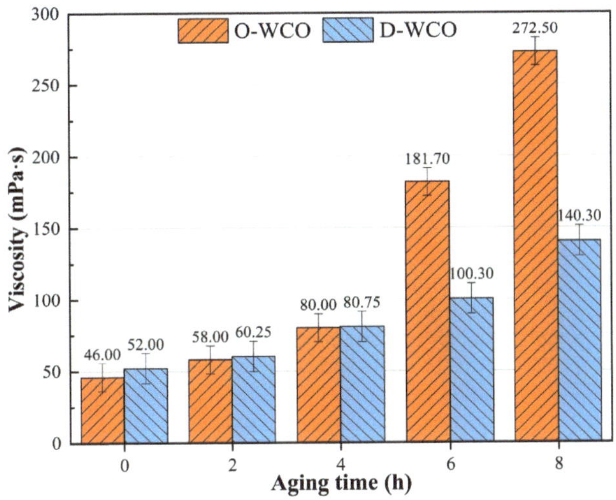

Figure 4. Variation of WCO viscosity with varying esterification levels.

3.1.1. Effect of Thermo-Oxidative Aging on Regenerants' Acid Value

The initial acid value of D-WCO was almost 96% lower than that of O-WCO, as evidenced by the data in Figure 2, suggesting that glycerol esterification significantly decreased the acid value of WCO. A modest increase in the acid value of D-WCO was seen with an increase in the thermo-oxidative aging duration. This could be attributed to the conversion of readily oxidizable functional groups in the oil to carboxyl groups during the thermo-oxidative aging process [43]. However, despite the increase in acid value, it was maintained at a low level compared to O-WCO, showing excellent performance stability.

The initial acid value for O-WCO was high, but as the duration of thermal oxygen treatment increased, it showed a slightly decreasing tendency. This may be due to the fact that it contained high levels of thermally less stable free fatty acids that decompose at high temperatures, and the amount of decomposition exceeds the amount of unsaturated fatty acids produced during the oxidation process, thus slightly lowering the acid value.

When consumed to a certain extent, the acid value slowly increased as the oxidation process proceeded.

3.1.2. Effect of Thermo-Oxidative Aging on Regenerants' Rate of Mass Loss

The mass loss of D-WCO and O-WCO during the thermal oxidative aging process are shown in Figure 3. According to the research, both oil samples had mass changes during thermal-oxidative aging; however, O-WCO shows a significantly higher mass loss rate than D-WCO. This indicates that during the glycerol esterification reaction, some of the compounds such as alcohols, aldehydes, ketones and heterocycles, which are readily volatile and decomposable under thermal oxidative conditions, were consumed, leading to a more stable performance of the pre-treated oil and effectively improving the rejuvenator's resistance to thermal–oxidative aging.

3.1.3. Effect of Thermo-Oxidative Aging on Regenerants' Viscosity

Figure 4 illustrates how the viscosity of both oil samples increased as the thermal oxidative aging period increased. This phenomenon was primarily attributed to the presence of a large number of unsaturated fatty acids (such as linoleic acid and alpha-linolenic acid) in the regenerants, whose α-position carbon–hydrogen bonds are easily attacked, promoting free radical chain reactions that generate oxidation and polymerization products, which caused a significant increase in viscosity. When exposed to thermal–oxidative conditions, D-WCO exhibited a more modest increase in viscosity than O-WCO, suggesting that the esterification reaction can effectively postpone the aging process of the WCO rejuvenator.

3.2. Pyrolysis Characteristics of WCO Regenerants with Different Degrees of Esterification

Good pyrolysis characteristics help ensure the stability of regenerant performance, as regenerants are prepared and used under high temperature conditions. The present investigation employed the thermogravimetric analysis (TGA) technique to examine the regenerant's thermal decomposition behavior. Additionally, thermogravimetric (TG) curves and derivative thermogravimetry (DTG) curves were plotted to visualize the variations in the WCO mass retention and weight loss rate with temperature during the warming process (Figure 5).

The thermal decomposition of WCO could be separated into three major stages, as evidenced by the trends in the mass retention and mass loss rates of the regenerant. The first stage (<350 °C) was the initial reaction phase, where the mass loss of the rejuvenator was primarily due to the volatilization of lighter components and small molecular oils in WCO under high-temperature conditions, resulting in a low weight loss rate. The second stage (350 °C–450 °C) was the reaction phase, in which the main components of the rejuvenator began to combust and decompose, reaching the highest weight loss rate. The third stage (>450 °C) was the end of the reaction phase; during this stage, both the mass retention and weight loss rate curves of the rejuvenator exhibited a more stable trend, indicating that the main components of the rejuvenator were largely burnt out.

The main differences between O-WCO and D-WCO's thermal breakdown characteristics were evident in the first stage, with the second and third stages showing significant similarities. O-WCO began to decompose at 235 °C, whereas D-WCO's initial decomposition temperature was boosted to more than 340 °C, representing a 48.81% increase. In comparison to D-WCO, the DTG curve of O-WCO had one small mass loss rate peak in the first stage, which indicates that the glycerol esterification almost completely consumed the components of WCO that were susceptible to volatilization under pyrolysis conditions and effectively improved the thermal stability of the WCO regenerant below 350 °C.

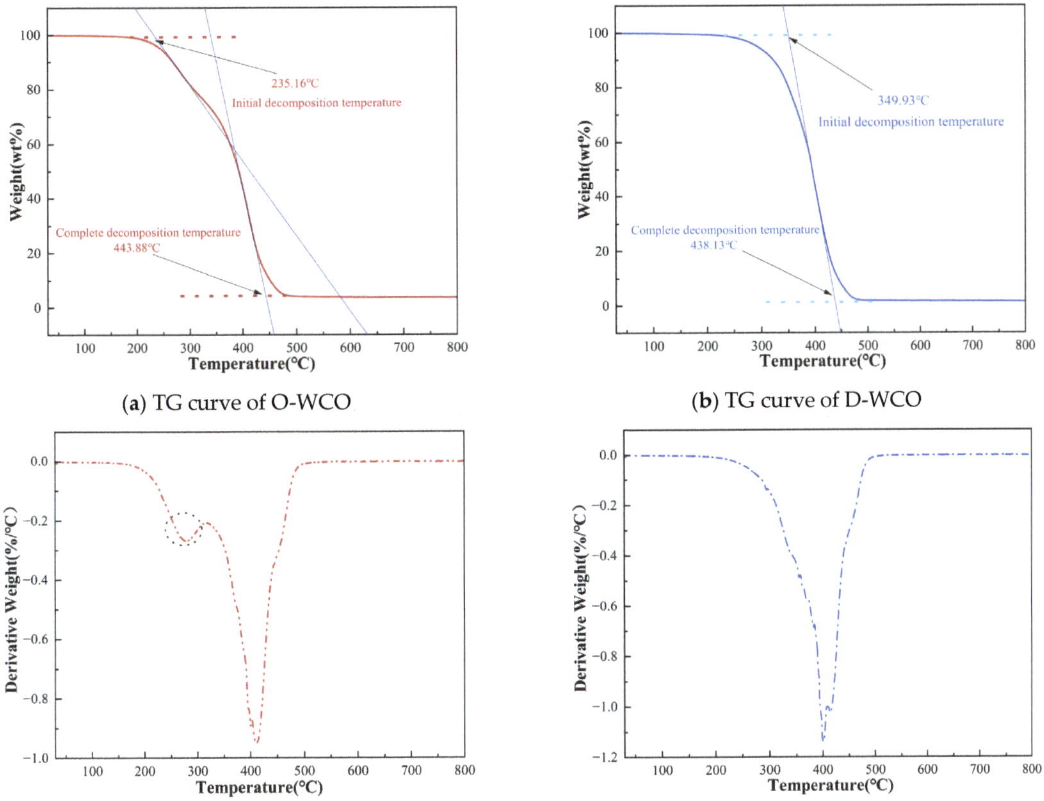

(**a**) TG curve of O-WCO (**b**) TG curve of D-WCO

(**c**) DTG curve of O-WCO (**d**) DTG curve of D-WCO

Figure 5. TG and DTG curves of WCO at varying esterification degrees.

Based on kinetic equations and the Coats–Redfern method, a kinetic model of the thermal decomposition reaction of two oil samples was created. Kinetic parameters, such as the activation energy E and fingering front factor A, were then calculated to serve as a reference for the quantitative investigation of the regenerant's thermal stability [44]. The outcomes are displayed in Table 5.

Table 5. Reaction mechanism model fitting and kinetic parameterization of regenerant's thermal decomposition process under nitrogen atmosphere.

Sample Name	Reaction Model	Simultaneous Equations	Correlation Coefficient	E	A	ΔH	ΔG	ΔS
D-WCO	D1	y = −3455.9x − 4.9965	0.9611	28,732	234	25,432	108,906	−210
O-WCO	F1	y = −601.9x − 12.071	0.9907	5004	0.0344	1577	118,599	−283

From the perspective of activation energy, D-WCO exhibited a higher activation energy E, indicating that the energy barrier for the thermal decomposition reaction was greater, making the decomposition process less likely to occur compared to O-WCO. Considering the change in enthalpy, the enthalpy change value of D-WCO was significantly greater than that of the untreated oil, suggesting that a larger amount of reaction heat was required for the thermochemical decomposition of WCO after glycerol esterification treatment, which reinforces the conclusion that the thermal decomposition reactions were less likely to occur.

This supports the findings of the activation energy analysis and provides more evidence that the esterification reaction can significantly improve the thermal stability of WCO.

3.3. *Performance Analysis of WCO Recycled Asphalt with Different Acid Values*

3.3.1. High-Temperature Stability

A multiple stress creep recovery (MSCR) test was applied to investigate the rheological properties of the recycled asphalts, calculate the stress sensitivity coefficients R_{diff} and $J_{nr\text{-}diff}$ of the various recycled asphalts, and analyze the impact of the esterification degree of waste cooking oil (WCO) on the high-temperature stability of the recycled asphalt. From Figure 6, it can be seen that the R_{diff} and $J_{nr\text{-}diff}$ values for the three types of recycled asphalts at 64 °C were all higher than those at 58 °C, indicating greater viscoelastic changes and higher stress sensitivity at elevated temperatures; this observation is consistent with the phenomenon of rutting in asphalt under high-temperature service conditions. Comparing the recycled asphalt prepared with the same dosage of three different esterification levels of reclaimers, D-RA showed the lowest stress sensitivity at the same temperature, indicating that the esterification treatment can improve the high temperature stability performance of this recycled asphalt.

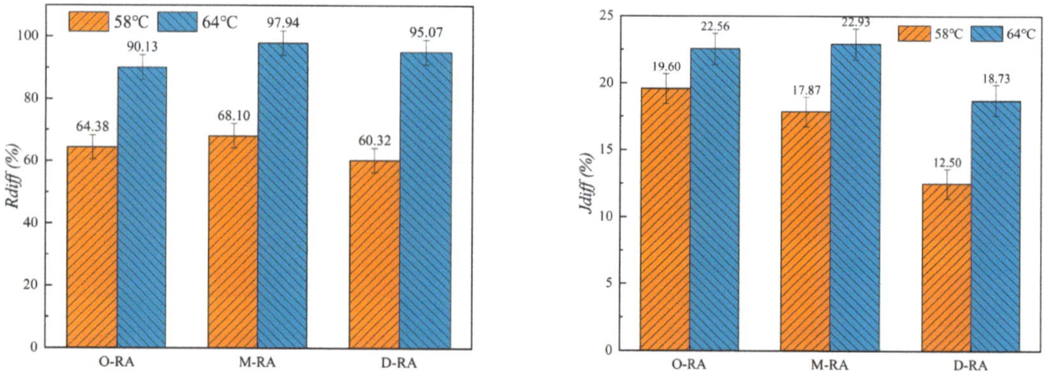

(**a**) R_{diff} (**b**) $J_{nr\text{-}diff}$

Figure 6. Effects of regenerants with different esterification degrees on stress sensitivity coefficients.

3.3.2. Low-Temperature Crack Resistance

The viscoelastic behavior of the asphalts at varying loading frequencies was investigated with the dynamic shear rheometer's frequency scanning mode. The asphalt storage modulus G′ and loss modulus G″ occurring in the high frequency range were determined utilizing the time–temperature equivalency concept, and the corresponding logarithmic curves were plotted. When the two lines connect, the cross-modulus and cross-frequency can be found, allowing for the evaluation of the asphalt material's crack resistance [45,46], as illustrated in Figure 7. Higher crossover frequencies were associated with lower asphaltene content, as the values demonstrated a strong association with asphaltene content [47]. D-RA had the highest crossover frequency, followed by M-RA and O-RA. This demonstrates that D-RA has better low-temperature cracking resistance, since it has lighter components and less asphaltene. The low-temperature cracking resistance of WCO-recycled asphalt improved as the esterification degree rose, according to an analysis of the crossover frequency trend.

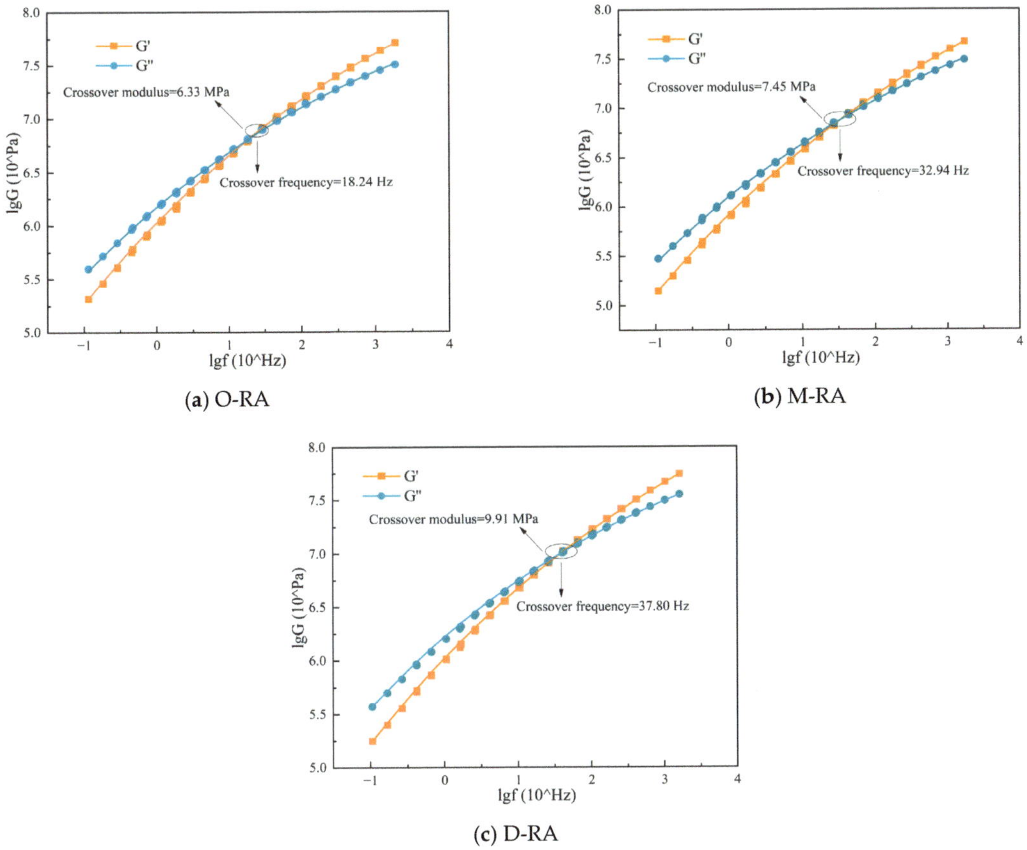

Figure 7. Cross-modulus and cross-frequency values of WCO recycled asphalts with different esterification degrees.

3.3.3. Fatigue Property

To assess the fatigue performance levels of the asphalts, the Linear Amplitude Sweep (LAS) test method, as outlined in AASHTO TP 101-12, was utilized in this study. Figure 8 displays the stress–strain curves of the recycled asphalts with varying acid values. When compared to the other two rejuvenators, D-RA showed the highest peak shear stress and maximum yield stress, indicating greater fatigue performance. Additionally, D-RA exhibited a significantly broader half peak than M-RA and O-RA, which indicates that it has the longest fatigue life. In conclusion, the oil components treated with esterification positively influenced the fatigue performance of recycled asphalt, improving the yield stress and extending the fatigue life.

3.3.4. Thermo-Oxidative Aging Resistance

Following short-term and long-term aging of the base asphalt and three types of recycled asphalts, temperature scans of the asphalt specimens were conducted. The complex modulus aging index (CAI) and phase angle aging index (PAI) were then utilized as the evaluation indexes of the asphalts' resistance to aging. For clarity of presentation, the short- and long-term aging resistance indexes of the complex modulus were denoted as R-CAI and P-CAI, and the short- and long-term aging resistance indexes of the phase angles as R-PAI and P-PAI. As shown in Figures 9 and 10, the complex modulus values rose, while the phase angles decreased as the asphalt aged, so the lower CAI and higher PAI values indicate better asphalt resistance to aging.

From the perspective of CAI, the R-CAI values of the WCO recycled asphalt with different degrees of esterification were quite similar and slightly higher than the short-term anti-aging index values of the base asphalt. This indicates that pretreatment has less of an effect on recycled asphalt's resistance to short-term aging. This could be as a result of less volatilization of the lightweight components under short-term aging circumstances of 163 °C and 85 min, leading to a closer change in the complex modulus. In contrast, the base asphalt has fewer lightweight components and a smaller molecule oil content than recycled asphalt, so it showed better resistance to short-term aging. D-RA had the lowest P-CAI among the three types of recycled asphalt, which was comparable to the base asphalt according to the trend of P-CAI changes in the recycled asphalt. This suggests that glycerol esterification treatment can enhance the thermal stability of the reclaimer, maintain the proportion of lightweight components of the reclaimed asphalt, and improve the resistance to long-term aging.

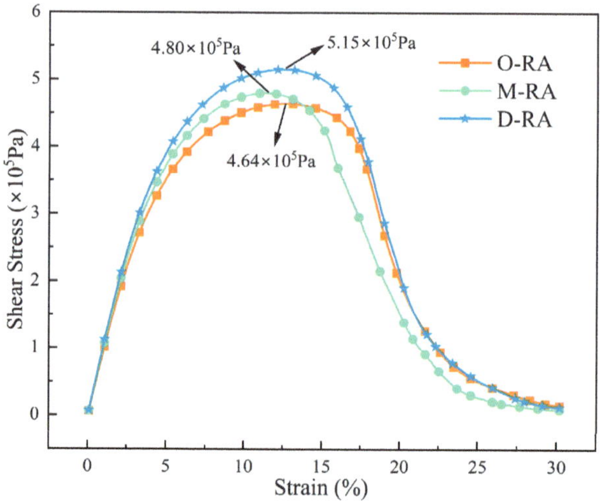

Figure 8. The stress–strain curves of WCO recycled asphalt with different degrees of esterification.

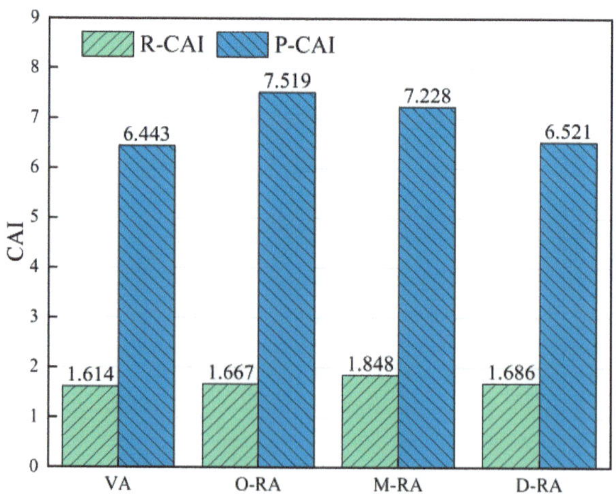

Figure 9. Complex modulus values anti-aging index.

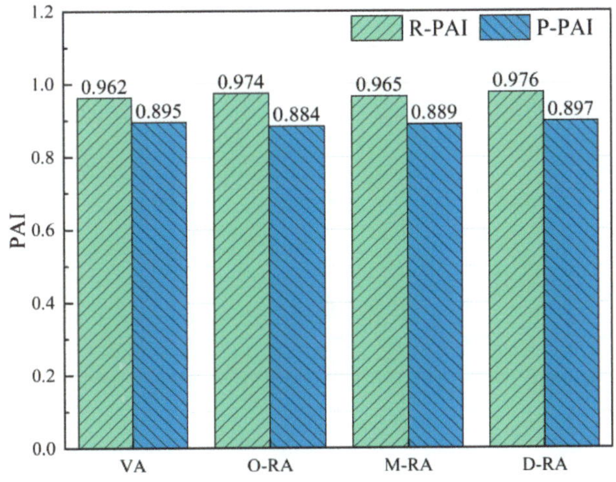

Figure 10. Phase angle values anti-aging index.

The effect of aging on the percentage of viscoelastic materials in asphalt was represented by the PAI values. D-RA exhibited the lowest phase angle loss during age under the identical regenerant dosing conditions, with M-RA and O-RA showing the greatest phase angle reductions in that order. This indicates that among rejuvenators with varying degrees of esterification, the deeply esterified rejuvenator has the most significant effect on slowing down the changes in the proportion of viscoelastic components in asphalt during the aging process. In addition to providing support for the lightweight components, the pre-treated rejuvenator also improved the thermo-oxidative stability of the recycled asphalt, reduced the loss of volatile components during thermo-oxidative aging, and increased the performance of the recycled asphalt retention capacity.

3.4. Microstructural and Mechanistic Analysis of Recycled Asphalt

3.4.1. Analysis of WCO Recycled Asphalt's Functional Group Changes and Antioxidant Properties

The infrared spectral tests were conducted on long-term aged bade asphalt, recycled asphalt, long0term aged recycled asphalt, and D-WCO using a Fourier transform infrared spectrometer (Nicolet iS50) with a wavelength test range from 500 cm^{-1} to 4000 cm^{-1}, and the infrared spectrogram is shown in Figure 11. LA stands for long-term aged matrix asphalt, RA denotes recycled asphalt, and RA-P refers to long-term aged recycled asphalt.

Stretching vibration peaks were seen in the spectra for methyl groups (2093 cm^{-3}), methylene groups (2850 cm^{-3}), and benzene rings (1700 cm^{-3}) for LA, RA, RA-P, and D-WCO. Furthermore, deformation vibration peaks, which belong to methylene groups, were visible at 1375 and 1450 cm^{-1}. Additionally, D-WCO showed characteristic peaks at 1740 cm^{-1}, 1375 cm^{-1}, and 1450 cm^{-1}, indicating that the functional group contains an ester group. When D-WCO was added to LA, a new ester carbonyl (-C=O-) absorption peak appeared in RA. In the recycled asphalt, besides the classic peaks of D-WCO and LA, no additional new absorption peaks were identified. This indicates that the physical co-mingling between WCO and asphalt was predominant and that no chemical reaction occurred or that the reaction was weak. The strength of the sulfoxide (-S=O-) absorption peak at 1030 cm^{-3} in LA decreased from 0.06 to around 0.04 after D-WCO was added, and then it increased to approximately 0.05 after long-term aging, which was still less than the initial value. This indicates that the sulfoxide intensity was effectively reduced by the addition of D-WCO and that the esterification reaction also somewhat increased the oxidative resistance of the D-WCO recycled asphalt.

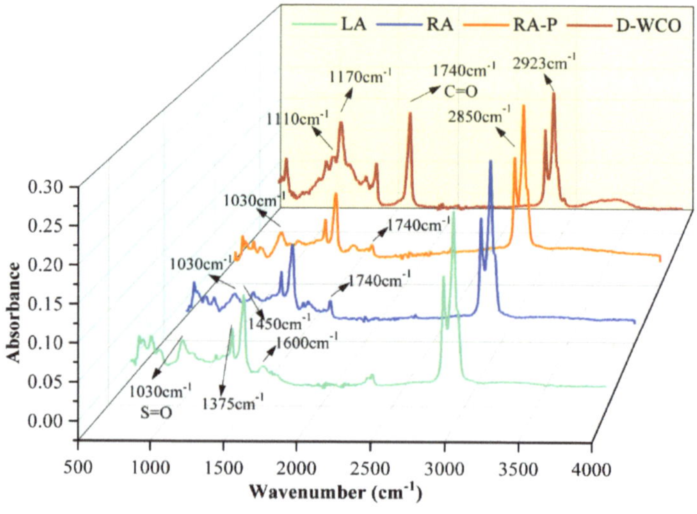

Figure 11. Infrared spectrogram of D-WCO and different asphalts.

3.4.2. Laws of Molecular Evolution during Aging of Base Asphalt and Recycled Asphalt

The molecular weight and distribution properties of base asphalt, short-term aged asphalt, long-term aged asphalt, recycled asphalt, and recycled asphalt with varying degrees of aging were examined in this study using gel chromatography (PL-GPC50). The relative proportions of molecular weights are displayed in Figure 12.

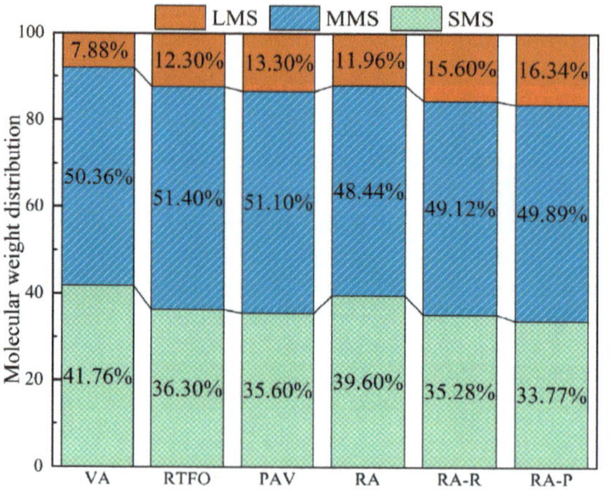

Figure 12. Effect of aging on molecular weight distribution.

By introducing the rejuvenator, the proportion of large molecular weight substances (LMSs) could be reduced while diluting the aged asphalt according to its rich light components, thereby adjusting the molecular weight distribution. In addition to the diluting effect, the depolymerization of asphalt dimers was also included in the rejuvenator's mechanism of action, which further reduces the aggregation of large molecules [48]. By reducing the proportion of large molecular weight compounds (LMSs) and boosting the amounts of small molecular weight substances (SMSs) and medium molecular weight molecules (MMSs), this process efficiently optimizes the molecular composition of aged asphalt. During the aging process of the asphalt materials, the proportion of LMSs increased while the proportion of SMSs decreased, revealing the impact of aging on the molecular composition of the asphalt. This phenomenon could be attributed to the volatilization and oxidation of light small molecular components during the aging stage, leading to a decrease in small molecular quantities or their transformation into medium and large molecular weight substances [49]. Since WCO rejuvenators primarily achieve the regeneration of aged asphalt by supplementing a large number of light components, the aging process of rejuvenated asphalt is also marked by the molecular change phenomena.

The number-average molecular weight (Mn), weight-average molecular weight (Mw), and polydispersity index (PDI) of asphalt at different aging degrees are listed in Table 6. The volatilization of oil components after aging leads to an increase in the Mn and a decrease in the Mw of the composite rejuvenated asphalt; however, the data presented show a trend of increasing Mw, indicating that aggregation occurs between asphaltene molecules after asphalt aging, resulting in an increase in the large molecular content [50].

Table 6. Effect of aging on molecular weight distribution parameters of base and recycled asphalt.

Molecular Mass	Mn (g/mol)			Mw (g/mol)			PDI (Mw/Mn)		
Aging degree	Unaged	RTFO	PAV	Unaged	RTFO	PAV	Unaged	RTFO	PAV
Base asphalt	822	880	894	1704	2230	2570	2.07	2.48	2.59
Recycled asphalt	864	980	991	2111	2463	2647	2.44	2.51	2.67

The polydispersity index (PDI) in the GPC results of asphalt were correlated with the high-temperature performance outcomes of the asphalts. As the PDI increased, the high-temperature performance improved [51]. With the increase in the aging degree, the PDIs of the asphalts also increased, and the PDIs of the recycled asphalts at each stage were greater than that of the base asphalt. This indicates that following pretreatment, the WCO eliminated some tiny molecules with poor thermal stability, producing a more stable and effective light component supplement for the aged asphalt that reduced volatility and improves the recycled asphalt's performance at high temperatures.

4. Conclusions

(1) After 2, 4, 6, and 8 h of thermal oxidation aging, the mass, viscosity, and acid value changes of D-WCO were all smaller than those of O-WCO, and no thermal decomposition happened below 350 °C. This suggests that glycerol esterification can effectively improve WCO rejuvenators' resistance to thermal–oxidative aging and improve the stability of their performance.

(2) The rheological performance test results indicate that D-RA is characterized by the lowest temperature sensitivity, a higher content of light components, superior high-temperature stability, and low-temperature crack resistance.

(3) Strong evidence for improving the durability of the recycled asphalt pavement is provided by the esterification treatment, which makes sure that the small molecular components added to the aged asphalt during the incorporation of WCO are more stable. This reduces the volatilization and oxidation during the thermal oxidation aging of the recycled asphalt.

Author Contributions: Conceptualization, J.W. and Q.Y.; methodology, Q.Y.; software, L.F.; validation, J.W., L.F. and C.X.; formal analysis, L.F.; investigation, J.W., L.F., C.X. and H.L.; resources, Q.Y., C.X. and H.L.; data curation, C.X. and H.L.; writing—original draft preparation, J.W.; writing—review and editing, J.W.; visualization, H.L.; supervision, Q.Y.; project administration, Q.Y.; funding acquisition, Q.Y. All authors have read and agreed to the published version of the manuscript.

Funding: This research was funded by the National Natural Science Foundation of China (No. 52378437) and the Ministry of Transport of the People's Republic of China (No. 2020-MS1-005).

Institutional Review Board Statement: Not applicable.

Informed Consent Statement: Not applicable.

Data Availability Statement: The data presented in this study are available on request from the corresponding author.

Conflicts of Interest: Authors Cheng Xie and Haobin Liu was employed by the company Guangxi Beitou Transportation Maintenance Technology Group Co. Ltd. The remaining authors declare that the research was conducted in the absence of any commercial or financial relationships that could be construed as a potential conflict of interest.

References

1. Liu, L.; Sun, L.; Xu, J.; Li, M.; Xing, C.; Zhang, Y. Effect of RAP's Preheating Temperature on the Secondary Aging and Performance of Recycled Asphalt Mixtures Containing High RAP Content. *Constr. Build. Mater.* **2024**, *411*, 134719. [CrossRef]
2. Yang, C.; Zhang, J.; Yang, F.; Cheng, M.; Wang, Y.; Amirkhanian, S.; Wu, S.; Wei, M.; Xie, J. Multi-Scale Performance Evaluation and Correlation Analysis of Blended Asphalt and Recycled Asphalt Mixtures Incorporating High RAP Content. *J. Clean. Prod.* **2021**, *317*, 128278. [CrossRef]
3. Jiang, X.; Zhang, F.; Huang, B.; Titi, H.; Polaczyk, P.; Ma, Y.; Wang, Y.; Cheng, Z. Full-Scale Accelerated Testing of Geogrid-Reinforced Inverted Pavements. *Geotext. Geomembr.* **2024**, *52*, 511–525. [CrossRef]
4. Zaumanis, M.; Mallick, R.B.; Frank, R. 100% Recycled Hot Mix Asphalt: A Review and Analysis. *Resour. Conserv. Recycl.* **2014**, *92*, 230–245. [CrossRef]
5. Nosetti, A.; Pérez-Madrigal, D.; Pérez-Jiménez, F.; Martínez, A.H. Effect of the Recycling Process and Binder Type on Bituminous Mixtures with 100% Reclaimed Asphalt Pavement. *Constr. Build. Mater.* **2018**, *167*, 440–448. [CrossRef]
6. Fakhri, M.; Saberi, K.F. The Effect of Waste Rubber Particles and Silica Fume on the Mechanical Properties of Roller Compacted Concrete Pavement. *J. Clean. Prod.* **2016**, *129*, 521–530. [CrossRef]
7. Fakhri, M.; Amoosoltani, E.; Aliha, M.R.M. Crack Behavior Analysis of Roller Compacted Concrete Mixtures Containing Reclaimed Asphalt Pavement and Crumb Rubber. *Eng. Fract. Mech.* **2017**, *180*, 43–59. [CrossRef]
8. Anthonissen, J.; Van Den Bergh, W.; Braet, J. Review and Environmental Impact Assessment of Green Technologies for Base Courses in Bituminous Pavements. *Environ. Impact Assess. Rev.* **2016**, *60*, 139–147. [CrossRef]
9. Zhao, W.; Yang, Q. Life-Cycle Assessment of Sustainable Pavement Based on the Coordinated Application of Recycled Asphalt Pavement and Solid Waste: Environment and Economy. *J. Clean. Prod.* **2024**, *434*, 140203. [CrossRef]
10. Li, D.; Leng, Z.; Yao, L.; Cao, R.; Zou, F.; Li, G.; Wang, H.; Wang, H. Mechanical, Economic, and Environmental Assessment of Recycling Reclaimed Asphalt Rubber Pavement Using Different Rejuvenation Schemes. *Resour. Conserv. Recycl.* **2024**, *204*, 107534. [CrossRef]
11. Zaumanis, M.; Poulikakos, L.; Arraigada, M.; Kunz, B.; Schellenberg, U.; Gassmann, C. Asphalt Recycling in Polymer Modified Pavement: A Test Section and Recommendations. *Constr. Build. Mater.* **2023**, *409*, 134005. [CrossRef]
12. Lu, D.; Jiang, X.; Leng, Z. Sustainable Microwave-Heating Healing Asphalt Concrete Fabricated with Waste Microwave-Sensitive Fillers. *J. Clean. Prod.* **2024**, *434*, 140343. [CrossRef]
13. Zaumanis, M.; Mallick, R.B. Review of Very High-Content Reclaimed Asphalt Use in Plant-Produced Pavements: State of the Art. *Int. J. Pavement Eng.* **2015**, *16*, 39–55. [CrossRef]
14. Zaumanis, M.; Mallick, R.B.; Frank, R. 100% Hot Mix Asphalt Recycling: Challenges and Benefits. *Transp. Res. Procedia* **2016**, *14*, 3493–3502. [CrossRef]
15. Lin, M.; Shuai, J.; Li, P.; Kang, X.; Lei, Y. Analysis of Rheological Properties and Micro-Mechanism of Aged and Reclaimed Asphalt Based on Multi-Scales. *Constr. Build. Mater.* **2022**, *321*, 126290. [CrossRef]
16. Li, H.; Dong, B.; Wang, W.; Zhao, G.; Guo, P.; Ma, Q. Effect of Waste Engine Oil and Waste Cooking Oil on Performance Improvement of Aged Asphalt. *Appl. Sci.* **2019**, *9*, 1767. [CrossRef]
17. Zadshir, M.; Oldham, D.J.; Hosseinnezhad, S.; Fini, E.H. Investigating Bio-Rejuvenation Mechanisms in Asphalt Binder via Laboratory Experiments and Molecular Dynamics Simulation. *Constr. Build. Mater.* **2018**, *190*, 392–402. [CrossRef]
18. Zhou, C.L.; Zheng, C.C.; Cheng, D.X. Evaluation of Environmental Pollution from Asphalt Recycling Technology. *AMR* **2014**, *898*, 482–485. [CrossRef]
19. Zargar, M.; Ahmadinia, E.; Asli, H.; Karim, M.R. Investigation of the Possibility of Using Waste Cooking Oil as a Rejuvenating Agent for Aged Bitumen. *J. Hazard. Mater.* **2012**, *233*, 254–258. [CrossRef]
20. Asli, H.; Ahmadinia, E.; Zargar, M.; Karim, M.R. Investigation on Physical Properties of Waste Cooking Oil—Rejuvenated Bitumen Binder. *Constr. Build. Mater.* **2012**, *37*, 398–405. [CrossRef]
21. Zhao, Y.; Chen, M.; Zhang, X.; Wu, S.; Zhou, X.; Jiang, Q. Effect of Chemical Component Characteristics of Waste Cooking Oil on Physicochemical Properties of Aging Asphalt. *Constr. Build. Mater.* **2022**, *344*, 128236. [CrossRef]
22. Behnood, A. Application of Rejuvenators to Improve the Rheological and Mechanical Properties of Asphalt Binders and Mixtures: A Review. *J. Clean. Prod.* **2019**, *231*, 171–182. [CrossRef]
23. Ma, J.; Hu, M.; Sun, D.; Lu, T.; Sun, G.; Ling, S.; Xu, L. Understanding the Role of Waste Cooking Oil Residue during the Preparation of Rubber Asphalt. *Resour. Conserv. Recycl.* **2021**, *167*, 105235. [CrossRef]
24. Ma, J.; Sun, D.; Pang, Q.; Sun, G.; Hu, M.; Lu, T. Potential of Recycled Concrete Aggregate Pretreated with Waste Cooking Oil Residue for Hot Mix Asphalt. *J. Clean. Prod.* **2019**, *221*, 469–479. [CrossRef]

25. Yan, K.; Liu, W.; You, L.; Ou, J.; Zhang, M. Evaluation of Waste Cooling Oil and European Rock Asphalt Modified Asphalt with Laboratory Tests and Economic Cost Comparison. *J. Clean. Prod.* **2021**, *310*, 127364. [CrossRef]
26. Yan, S.; Dong, Q.; Chen, X.; Zhou, C.; Dong, S.; Gu, X. Application of Waste Oil in Asphalt Rejuvenation and Modification: A Comprehensive Review. *Constr. Build. Mater.* **2022**, *340*, 127784. [CrossRef]
27. El-Shorbagy, A.M.; El-Badawy, S.M.; Gabr, A.R. Investigation of Waste Oils as Rejuvenators of Aged Bitumen for Sustainable Pavement. *Constr. Build. Mater.* **2019**, *220*, 228–237. [CrossRef]
28. Cao, X.; Cao, X.; Tang, B.; Wang, Y.; Li, X. Investigation on Possibility of Waste Vegetable Oil Rejuvenating Aged Asphalt. *Appl. Sci.* **2018**, *8*, 765. [CrossRef]
29. Cao, Z.; Chen, M.; Han, X.; Wang, R.; Yu, J.; Xu, X.; Xue, L. Influence of Characteristics of Recycling Agent on the Early and Long-Term Performance of Regenerated SBS Modified Bitumen. *Constr. Build. Mater.* **2020**, *237*, 117631. [CrossRef]
30. Zhang, R.; Wang, H.; You, Z.; Jiang, X.; Yang, X. Optimization of Bio-Asphalt Using Bio-Oil and Distilled Water. *J. Clean. Prod.* **2017**, *165*, 281–289. [CrossRef]
31. *ASTM D5*; Standard Test Method for Penetration of Bituminous Materials. ASTM International: West Conshohocken, PA, USA, 2006.
32. *ASTM D36*; Standard Test Method for Softening Point of Bitumen (Ring-and-Ball Apparatus). ASTM International: West Conshohocken, PA, USA, 2020.
33. *ASTM D113*; Standard Test Method for Ductility of Asphalt Materials. ASTM International: West Conshohocken, PA, USA, 2017.
34. *ASTM D4402*; Standard Test Method for Viscosity Determination of Asphalt at Elevated Temperatures Using a Rotational Viscometer. ASTM International: West Conshohocken, PA, USA, 2006.
35. *ASTM D92*; Standard Test Method for Flash and Fire Points by Cleveland Open Cup Tester. ASTM International: West Conshohocken, PA, USA, 2018.
36. *ASTM D1298*; Standard Test Method for Density, Relative Density, or API Gravity of Crude Petroleum and Liquid Petroleum Products by Hydrometer Method. ASTM International: West Conshohocken, PA, USA, 2017.
37. *ASTM D445*; Standard Test Method for Kinematic Viscosity of Transparent and Opaque Liquids (and Calculation of Dynamic Viscosity). ASTM International: West Conshohocken, PA, USA, 2024.
38. *ASTM D974*; Standard Test Method for Acid and Base Number by Color-Indicator Titration. ASTM International: West Conshohocken, PA, USA, 2022.
39. *ASTM D5558*; Standard Test Method for Determination of the Saponification Value of Fats and Oils. ASTM International: West Conshohocken, PA, USA, 2023.
40. *ASTM D2872*; Standard Test Method for Effect of Heat and Air on a Moving Film of Asphalt Binder (Rolling Thin-Film Oven Test). ASTM International: West Conshohocken, PA, USA, 2022.
41. *ASTM D6521*; Standard Practice for Accelerated Aging of Asphalt Binder Using a Pressurized Aging Vessel (PAV). ASTM International: West Conshohocken, PA, USA, 2022.
42. Fang, Y.; Zhang, Z.; Yang, J.; Li, X. Comprehensive Review on the Application of Bio-Rejuvenator in the Regeneration of Waste Asphalt Materials. *Constr. Build. Mater.* **2021**, *295*, 123631. [CrossRef]
43. Li, K.; Chen, J.; Nie, X. Synthesis of Biodiesel from Waste Oil by Glycerol Pre-Esterification. *Energy Rep.* **2023**, *10*, 3223–3228. [CrossRef]
44. Fan, Y.; Li, L.; Tippayawong, N.; Xia, S.; Cao, F.; Yang, X.; Zheng, A.; Zhao, Z.; Li, H. Quantitative Structure-Reactivity Relationships for Pyrolysis and Gasification of Torrefied Xylan. *Energy* **2019**, *188*, 116119. [CrossRef]
45. Asadi, B.; Tabatabaee, N.; Hajj, R. Use of Linear Amplitude Sweep Test as a Damage Tolerance or Fracture Test to Determine the Optimum Content of Asphalt Rejuvenator. *Constr. Build. Mater.* **2021**, *300*, 123983. [CrossRef]
46. Samieadel, A.; Islam Rajib, A.; Phani Raj Dandamudi, K.; Deng, S.; Fini, E.H. Improving Recycled Asphalt Using Sustainable Hybrid Rejuvenators with Enhanced Intercalation into Oxidized Asphaltenes Nanoaggregates. *Constr. Build. Mater.* **2020**, *262*, 120090. [CrossRef]
47. Pahlavan, F.; Hung, A.; Fini, E.H. Evolution of Molecular Packing and Rheology in Asphalt Binder during Rejuvenation. *Fuel* **2018**, *222*, 457–464. [CrossRef]
48. Fini, E.; Rajib, A.I.; Oldham, D.; Samieadel, A.; Hosseinnezhad, S. Role of Chemical Composition of Recycling Agents in Their Interactions with Oxidized Asphaltene Molecules. *J. Mater. Civ. Eng.* **2020**, *32*, 04020268. [CrossRef]
49. Liu, F.; Zhou, Z.; Zhang, X. Construction of Complex Shear Modulus and Phase Angle Master Curves for Aging Asphalt Binders. *Int. J. Pavement Eng.* **2022**, *23*, 536–544. [CrossRef]
50. Bocci, E.; Prosperi, E.; Mair, V.; Bocci, M. Ageing and Cooling of Hot-Mix-Asphalt during Hauling and Paving—A Laboratory and Site Study. *Sustainability* **2020**, *12*, 8612. [CrossRef]
51. Ye, W.; Jiang, W.; Li, P.; Yuan, D.; Shan, J.; Xiao, J. Analysis of Mechanism and Time-Temperature Equivalent Effects of Asphalt Binder in Short-Term Aging. *Constr. Build. Mater.* **2019**, *215*, 823–838. [CrossRef]

materials

MDPI

Article

Evaluating Impact of Thermo-Oxidative and Ultraviolet Aging on Performance of Hot In-Place Recycled Asphalt Mixtures

Yao Guan [1], Yao Zhang [1,2,*], Tianyi Sang [1], Yifeng Ding [1], Zichao Yan [1] and Aihong Kang [1,2]

[1] College of Civil Science and Engineering, Yangzhou University, Yangzhou 225127, China; yaogwhx@163.com (Y.G.); stytt7@163.com (T.S.); goethe0411@126.com (Y.D.); 18156736818@163.com (Z.Y.); 003129@yzu.edu.cn (A.K.)

[2] Research Center for Basalt Fiber Composite Construction Materials, Yangzhou 225127, China

* Correspondence: yaozhang@yzu.edu.cn

Abstract: Hot in-place recycling (HIR) is a sustainable pavement rehabilitation method. However, it is susceptible to aging processes that can compromise its mechanical properties and long-term performance. This study investigates the effects of thermo-oxidative (TO) and ultraviolet (UV) aging on HIR mixtures. Basic performance tests were conducted on the aggregate gradation, moisture content, and asphalt content of the reclaimed asphalt pavement (RAP) to assess the aging level. Simulations of long-term and short-term oxidative aging of the HIR mixture, along with 12 months of UV irradiation, were performed to evaluate its high-temperature stability, low-temperature crack resistance, and water stability. The Verhulst model was employed to establish a predictive equation for performance attenuation under UV aging. To quantify the photoaging effect, indicators for UV aging degree were proposed to characterize the road performance of the HIR mixture, including the aging rate and the aging residual index. Results indicate that the improvement in high-temperature performance after aging is limited, but cracking resistance decreases substantially. Notably, the flexural tensile strain was reduced by 129.25 με for 10 years of TO aging compared to 12 months of UV exposure, underscoring the importance of considering environmental factors in performance predictions. This study emphasizes the need for enhanced aging mitigation strategies to improve the sustainability and reliability of HIR mixtures in practical applications.

Keywords: asphalt pavement; hot in-place recycling technology; ultraviolet aging; pavement performance; Verhulst model; performance prediction

Citation: Guan, Y.; Zhang, Y.; Sang, T.; Ding, Y.; Yan, Z.; Kang, A. Evaluating Impact of Thermo-Oxidative and Ultraviolet Aging on Performance of Hot In-Place Recycled Asphalt Mixtures. *Materials* **2024**, *17*, 5813. https://doi.org/10.3390/ma17235813

Academic Editor: Giovanni Polacco

Received: 21 October 2024
Revised: 22 November 2024
Accepted: 25 November 2024
Published: 27 November 2024

1. Introduction

HIR pavement technology has become one of the most widely used methods in road maintenance and repair projects due to its advantages in energy efficiency, environmental sustainability, and cost reduction [1,2]. However, reclaimed asphalt pavement (RAP) deteriorates significantly during long-term service [3]. While the viscoelastic properties of asphalt can be partially restored with the addition of rejuvenators [4], recycled asphalt mixtures degrade more rapidly than new asphalt mixtures when exposed to high-temperature construction, traffic loads, and environmental factors [5,6]. These factors lead to an increased frequency of maintenance, which hampers the broader adoption and application of HIR pavement technology. Most current research focuses on the effects of aging on new asphalt mixtures and the performance enhancement of recycled asphalt mixtures. However, the impact of aging during the continued service life of recycled asphalt mixtures is seldom addressed, and the underlying aging mechanisms still need to be clarified.

During the service life of asphalt mixtures, the light fractions of asphalt undergo volatilization and cracking, leading to increased stiffness and reduced flexibility [7]. To address this, admixtures are often added during RAP regeneration to restore or improve the crack resistance of the asphalt mixture. Rejuvenators are used to replenish the light components in RAP, while fibers are incorporated to enhance the network structure of the asphalt binder [8,9].

Currently, the most commonly used recycling methods in engineering are HIR and hot mix plant recycling [10]. With advancements in hot mix plant recycling technology, the proportion of RAP in these applications is gradually increasing. However, compared to HIR, which allows for RAP blending ratios of over 70%, the economic benefits of hot mix plant recycling are limited when the RAP content is less than 50%. Although HIR offers advantages such as high energy efficiency and short construction periods [9], the performance of recycled asphalt mixtures after HIR is significantly affected by aging, primarily due to the high RAP content and secondary aging in thermal regeneration construction.

Many researchers have studied the performance effects of aging on asphalt mixtures and simulated TO and UV aging in natural engineering environments using indoor artificial accelerated aging tests [11,12]. The carbon (C) and sulfur (S) elements in asphalt undergo oxidation reactions with oxygen, and elevated temperatures accelerate these reactions. As a result, TO aging is commonly used to investigate the effects of aging on the thermal stability of styrene-butadiene-styrene (SBS)-modified asphalt [13]. Pei Z conducted TO and UV aging tests on asphalt using a pressurized aging vessel and custom UV aging equipment, revealing significant differences in the microstructure of aged asphalt [14,15]. Additionally, UV aging is a critical factor contributing to the deterioration of asphalt pavements. UV radiation severely damages the network structure formed in SBS-modified asphalt, exacerbating the degradation of SBS and causing significant changes in rheological properties following UV aging [16]. Chen Z et al. found that as UV aging progresses, asphalt films exhibit vertical gradients in molecular size and oxidation product content [17]. Some researchers have examined the combined effects of TO and UV aging on asphalt mixtures, observing that UV aging weakens asphalt cohesion and interfacial adhesion [18]. Xu Y et al. investigated the aging mechanisms of asphalt under both TO and UV conditions, concluding that composite aging leads to more severe stratification compared to single aging conditions [19].

The development of aging models is crucial for understanding the effects of aging on asphalt mixture performance and for minimizing resource waste associated with repetitive and time-consuming performance tests [20]. Nonlinear modeling has proven effective in accurately characterizing the aging rate and degree of asphalt binders [21]. Guan M et al. conducted multifactor coupling aging tests on asphalt binders using an accelerated environmental aging oven, incorporating UV radiation, temperature, and moisture. The findings indicate that the Glover-Rowe parameter serves as a reliable, comprehensive index for evaluating the rheological properties of asphalt based on correlation analysis [22]. Aging in asphalt tends to follow a gradient from the surface layer to the deeper layers of the pavement, and the Verhulst model has been shown to effectively explain the aging behavior at different nanoscale depths [23].

In this study, TO and UV aging of hot in-place recycled mixtures were simulated to reflect the actual environmental conditions of the in-serviced pavement. The high-temperature stability, low-temperature crack resistance, and water stability of mixtures after secondary aging were then evaluated. The Verhulst model was used to establish a predictive model for performance attenuation under UV aging. Additionally, UV aging indicators were proposed, including the aging rate and aging residual index, to quantify the photoaging effects on the road performance of HIR asphalt mixtures.

2. Raw Materials and Experimental Design

2.1. Overview of the Original Pavement Project

The RAP used was sourced from the HIR project on the Baoying section of the G233 national and provincial trunk road, which has a pavement design traffic load rating for heavy traffic. The project was fully completed and opened to traffic in September 2014, and the RAP had been in service for nine years at the time of sampling.

(1) Introduction of pavement distress

Figure 1 illustrates the significant rutting distress observed in the original pavement section. The primary distress type identified was transverse cracking, which accounted for 49.67% of the total affected area and exhibited a relatively dense distribution. Longitudinal cracking, comprising 15.97% of the affected area, was found across multiple sections. Field investigations indicated that substandard repair methods for localized pavement issues led to secondary problems, such as potholes and slurry formation under traffic loads. These defects not only degrade overall driving quality but also pose risks to pavement safety.

(a) (b) (c) (d)

Figure 1. Distress types on the original pavement: (**a**) Rutting, (**b**) Transverse cracks, (**c**) Longitudinal cracks, (**d**) Pavement patching.

(2) Asphalt content and gradation of the original pavement

The RAP was obtained from the original pavement by heated milling, which does not broadly break the coarse aggregate particles, and the original gradation is not overly refined. Therefore, the RAP's gradation is within the limits of the specification. Trichloroethylene was used as a solvent to remove the asphalt from the RAP. The gradation was determined using the water sieve method, as shown in Table 1. Following the "Technical Specification for Highway Asphalt Pavement Recycling", moisture and asphalt content tests were conducted on the RAP sample. The moisture content of the coarse and fine aggregates in the RAP material was approximately 2.31% and 4.01%, respectively. The asphalt content of the reclaimed pavement was around 5.45%.

Table 1. Original pavement gradation.

	Percentage Passing Through the Aggregate Sieves/%									
Sieve Size (mm)	16.0	13.2	9.5	4.75	2.36	1.18	0.6	0.3	0.15	0.075
Original gradation	100.0	96.2	77.4	40.3	27.3	21.1	15.8	11.0	8.2	4.9
Upper limit	100.0	100.0	85.0	68.0	50.0	38.0	28.0	20.0	15.0	8.0
Median	100.0	95.0	76.5	53.0	37.0	26.5	19.0	13.5	10.0	6.0
Lower limit	100.0	90.0	68.0	38.0	24.0	15.0	10.0	7.0	5.0	4.0

2.2. *Raw Materials and Fundamental Properties*

The raw materials primarily consist of reclaimed asphalt and aggregates from the RAP, virgin SBS-modified asphalt, basalt aggregate, and a rejuvenator sourced from Jiangsu Subote Company (Ninjing, China).

(1) Virgin and reclaimed asphalt

The reclaimed asphalt was recycled using the distillation separation method, and the technical properties of both the virgin and reclaimed asphalt were tested. The asphalt extraction instruments are shown in Figures 2 and 3, and the results are presented in Table 2. The new asphalt selected SBS-modified asphalt, with a softening point, 25 °C penetration, and 5 °C ductility of 77.1 °C, 50.1 (0.1 mm), and 29.4 cm, respectively. The performance of reclaimed asphalt is inferior to that of virgin asphalt, with a 5 °C ductility of only 8.6 cm.

Figure 2. Centrifugal extraction test chart.

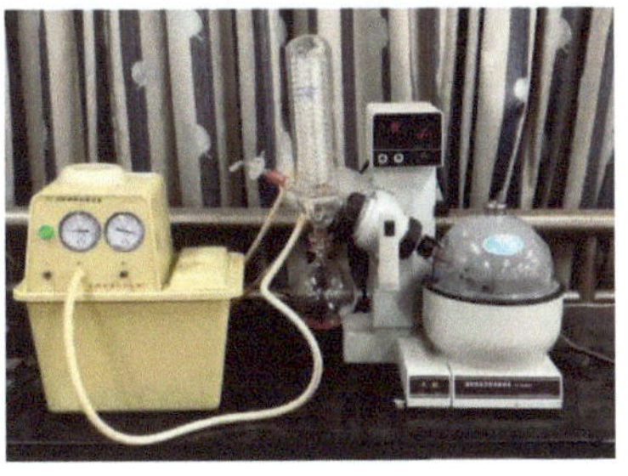

Figure 3. Rotational evaporation test diagram.

Table 2. Asphalt performance test results.

Technical Indicators	Reclaimed Asphalt Binder	SBS-Modified Asphalt Binder	Requirement	Test Method
Penetration (25 °C,100 g, 5 s)/0.1 mm	42.1	50.1	40~60	T0604
Ductility (5 cm/min, 5 °C)/cm	8.6	29.4	≥20	T0605
Softening point/°C	56.3	77.1	≥60	T0606
Rotational viscosity (135 °C)/Pa·s	2.62	2.75	≤3	T0613

According to Table 2, the aging grade of RAP is determined based on needle penetration and the viscosity index. As shown in Table 3, the RAP used in this study belongs to aging grade II, and its performance can be partially improved by adding the rejuvenator.

Table 3. Basis for classification of reclaimed asphalt aging grades.

Used Material Category	SBS Asphalt Pavement Binder					
Viscosity (Pa·s)	$\eta \leq 1.6$		$1.6 < \eta \leq 3$		$\eta > 3$	
Penetration (0.1 mm)	$p > 30$	$p > 30$	$20 < p \leq 30$	$10 < p \leq 20$	$20 < p \leq 30$	$10 < p \leq 20$
Aging grade	I	II	III	IV	V	VI

(2) Rejuvenator

The rejuvenator (RA-102) was selected from Jiangsu Subote Company. It features a flash point of 252 °C and a 90 °C viscosity of 4000 cP. The viscosity ratio before and after the rolling thin film oven test (RTFOT) is 1.64, and the change in mass before and after the RTFOT is 1.14%. The technical properties and specification requirements are presented in Table 4 and satisfy the standard request.

Table 4. Technical indexes of the RA-102 rejuvenator.

Technical Indicators	RA-102	Requirement	Test Method
90 °C viscosity/cP	4000	/	T0619
Flash point/°C	252	≥220	T0633
Saturation content/%	23.4	≤30	T0618
Aromatic content/%	49	≥30	T0618
The viscosity ratio before and after the RTFOT	1.64	≤3	T0610
Change in mass before and after the RTFOT/%	1.14	≤4%	T0603

(3) Mineral properties

Both the coarse and fine aggregates are basalt. The coarse aggregates are clean, rough, and free of impurities, with their physical and mechanical properties meeting the requirements and relative density, as shown in Table 5. The particle size distribution of the fine aggregate must adhere to the design specifications, and it should be clean and free from impurities such as soil, clay, and organic matter. The relative density of fine aggregates is provided in Table 6.

Table 5. Relative density of coarse aggregates.

Relative Density	Sieve Size/mm			
	13.2~16.0	9.5~13.2	4.75~9.5	2.36~4.75
Apparent relative density	2.732	2.734	2.736	2.735
Relative density of surface drying	2.704	2.705	2.711	2.712
Gross volume relative density	2.691	2.692	2.696	2.699

Table 6. Relative density of fine aggregates.

Relative Density	Sieve Size/mm					
	1.18~2.36	0.6~1.18	0.3~0.6	0.15~0.3	0.075~0.1	0~0.075
Apparent relative density	2.749	2.757	2.763	2.771	2.775	2.786

Limestone mineral powder was used as filler, and the performance indexes are shown in Table 7. The water content is 0.38%, the relative density is 2.769 kg/m^3, and the particle size range satisfies the standard request.

Table 7. Technical properties of fillers.

Test Metrics		Test Results	Requirement	Test Method
Water content/%		0.38	≤1	Drying method
Relative density		2.769	≥2.5	T0352
Particle size range	<0.6 mm	100	100	T0351
	<0.15 mm	92.9	90~100	
	<0.075 mm	91.2	75~100	

2.3. Mix Design of AC-13 Hot In-Place Recycling Mixture

The original pavement gradation used a nominal maximum particle size of 13.2 mm (AC-13), which was selected for the gradation design. Based on actual engineering conditions and the Marshall test, the optimum asphalt content for the HIR mixture was determined.

(1) Determination of rejuvenator content

The rejuvenator was added to the reclaimed asphalt at 2%, 4%, and 6% by weight of the reclaimed asphalt, and tests on penetration, softening point, and ductility were performed. The test results are presented in Figure 4.

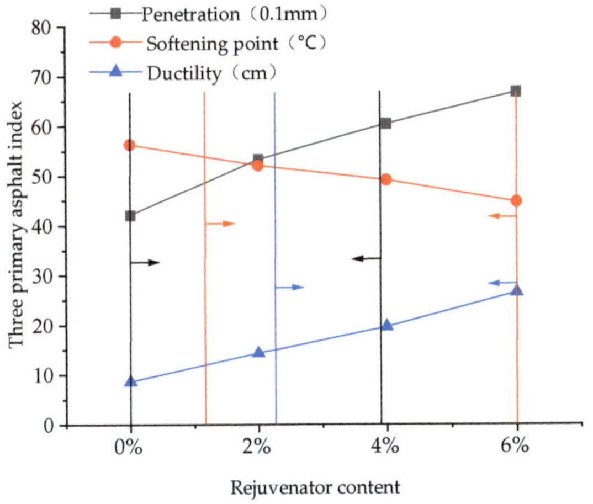

Figure 4. Effect of rejuvenator dosage on asphalt properties.

According to the specification requirements, due to the process characteristics of hot recycling, the virgin asphalt grade of the pavement is typically reduced by one level to determine the target grade for reclaimed asphalt. This corresponds to a penetration index of 40–60 (0.1 mm), which is associated with a rejuvenator content of 0–3.9%; a softening point of no less than 49 °C, corresponding to a rejuvenator dosage of 1.2–6%; and ductility of no less than 15 cm, corresponding to a rejuvenator dosage of 2.3–6%. To meet these performance requirements, a rejuvenator content range of 2.3% to 3.9% was selected. Considering the high-temperature performance and operability of rejuvenated asphalt, a rejuvenator content of 3% was ultimately selected for subsequent testing in this study.

(2) Rejuvenated asphalt mixture mix design

Based on the actual engineering conditions, the original pavement in the section exhibited significant rutting. Therefore, the gradation at the lower limit of the specification was chosen as the final composite gradation, and the gradation diagram is shown in Figure 5. The Marshall test determined that the optimum asphalt content for the HIR mixture was 4.67%.

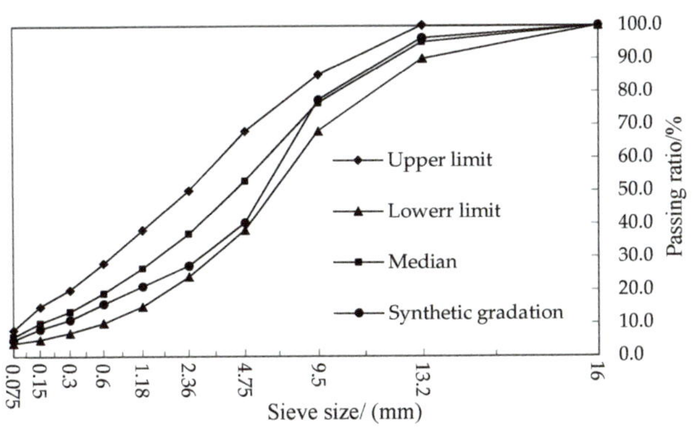

Figure 5. Mixture gradation curve.

2.4. Design of Thermo-Oxidative and Ultraviolet Aging Tests

Pavement aging typically involves two mechanisms: TO aging and UV aging. TO aging occurs when asphalt pavement is exposed to high temperatures and oxygen over extended periods. This leads to the volatilization of lightweight components and changes in the molecular structure due to the combined effects of heat and oxygen. These changes increase the viscosity and elastic modulus of the asphalt while reducing its ductility, ultimately making the mixture harder and more brittle [24]. UV aging, on the other hand, results from prolonged exposure to sunlight. Intense UV irradiation causes the asphalt to become brittle and reduces its viscosity [18,25], which eventually contributes to pavement distress.

The power of the UV lamp primarily determines the intensity of UV radiation, while the duration of exposure directly impacts the accuracy of the test results [26]. To simulate long-term exposure to sunlight, the UV radiation produced by the equipment should be at least ten times stronger than that of the outdoor environment [27]. In this study, an ultraviolet aging chamber was used to irradiate the samples, providing a more accurate simulation of UV aging under natural environmental conditions. The maximum wavelength is 365 nm, the minimum is 313 nm, and the power consists of one 300 W lamp, two 80 W lamps, and two 40 W lamps. The UV aging simulations are shown in Figures 6 and 7.

Figure 6. Simulation method of UV aging.

Figure 7. UV aging chamber.

The laboratory aging time to simulate actual radiation exposure was calculated using the equivalent conversion method for radiation exposure [28], as detailed in Equation (1).

$$t = \frac{F \times b}{3600 \times Q} \quad (1)$$

where F is the total annual sunlight radiation (J/m$^{2)}$; b is the proportion of UV rays in sunlight, which is taken as 7%; Q is the radiation intensity of the UV lamp (W/m$^{2)}$; and t is the adequate time for the conversion of natural radiation to indoor radiation, the unit is h.

Previous studies have shown that the effect of UV aging on asphalt pavement is most pronounced during the early service period [16]. In this study, the HIR mixture specimens were aged for 27, 54, 81, 108, 135, and 162 h to simulate UV exposure corresponding to

2, 4, 6, 8, 10, and 12 months of service in the city of Yangzhou, Jiangsu Province, China, respectively, as detailed in Table 8.

Table 8. Comparison of UV aging test with actual pavement radiation duration.

Lab Aging Time (h)	Simulated Pavement Aging Time (Months)	Lab UV Radiation (MJ/m^2)
27	2	54.4
54	4	108.8
81	6	163.2
108	8	217.6
135	10	272.0
162	12	326.4

Hence, this study simulates the actual aging of rejuvenated pavement by varying both aging methods and duration. Among them, short-term aging at 135 °C and 155 °C was used to simulate multiple high-temperature heating phases of HIR mixtures during the construction process, and long-term aging for 120 h and 240 h simulated natural aging for 5 years and 10 years, respectively [29]. A control group without aging and another with eight different aging schemes were established, as shown in Table 9.

Table 9. TO and UV aging test plan.

Group Name	Aging Schemes
N_{on}	Blank control group—unaged HIR mixture control group.
D_{135}	Short-term aging I—Before molding, the specimen was heated at 135 °C for 4 h.
D_{155}	Short-term aging II—Before molding, the specimen was heated at 155 °C for 4 h.
C_{120}	Long-term aging I—After short-term aging I, the specimen was heated at 85 °C for 120 h.
C_{240}	Long-term aging II—After short-term aging I, the specimen was heated at 85 °C for 240 h.
Z_{27}	UV aging I—After short-term aging I, the specimen was UV aged at 50 °C, 556 W/m^2 radiation for 27 h.
Z_{54}	UV aging II—After short-term aging I, the specimen was UV aged at 50 °C, 556 W/m^2 radiation for 54 h.
Z_{81}	UV aging III—After short-term aging I, the specimen was UV aged at 50 °C, 556 W/m^2 radiation for 81 h.
Z_{108}	UV aging IV—After short-term aging I, the specimen was UV aged at 50 °C, 556 W/m^2 radiation for 108 h.
Z_{135}	UV aging V—After short-term aging I, the specimen was UV aged at 50 °C, 556 W/m^2 radiation for 135 h.
Z_{162}	UV aging VI—After short-term aging I, the specimen was UV aged at 50 °C, 556 W/m^2 radiation for 162 h.

3. Performance Tests—Analysis and Prediction

To evaluate the changes in the HIR mixture following TO and UV aging, this study systematically investigated their high-temperature, low-temperature, and moisture damage properties. The evaluation was conducted through a series of performance tests, including the wheel track rutting test, uniaxial penetration test, low-temperature beam bending test, immersion Marshall test, and freeze–thaw splitting test.

3.1. Wheel Track Rutting Test

Following the "Highway Engineering Asphalt and Asphalt Mixture Test Regulations", wheel track rutting tests were conducted on the 100% RAP and HIR mixture under various aging conditions, using dynamic stability as the evaluation index. Rutting experiments were carried out using a Wheel Track Rutting Instrument from Shengshi Huike Testing Equipment Co., Ltd. (Shanghai, China), a plate specimen sized 300 mm × 300 mm × 50 mm, an experimental temperature of 60 °C, and a wheel pressure of 0.7 MPa. The test results are presented in Figure 8.

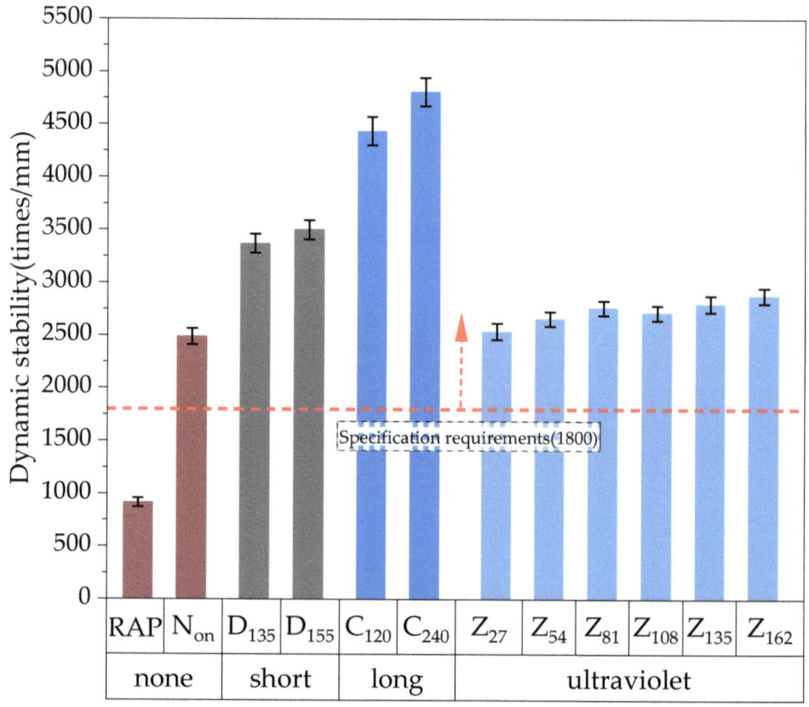

Figure 8. Results of high-temperature dynamic stability.

Pavement gradation and asphalt content are critical factors influencing the high-temperature stability of the mixture. The results of the original pavement and the composite mixture after rutting tests are shown in Figures 9 and 10. As shown in Figure 9, the original pavement, with a high asphalt content of 5.45%, exhibits a more significant presence of surface fines and free asphalt, resulting in noticeable binder bleeding and increased mobility. After the rutting test, deeper rutting grooves were observed, likely caused by the excessive asphalt binder content, which reduces the interlocking and occlusion effect between the coarse aggregates. This weakening of the internal skeleton structure makes the mixture more susceptible to deformation under repeated wheel loading. Furthermore, in the fine-graded mixture, a higher asphalt content thickens the asphalt film, coating the mineral material and increasing the mixture's mobility and temperature sensitivity at high temperatures while reducing its overall stiffness.

Figure 9. Rutting depth of the original pavement.

Figure 10. Rutting depth of the composite mixture.

As shown in Figure 9, the original pavement exhibited a significant rutting depth, with a dynamic stability (*DS*) of only 913 cycles/mm, indicating poor high-temperature performance. By optimizing the aggregate gradation and controlling the asphalt content, the *DS* increased to 2490 cycles/mm, representing a 172.7% improvement compared to the 100% RAP control group, thus meeting specification requirements. The dynamic stability of the specimen further improved after TO aging. Compared to the non-aged group, the *DS* after short-term aging increased by 35.3%. When the short-term aging temperature reached 155 °C, the *DS* further increased to 3500 cycles/mm. Following long-term aging after short-term aging, the *DS* of the HIR mixture improved significantly. After 240 h of long-term aging, the *DS* reached a maximum of 4809 cycles/mm, a 93.1% increase compared to the non-aged group.

3.2. Uniaxial Penetration Test

The uniaxial penetration test was conducted on the HIR mixture using a UTM-25 universal testing machine from Australia IPC Company (Guangzhou, China) to evaluate its high-temperature shear capacity through penetration stress. Marshall specimens were penetrated using a 50 mm long metal rod with a diameter of 28.5 mm, at a penetration rate of 1 mm/min. Specimens should be held in an oven at 60 °C for 6 h to ensure that the test is carried out at the specified temperature. The test results are shown in Figure 11.

As shown in Figure 11, the results of the uniaxial penetration test exhibit a similar increasing trend to those of the rutting test. Both the penetration stress and the penetration strength of the HIR mixture with the optimized gradation were increased by 19.4% compared to the 100% RAP control group. These findings indicate that the high-temperature performance of the HIR mixture was significantly enhanced by incorporating the virgin mixture.

The performance of the HIR mixture was further evaluated under various aging conditions. Compared to the unaged control group, the penetration strength of specimens increased markedly after TO aging. Short-term aging at 135 °C and 155 °C increased penetration strength by 5.0% and 5.8%, respectively, while long-term aging for 120 h and 240 h led to increases of 11.6% and 13.6%, respectively. The high-temperature performance indices of the specimens also improved after UV aging, although the effect was modest.

The most notable improvement was observed after 162 h of UV aging, with penetration strength reaching 2.434 MPa.

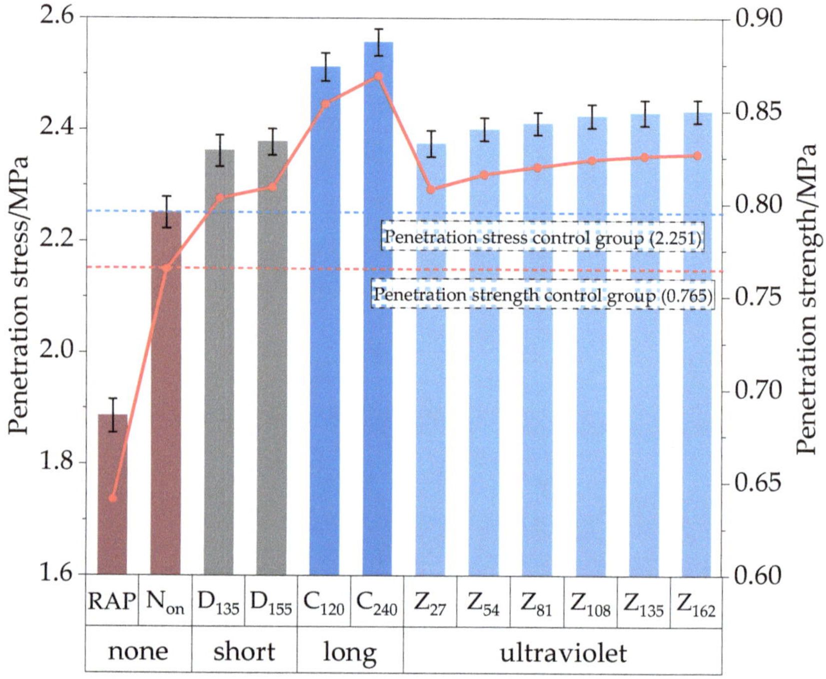

Figure 11. Uniaxial penetration test results.

3.3. *Low-Temperature Cracking Test*

The trabecular bending test was used to evaluate the low-temperature cracking resistance of the HIR mixture under various aging conditions. The specimens were prepared according to specification requirements. They are rectangular and 100 mm × 10 mm × 10 mm. After being conditioned at −10 °C for 6 h, the specimens were loaded to failure using UTM equipment at a loading rate of 50 mm/min. The results are presented in Figure 12.

As shown in Figure 12, the flexural tensile strength (R_B) and maximum flexural strain (ε_B) of the mixture decreases with increased aging. At the same time, the bending stiffness modulus (S_B) exhibits a corresponding increase. Comparing Group N_{on} to Group D_{135}, after short-term aging at 135 °C, the change in R_B was negligible. However, ε_B decreased by 3.4%. The primary distinction between Group D_{135} and Group D_{155} lies in the short-term aging temperature. As the temperature increased, R_B decreased by 1.5%, ε_B dropped to 1896.56 με, and S_B increased by 8.4%. This indicates that elevated temperatures during the construction process negatively impact the low-temperature performance of the mixture, as excessive heating leads to hardening and embrittlement of the HIR mixture, thus reducing its low-temperature performance.

Compared to Group D_{135}, Group C_{120} experienced a 5.5% reduction in R_B, and ε_B decreased to 1708.30 με after long-term aging. Comparing Group C_{240} with Group C_{120}, after extending the long-term aging duration to 240 h, R_B decreased from 12.36 MPa to 11.98 MPa, while ε_B dropped from 1708.30 με to 1538.98 με. This trend indicates that the low-temperature performance of the HIR mixture declined with prolonged TO aging, while S_B continued to increase.

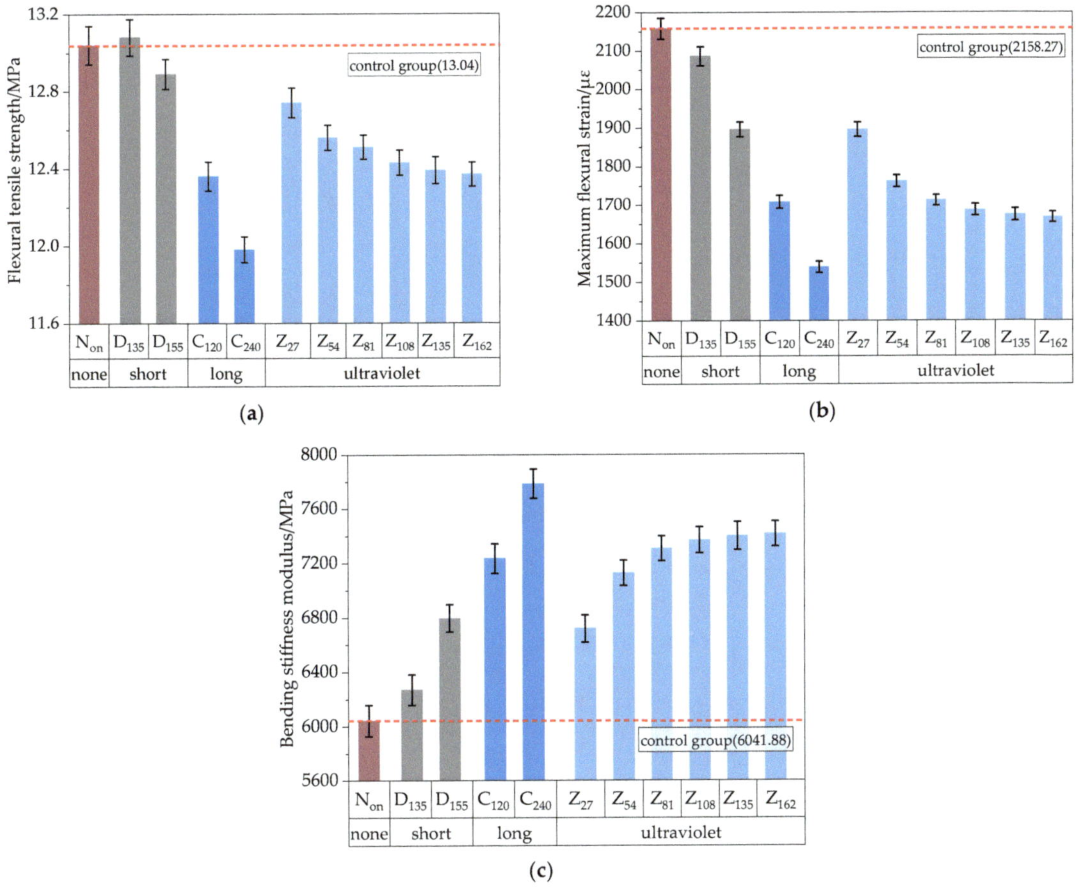

Figure 12. Results of low-temperature beam bending test. (**a**) Flexural tensile strength, (**b**) Maximum flexural strain, (**c**) Bending stiffness modulus.

Comparing Group Z_{27} with Group D_{135}, after 27 h of UV aging, the R_B of the HIR mixture decreased by 2.6%, while ε_B declined by 9.1%. The primary distinction among the UV aging control groups lies in the duration of UV exposure. The influence of aging duration on mixture performance becomes evident when comparing Group Z_{27} to Group Z_{162}. Specifically, Group Z_{54} exhibited a 1.4% reduction in R_B and a 7.1% reduction in ε_B compared to Group Z_{27}. As the UV aging duration increased, Group Z_{81} showed a smaller performance decline compared to Group Z_{54}, with a 0.4% reduction in R_B and a 3.5% reduction in ε_B. Furthermore, the R_B and ε_B of Group Z_{162} decreased to 12.37 MPa and 1668.23 με, respectively. The results indicate that prolonged UV aging gradually reduced R_B and ε_B in the HIR mixture. However, after 54 h of aging, the rate of change in performance indices slowed, suggesting that the impact of UV aging is most pronounced in the early stages. Meanwhile, the increasing trend in S_B mirrored the changes observed in R_B and ε_B.

3.4. Water Stability Performance

The water stability performance of the mixture is typically assessed using the immersion Marshall test and freeze–thaw splitting test. In this work, the water stability performance of the asphalt mixture after TO and UV aging was examined.

(1) Immersing Marshall test

The water stability of the HIR mixture under various aging conditions was tested using the immersion Marshall test. To determine its stability, the Marshall specimen was placed in a thermostat tank at 60 °C for 48 h and then compared to the uninsulated specimen. The test results are presented in Figure 13.

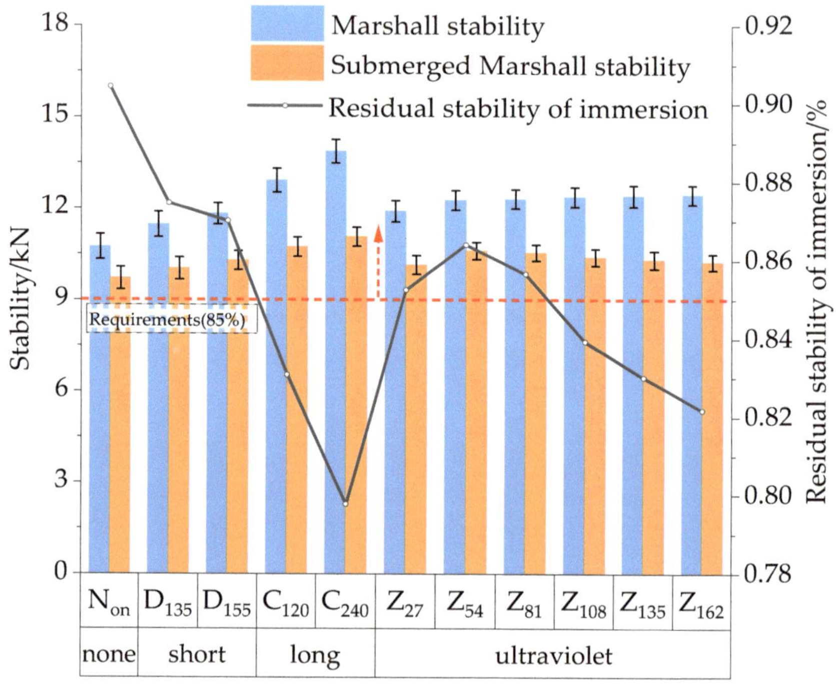

Figure 13. Results of the immersion Marshall test.

The data in Figure 13 indicate that the stability of the HIR mixture increases with aging time and temperature, whereas the residual stability of immersion (MS_0) exhibits a decreasing trend. A comparative analysis of Group Non and Group D_{135} shows that after short-term aging, stability increases by 6.8%. In comparison, MS_0 decreases by 3.0%, suggesting that short-term aging reduces the stability of the mixture to some extent. When the short-term aging temperature increases to 155 °C, the mixture stability further improves to 11.83 kN, with a minimal MS_0 decrease of only 0.3%. These results suggest that the impact of temperature on the water stability of the asphalt mixture during construction is minimal, causing only a slight reduction in resistance to water damage as the temperature rises.

Further analysis revealed that the stability of Group C_{120} increased from 11.47 kN after short-term aging to 12.93 kN following 120 h of long-term aging, while MS_0 decreased by 4.4%, falling short of the 85% specification. In Group C_{240}, where the long-term aging duration was extended to 240 h, stability improved by 7.4%, while MS_0 decreased by 3.3% compared to Group C_{120}. This increase in stability can be attributed to the deepening of TO aging, which reduces the lightweight components in the asphalt and enhances the material's hardness and rigidity. However, the presence of internal defects in the HIR mixture, combined with severe aging, led to compaction difficulties, ultimately contributing to a reduction in MS_0.

A comparison between Group Z_{27} and Group D_{135} reveals that after UV aging, the stability of the specimens increased by 3.8%, while MS_0 decreased by 2.2%. When the UV radiation time was extended to 54 h, the specimens exhibited higher stability but lower MS_0. As UV radiation time increased to 81 and 108 h, the performance indices from the immersion Marshall tests stabilized. These results indicate that the performance changes in the mixture are more moderate under UV aging conditions, with the mixture showing higher sensitivity to TO aging in terms of water stability.

(2) Freeze–thaw splitting test

The freeze–thaw splitting test was conducted on the HIR mixture under various aging conditions. Marshall specimens were placed in a plastic bag with 10 mL of water and frozen in a refrigerator at −18 °C for 16 h. The specimens were removed and immersed in a thermostat tank at 25 °C for not less than 2 h. Finally, the specimens were loaded with a UTM tester at a rate of 50 mm/min and a temperature of 25 °C. The test procedure is shown in Figure 14.

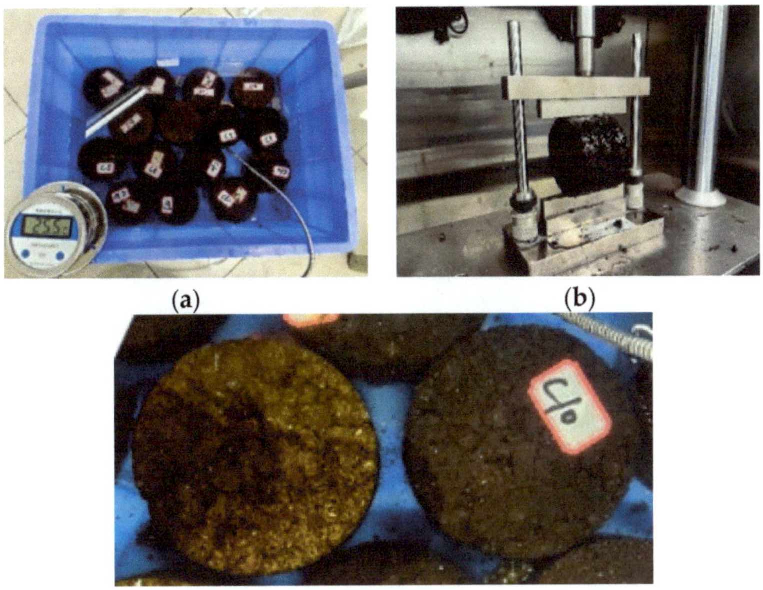

(a) (b) (c)

Figure 14. Freeze–thaw splitting setup. (**a**) Specimen insulation, (**b**) Freeze–thaw splitting test, (**c**) Freeze–thaw splitting specimens under UV aging.

The water damage resistance was evaluated using the freeze–thaw splitting tensile strength ratio (*TSR*) and the results are shown in Figure 15. As shown in Figure 15, the *TSR* of the aged mixture in the freeze–thaw test meets the specification requirement of not less than 80%. Overall, the splitting strength of the mixture increased, while the *TSR* decreased with progressive aging. Specifically, compared to the Non control group, the pre-freeze–thaw strength of Group D_{135} increased by 12.9%, and the post-freeze–thaw strength of Group D_{135} increased by 8.1%. However, the *TSR* decreased by 4.0%. When the short-term aging temperature rose from 135 °C to 155 °C, performance indices showed no significant changes. In contrast, the long-term aging group C_{120} demonstrated more pronounced changes compared to Group D_{135}, with a 13.9% increase in pre-freeze–thaw strength, a 12.2% increase in post-freeze–thaw strength, and a 1.4% decrease in *TSR*, which further declined with extended long-term aging.

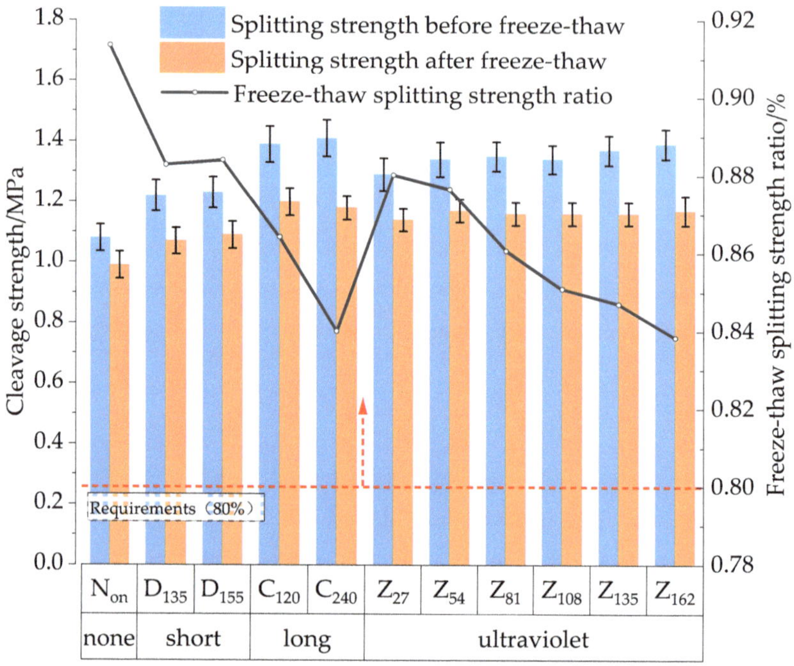

Figure 15. Results of freeze–thaw splitting test.

Regarding UV aging, Figure 14c shows that the surface of UV-aged specimens turned dark yellow after freeze–thaw cycles, likely due to changes in the chemical structure of the asphalt from prolonged UV radiation, oxidation, and temperature effects. Analyzing the test results, a comparison between Group Z_{27} and Group D_{135} reveals a 5.7% increase in splitting strength after UV aging, while *TSR* slightly decreased. As UV aging time increased, splitting strength gradually improved, but *TSR* continued to decline.

After more than 54 h of UV aging, the splitting strength exhibited no significant variation, while the *TSR* exhibited some fluctuations. This may be attributed to internal defects in the HIR mixture, affecting its homogeneity and performance stability during the UV aging process. Moreover, according to the time–temperature equivalence principle, the material's behavior over an extended period at low temperatures can be equivalent to its behavior over a shorter period at high temperatures, making it challenging to isolate the influence of the test temperature on long-term UV aging results. The findings indicate that both TO and UV aging significantly impact the water stability of the mixture. As the degree of aging intensifies, the water stability decreases. However, when UV aging reaches a specific duration, its effect on water stability tends to stabilize.

3.5. Performance Prediction Model of HIR Mixtures Under Ultraviolet Irradiation

The long-term UV aging test is complex and labor-intensive. In this work, only a one-year UV aging test was simulated. The effects of UV aging over one year remain unclear. Therefore, a model prediction was performed based on the one-year UV aging simulation to estimate the impacts of long-term UV aging. The Verhulst model [30], within the grey system theory, was applied to establish a prediction equation for the HIR mixture pavement performance index under varying UV irradiation durations. Initially, the performance index of the HIR mixture corresponding to different UV irradiation times was selected as the original sequence, as shown in Equation (2). Then, the whitening equation for the Verhulst model can be written as Equation (3). Using the method of separating variables, the general solution of the ordinary differential equation is Equation (4). When $t = 0$,

$y(0) = y_0$, let $k = \alpha/\beta$, we can obtain Equation (5). Equation (6) can be calculated by making $ARI_\infty = k/y_0$. When the time is infinite, the road performance index of the asphalt mixture after UV aging is calculated by Equation (7) [31].

$$y_i(t) = [y_i(t), y_i(t), \cdots, y_i(t)];\ t = 1, 2, \cdots, n;\ i = 1, 2, \cdots, m \tag{2}$$

$$\frac{dy(t)}{dt} = \alpha y(t) - \beta y^2(t) \tag{3}$$

$$y(t) = \frac{\alpha}{\beta(1 + Ce^{-\alpha t})} \tag{4}$$

$$y(t) = \frac{k}{1 + e^{-\alpha t}(\frac{k}{y_0} - 1)} \tag{5}$$

$$y(t) = \frac{ARI_\infty y_0}{1 - e^{-\alpha t}(1 - ARI_\infty)} \tag{6}$$

$$\frac{y(t)_{t\to\infty}}{y_0} = ARI_\infty \tag{7}$$

where $y(t)$ is the performance evaluation index of the HIR mixture at the time of t, the unit of t is h; e is a transcendental number equal to approximately 2.71828; ARI_∞ is the Aging Residual Index; and α and β are fitting parameters.

As can be seen from the formula, ARI_∞ represents the ratio between the pavement performance evaluation index of the asphalt mixture after stabilization by UV aging and its initial value. ARI_t is calculated by Equation (8), which represents the ratio of road performance at a specific aging time to the initial performance and reflects the degree of performance change in the asphalt mixture over that aging period.

$$ARI_t = \frac{y(t)}{y_0} = \frac{ARI_\infty}{1 - e^{-\alpha t}(1 - ARI_\infty)} \tag{8}$$

Equation (8) characterizes the relationship between the pavement performance index of the HIR mixture and the time of UV irradiation, and the derivative of Equation (8) can be used to obtain Equation (9), which indicates the rate of change of the pavement performance index of the HIR mixture with the time of UV irradiation.

$$UAR_t = \frac{dy(t)}{dt} = \frac{\alpha ARI_\infty y_0(ARI_\infty - 1)e^{\alpha t}}{[ARI_\infty + e^{\alpha t} - 1]^2} \tag{9}$$

where UAR_t indicates the rate of change of the pavement performance index of the HIR mixture with the time of UV irradiation.

Equation (8) is applied to predict the performance of the HIR mixture under UV irradiation. The prediction results are presented in Figure 16.

As shown in Figure 16, the UV aging residual index model is well-established, with R-squared values for various pavement performance metrics all exceeding 0.92. Notably, the R-squared values for penetration strength and stiffness modulus even exceed 0.99. The analysis of the aging residual index reveals that dynamic stability, penetration strength, and stiffness modulus demonstrate an increasing trend, with rapid growth in the early stages followed by more moderate growth. As UV irradiation time increases, the aging residual index gradually decreases and eventually levels off. However, the leveling-off period is prolonged, with the final index value remaining relatively low at around 0.5.

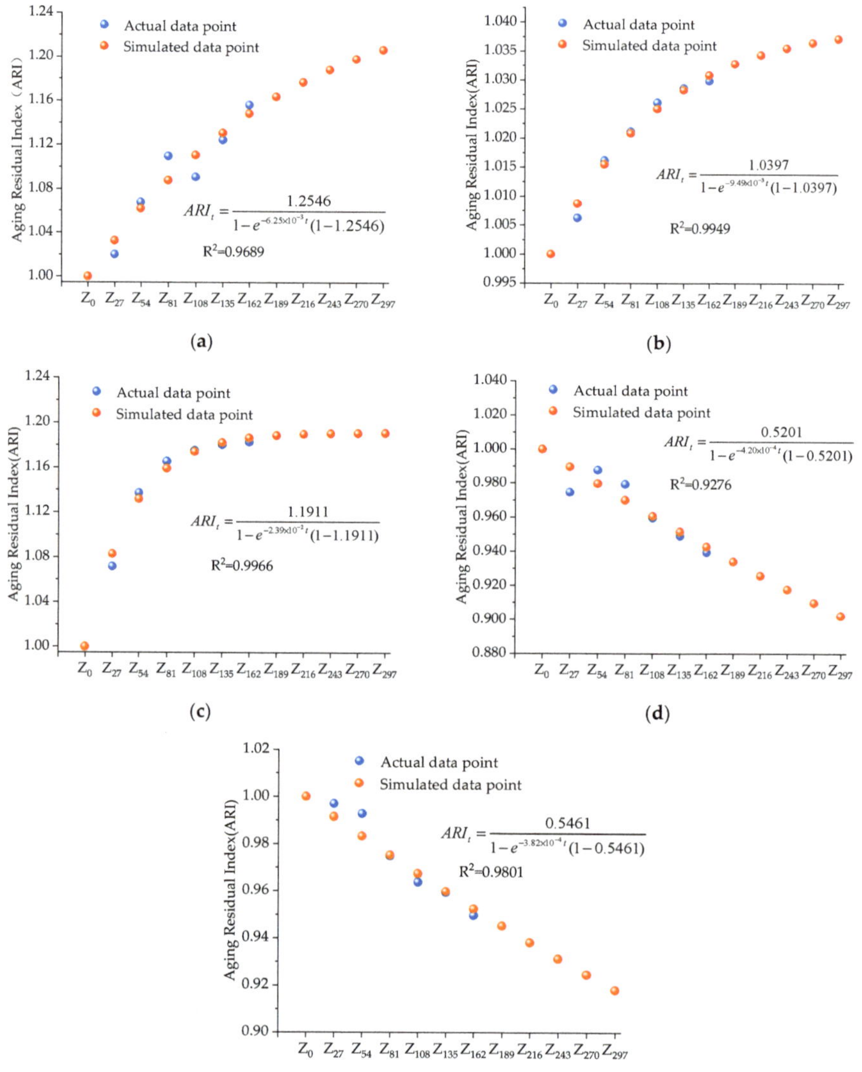

Figure 16. The performance index prediction accuracy of UV aging. (**a**) Dynamic stability, (**b**) Penetration strength, (**c**) Bending stiffness modulus, (**d**) Residual stability of immersion, and (**e**) Freeze–thaw splitting strength ratio.

As depicted in Figure 17, the UV aging residual index model demonstrates that, apart from water damage resistance, all pavement performance indices experience a rapid initial aging rate, evidenced by a steep curve in the early stages. Specifically, dynamic stability, penetration strength, and stiffness modulus increase sharply at first and then stabilize over time. This is based on the average annual sunshine hours in Yangzhou, Jiangsu Province, China, ranging from 2400 to 3500 h, and calculated using an average indoor UV irradiation intensity of 200 W/m^2. It is estimated that after approximately 194 h of indoor UV aging, the high- and low-temperature performance indicators of the HIR mixture exhibit minimal changes in the aging rate and residual index. This can be attributed to the presence of reactive double bonds in the asphalt mixtures, which are susceptible to photoreactions. These reactions with UV light result in increased hardness and stiffness, while deteriorating the low-temperature properties. As the irradiation time increases, most of the reactive double bonds on the surface of the asphalt material are consumed, and further changes in material properties slow down. Therefore, in the early stages of aging, strategies should be implemented to reduce the reaction between reactive double bonds and UV light, thus extending the duration of light-induced reactions and mitigating the effects of UV exposure on the road performance of asphalt mixtures.

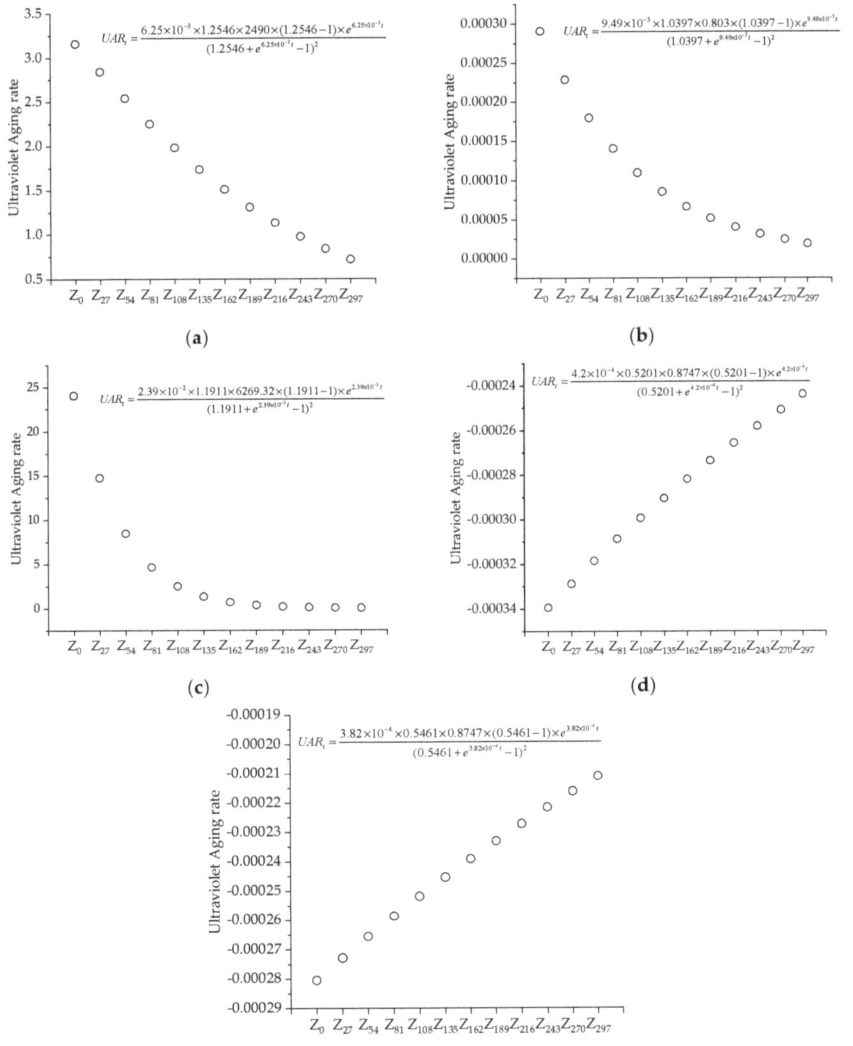

Figure 17. The rate of change of the road performance index with UV irradiation time. (**a**) Dynamic stability, (**b**) Penetration strength, (**c**) Stiffness modulus, (**d**) Residual stability of immersion, and (**e**) Freeze-thaw splitting strength ratio.

4. Discussion

To visually illustrate the effects of aging on high- and low-temperature performance test results, the data from the HIR mixture subjected to different aging methods and durations were plotted as biaxial line graphs, as shown in Figure 18. The figure demonstrates that the high- and low-temperature performance indices of the recycled mixture exhibit opposite trends as aging time increases. Regardless of whether subjected to TO or UV aging, the asphalt mixture shows an increase in high-temperature performance alongside a significant decrease in low-temperature performance. Therefore, greater emphasis should be placed on low-temperature cracking performance during the design process of the HIR mixture while also considering the balance of various properties.

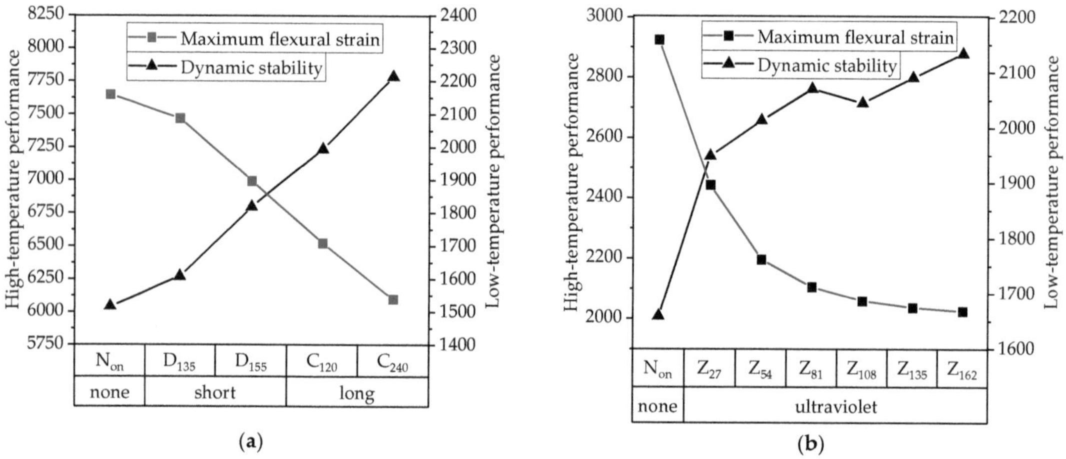

Figure 18. Effect of aging time on high- and low-temperature performance indicators of HIR mixtures. (**a**) TO aging, (**b**) UV aging.

This study primarily focused on recycled asphalt mixtures to examine the effects of aging in commonly used reclaimed materials. Under natural weather conditions, road asphalt materials undergo various effects of aging. The aging effects of TO and UV aging on recycled asphalt were investigated separately. However, the combined aging of TO and UV represents an important area of future research. Combining the effects of UV and TO aging on recycled asphalt mixtures could provide deeper insights into aging mechanisms, thereby improving the aging resistance of asphalt mixtures.

5. Conclusions

This study investigated the effects of thermo-oxidative (TO) and ultraviolet (UV) aging on hot in-place recycling (HIR) mixtures through simulations of long-term and short-term oxidative aging and 12 months of UV irradiation. The primary objective was to better understand how these aging processes influence the performance of recycled asphalt pavements, specifically in terms of high-temperature stability, low-temperature crack resistance, and water stability. To achieve this, performance indicators such as dynamic stability, low-temperature flexibility, and resistance to water damage were measured before and after aging, and the Verhulst model was applied to predict performance attenuation under UV aging.

The results showed that both TO and UV aging improved the high-temperature performance of the HIR mixture, with TO aging having a more significant impact. Specifically, 10 years of TO aging resulted in a 78.2% increase in dynamic stability, whereas 12 months of UV aging led to only a 15.6% increase. However, both aging processes led to a decline in low-temperature cracking resistance. Notably, flexural tensile strain decreased by 129.25 με after 10 years of TO aging, while the reduction was less pronounced under UV aging. These findings highlight the trade-off between improved high-temperature performance and reduced low-temperature cracking resistance.

The Verhulst model indicated that UV aging initially significantly impacts pavement performance, underlining the importance of mitigating UV-induced photochemical reactions early in the service life of pavement. Effective aging mitigation strategies are crucial to extending the service life of HIR pavements.

Overall, this study emphasizes the need for a balanced approach in the design of HIR mixtures, considering both high- and low-temperature properties. Future research should focus on improving low-temperature cracking resistance and enhancing aging resistance, particularly in areas with high UV exposure.

Author Contributions: Conceptualization, T.S.; Methodology, Y.G.; Formal analysis, T.S.; Investigation, Y.D.; Resources, Y.D. and Z.Y.; Data curation, Z.Y.; Writing—original draft, Y.G. and Y.Z.; Writing—review & editing, Y.Z.; Supervision, A.K.; Project administration, A.K.; Funding acquisition, Y.G. and Y.Z. All authors have read and agreed to the published version of the manuscript.

Funding: This study is financially supported by the National Natural Science Foundation of China (No. 52108422), the fellowship of China National Postdoctoral Program for Innovative Talents (BX20230302), the Municipal-School Cooperation Special Project of Yangzhou Science and Technology Plan (No. YZ2022176), the Jiangsu Postgraduate Practice Innovation Project (No. SJCX24_2230), and the Yangzhou Government-Yangzhou University Cooperative Platform Project for Science and Technology Innovation (No. YZ2020262).

Institutional Review Board Statement: Not applicable.

Informed Consent Statement: Not applicable.

Data Availability Statement: The original contributions presented in the study are included in the article, further inquiries can be directed to the corresponding author.

Conflicts of Interest: The authors declare no conflicts of interest.

References

1. Yao, Y.; Yang, J.; Gao, J.; Zheng, M.; Xu, J.; Zhang, W.; Song, L. Strategy for improving the effect of hot in-place recycling of asphalt pavement. *Constr. Build. Mater.* **2023**, *366*, 130054. [CrossRef]
2. Li, Q.; Sun, G.; Lu, Y.; Meng, Y.; Luo, S.; Gao, L. Effects of warm-mix asphalt technologies and modifiers on pavement performance of recycled asphalt binders. *J. Clean. Prod.* **2021**, *282*, 125435. [CrossRef]
3. Ji, Z.; Wu, X.; Zhang, Y.; Milani, G. Aging Behavior and Mechanism Evolution of Nano-Al_2O_3/Styrene-Butadiene-Styrene-Modified Asphalt under Thermal-Oxidative Aging. *Materials* **2023**, *16*, 5866. [CrossRef] [PubMed]
4. Pan, P.; Kuang, Y.; Hu, X.; Zhang, X. A comprehensive evaluation of rejuvenator on mechanical properties, durability, and dynamic characteristics of artificially aged asphalt mixture. *Materials* **2018**, *11*, 1554. [CrossRef] [PubMed]
5. Sabouri, M. Evaluation of performance-based mix design for asphalt mixtures containing Reclaimed Asphalt Pavement (RAP). *Constr. Build. Mater.* **2020**, *235*, 117545. [CrossRef]
6. Wang, H.; Zhu, Y.; Zhang, W.; Shen, S.; Wu, S.; Mohammad, L.N.; She, X. Effects of field aging on material properties and rutting performance of asphalt pavement. *Materials* **2022**, *16*, 225. [CrossRef]
7. Zhang, Y.; Wei, H.; Dai, Y. Influence of different aging environments on rheological behavior and structural properties of rubber asphalt. *Materials* **2020**, *13*, 3376. [CrossRef]
8. Yan, S.; Dong, Q.; Chen, X.; Zhou, C.; Dong, S.; Gu, X. Application of waste oil in asphalt rejuvenation and modification: A comprehensive review. *Constr. Build. Mater.* **2022**, *340*, 127784. [CrossRef]
9. Zhang, Y.; Wang, Y.; Kang, A.; Wu, Z.; Li, B.; Zhang, C.; Wu, Z. Analysis of rejuvenating fiber asphalt mixtures' performance and economic aspects in high-temperature moisture susceptibility. *Materials* **2022**, *15*, 7728. [CrossRef]
10. Zaumanis, M.; Mallick, R.B.; Frank, R. 100% recycled hot mix asphalt: A review and analysis. *Resour. Conserv. Recycl.* **2014**, *92*, 230–245. [CrossRef]
11. Han, X.; Mao, S.; Zeng, S.; Duan, H.; Liu, Q.; Xue, L.; Yu, J. Effect of reactive flexible rejuvenators on thermal-oxidative aging resistance of regenerated SBS modified asphalt. *J. Clean. Prod.* **2022**, *380*, 135027. [CrossRef]
12. Hu, Z.; Zhang, H.; Wang, S.; Xu, T. Thermal-oxidative aging mechanism of asphalt binder based on isothermal thermal analysis at the SARA level. *Constr. Build. Mater.* **2020**, *255*, 119349. [CrossRef]
13. Wang, L.; Xue, Z.; Zhang, L. Effect of Thermal-Oxygen Aging on Temperature Stability of Styrene-Butadiene-Styrene–Modified Asphalt. *J. Test. Eval.* **2022**, *50*, 1832–1843. [CrossRef]

14. Pei, Z.; Xu, M.; Cao, J.; Feng, D.; Hu, W.; Ren, J.; Jing, R.; Yi, J. Analysis of the micro-characteristics of different kinds of asphalt based on different aging conditions. *Mater. Struct.* **2022**, *55*, 250. [CrossRef]
15. Xiang, H.; He, Z.; Tang, H.; Chen, L. Effects of thermo-oxidative aging on the macro-physical properties and microstructure of asphalt. *Constr. Build. Mater.* **2023**, *395*, 132315. [CrossRef]
16. Yu, H.; Bai, X.; Qian, G.; Wei, H.; Gong, X.; Jin, J.; Li, Z. Impact of ultraviolet radiation on the aging properties of SBS-modified asphalt binders. *Polymers* **2019**, *11*, 1111. [CrossRef]
17. Chen, Z.; Zhang, H.; Duan, H. Investigation of ultraviolet radiation aging gradient in asphalt binder. *Constr. Build. Mater.* **2020**, *246*, 118501. [CrossRef]
18. Yang, J.; Zhang, Z.; Xue, J.; Lei, J.; Liu, Y.; Wang, Y.; Fang, Y. Variation and mechanism of asphalt-aggregate interface features under ultraviolet aging based on meso-and micro-observations. *Constr. Build. Mater.* **2023**, *400*, 132830. [CrossRef]
19. Xu, Y.; Niu, K.; Zhu, H.; Chen, R.; Ou, L.; Wang, J. Generation and evolution mechanisms of asphalt aging under the combined effects of intense UV radiation and large temperature variations. *Constr. Build. Mater.* **2024**, *438*, 136786. [CrossRef]
20. Shi, P.; Li, X.; Shen, J. Nonlinear modelling of selected micro-and macro-properties of weathered asphalt mixtures. *Constr. Build. Mater.* **2020**, *253*, 119097. [CrossRef]
21. Guan, M.; Guo, M.; Tan, Y.; Du, X. Study on the effect of aging on cracking resistance of virgin asphalt binder and its evolution model. *Constr. Build. Mater.* **2023**, *407*, 133443. [CrossRef]
22. Guan, M.; Guo, M.; Liu, X.; Du, X.; Tan, Y. Effect of Multifactor Coupling Aging on Rheological Properties of Asphalt Binders and Correlations between Various Rheological Indexes. *J. Mater. Civ. Eng.* **2024**, *36*, 04024088. [CrossRef]
23. Zhang, D.; Zhang, K.; Zhang, Y.; Zhou, Q.; Yuan, G.; Jiang, H.; Qian, G. Evolution of aging depth gradient distribution in SBS-modified bitumen during the aging of field. *Case Stud. Constr. Mater.* **2024**, *20*, e03260. [CrossRef]
24. Fernández-Gómez, W.D.; Rondón Quintana, H.; Reyes Lizcano, F. A review of asphalt and asphalt mixture aging: Una revisión. *Ing. Investig.* **2013**, *33*, 5–12. [CrossRef]
25. Zou, Y.; Pang, L.; Xu, S.; Wu, S.; Yuan, M.; Amirkhanian, S.; Xu, H.; Lv, Y.; Gao, X. Investigation of the rheological properties and chemical structure of asphalt under multiple aging conditions of heat, UV and aqueous solution. *Materials* **2022**, *15*, 5711. [CrossRef]
26. Yu, H.; Yao, D.; Qian, G.; Cai, J.; Gong, X.; Cheng, L. Effect of ultraviolet aging on dynamic mechanical properties of SBS modified asphalt mortar. *Constr. Build. Mater.* **2021**, *281*, 122328. [CrossRef]
27. Zou, L.; Zhang, Y.; Liu, B. Aging characteristics of asphalt binder under strong ultraviolet irradiation in northwest China. *Sustainability* **2021**, *13*, 10753. [CrossRef]
28. Wu, S.; Ye, Y.; Li, Y.; Li, C.; Song, W.; Li, H.; Li, C.; Shu, B.; Nie, S. The effect of UV irradiation on the chemical structure, mechanical and self-healing properties of asphalt mixture. *Materials* **2019**, *12*, 2424. [CrossRef]
29. Somé, S.C.; Barthélémy, J.-F.; Mouillet, V.; Hammoum, F.; Liu, G. Effect of thermo-oxidative ageing on the rheological properties of bituminous binders and mixes: Experimental study and multi-scale modeling. *Constr. Build. Mater.* **2022**, *344*, 128260. [CrossRef]
30. Verhulst, F. *Nonlinear Differential Equations and Dynamical Systems*; Springer Science & Business Media: Berlin/Heidelberg, Germany, 2012.
31. Alnadish, A.M.; Aman, M.Y.; Katman, H.Y.B.; Ibrahim, M.R. Influence of the long-term oven aging on the performance of the reinforced asphalt mixtures. *Coatings* **2020**, *10*, 953. [CrossRef]

materials

MDPI

Article

Separation of Macro- and Micro-Texture to Characterize Skid Resistance of Asphalt Pavement

Tao Xie [1,2], Enhui Yang [3,4], Qiang Chen [3,4], Junying Rao [5], Haopeng Zhang [3,4,*] and Yanjun Qiu [3,4]

1 College of Civil Engineering, Guizhou University, Guiyang 550025, China; xiet@gzu.edu.cn
2 Guizhou University Survey and Design Institute Co., Ltd., Guizhou University, Guiyang 550025, China
3 School of Civil Engineering, Southwest Jiaotong University, Chengdu 610031, China; ehyang@swjtu.edu.cn (E.Y.); qchen@swjtu.edu.cn (Q.C.); yjqiu@swjtu.edu.cn (Y.Q.)
4 Highway Engineering Key Laboratory of Sichuan Province, Southwest Jiaotong University, Chengdu 610031, China
5 Research Center of Space Structure, Guizhou University, Guiyang 550025, China; jyrao@gzu.edu.cn
* Correspondence: haopeng.zhang@my.swjtu.edu.cn

Abstract: The skid resistance of asphalt pavement is an important factor affecting road safety. However, few studies have characterized the contribution of the macro- and micro-texture to the skid resistance of asphalt pavement. In this paper, the generalized extreme studentized deviate (GESD) and neighboring-region interpolation algorithm (NRIA) were used to identify and replace outliers, and median filters were used to suppress noise in texture data to reconstruct textures. On this basis, the separation of the macro- and micro-texture and the Monte Carlo algorithm were used to characterize the skid resistance of asphalt pavement. The results show that the GESD method can accurately identify outliers in the texture, and the median filtering can eliminate burrs in texture data while retaining more original detail information. The contribution of the macro-texture on the skid resistance is mainly attributed to the frictional resistance caused by the adhesion and elastic hysteresis, and the main contribution of the micro-texture is a micro-bulge cutting part in the friction mechanism. This investigation can provide inspiration for the interior mechanism and the specific relationship between the pavement textures and the skid resistance of asphalt pavement.

Keywords: asphalt pavement; texture separation; macro-texture characteristics; micro-texture features; Monte Carlo algorithm; skid resistance

Citation: Xie, T.; Yang, E.; Chen, Q.; Rao, J.; Zhang, H.; Qiu, Y. Separation of Macro- and Micro-Texture to Characterize Skid Resistance of Asphalt Pavement. *Materials* **2024**, *17*, 4961. https://doi.org/10.3390/ma17204961

Academic Editor: Cesare Sangiorgi

Received: 13 September 2024
Revised: 27 September 2024
Accepted: 30 September 2024
Published: 11 October 2024

1. Introduction

The friction between the tire and the pavement surface can lead to sliding resistance and energy loss, and it depends on the contact state between the tire and the pavement surface, which starts from the top of a few rough micro-bulges on the two surfaces [1,2]. The contact pressure can cause plastic deformation at the contact point [3–5], and the contact points between the tire and the road surface are not continuous, but discrete. The contact state determines the friction between the tire and the pavement, which mainly includes intermolecular, adhesion, elastic hysteresis, and micro-bulge cutting interactions [6]. In addition, some types of aggregate could be too abrasive and create ‘too good adhesion’, which can contribute to accelerated tire wear and the formation of micro-rubber particles [7]. Therefore, the skid resistance of the pavement is affected by several factors such as pavement texture, material properties, vehicle operating status, tire pressure, and road pollutants [8].

Although there are many factors needed to be considered, it is generally believed that the pavement textures play a key role in the skid resistance of asphalt pavement [9–11]. On the one hand, the textures affect the actual contact area between the tire and the pavement surface and have an important influence on the intermolecular force and adhesion in friction [10]. On the other hand, the texture directly participates in the elastic hysteresis and cutting process of micro-bulges, which influences the size of these two components in the friction [9,11]. Therefore, it is very important to accurately collect and comprehensively evaluate the pavement texture topography for the investigation of friction characteristics between the tire and the pavement surface.

The deviation between the surface and the horizontal plane of the pavement is defined as the pavement texture, and the wavelength is the minimum interval of a periodically repeated part, which can determine different scale ranges and pavement textures [12,13]. The Permanent International Association of Road Congress (PIARC) divides pavement texture into four categories according to the wavelength range, including unevenness, macrostructure, macro-texture, and micro-texture, where the micro-texture is less than 0.5 mm, the macro-texture is in the range of 0.5–50.0 mm, the macrostructure is between 50 and 500 mm, and the uneven texture is more than 500 mm [14,15]. On this basis, some researchers try to measure the macro-texture parameters of asphalt pavement and establish a prediction model of the skid resistance for asphalt pavement. Then, the parameters of macro-texture can be termed as a direct input parameter of the prediction model [16,17]. Although these models can make full use of laboratory data to avoid some troubles caused by field measurements to predict the skid resistance of asphalt pavement at a certain extent, the correlation between the results of the prediction model and field measured data is not good according to the standard of correlation coefficients [18–20].

Based on the above literature, scholars mainly obtain macro-texture topography through various methods to evaluate the skid resistance of asphalt pavement. However, there are few studies on comprehensively characterizing the skid resistance of asphalt pavement through the separation of the macro- and micro-texture. In this paper, three-dimensional laser image technology was used to extract and separate high-precision feature information of the macro- and micro-texture for asphalt pavement, and the Monte Carlo expectation method was performed to calculate the average structure depth. On this basis, the prediction relationship between pavement texture features and the skid resistance is also established. This study can provide some inspirations for predicting the skid resistance of asphalt pavement and the corresponding mechanism.

2. Research Objectives

This paper is to comprehensively investigate contributions of macro- and micro-texture to the skid resistance of asphalt pavement through the separation of macro- and micro-texture and the Monte Carlo algorithm, which can reveal the interior mechanism and the specific relationship between pavement textures and the skid resistance of asphalt pavement. The diagram of research objectives and the method are shown in Figure 1, and there are four objectives in this paper as follows.

(1) To identify and replace the outliers using the generalized extreme studentized deviate (GESD) and neighboring-region interpolation algorithm (NRIA).
(2) To suppress noise in the pavement textures data and reconstruct the road texture through the median filter.
(3) To separate the macro- and micro-texture in asphalt pavement according to their corresponding frequency.
(4) To establish the relationship between the characteristics of the macro- and micro-texture and the skid resistance.

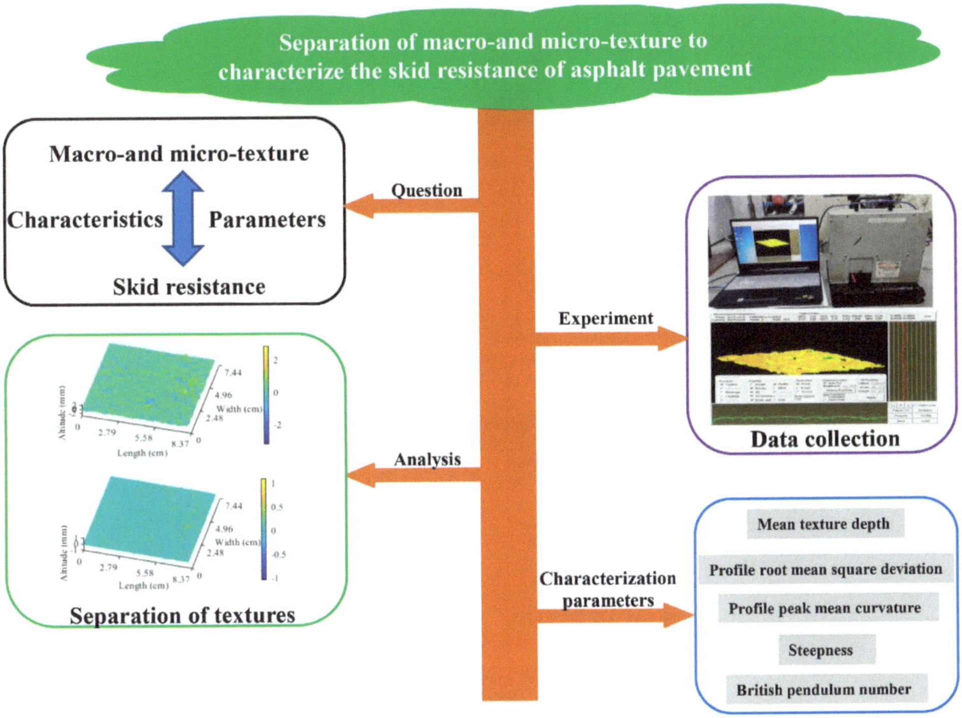

Figure 1. Diagram of research objectives and methods.

3. Materials and Methods

3.1. Raw Materials

The origin bitumen used in this paper originated from the Mianyang Bitumen Company, and the basic physical properties of the origin bitumen are shown in Table 1. In addition, the styrene–butadiene–styrene (SBS) was also purchased from the Mianyang Bitumen Company. In order to investigate the changes of the macro- and micro-texture characteristics and the skid resistance of the asphalt pavement, three typical-graded asphalt mixtures including dense-graded asphalt concrete (AC), stone mastic asphalt (SMA), and open-graded friction course (OGFC) were selected, and the rutting slabs with different maximum nominal particle sizes were made for the experiment. The maximum nominal particle size of AC and SMA was 16 mm, 13 mm, and 10 mm, and the maximum nominal particle size of OGFC was 13 mm and 10 mm. The gradation design of the mixture is shown in Table 2. The size of the rutting plate specimen was 300 mm × 300 mm × 50 mm, and the lignocellulose was doped into the asphalt mixture, and the added content was 0.3% of weight of the asphalt mixture.

Table 1. Basic physical properties of the origin bitumen.

Test Parameters	Value	Test Methods
Density, g/cm^3	1.004	ASTM D70 [21]
Ductility at 25 °C, cm	88	ASTM D113-99 [22]
Viscosity at 135 °C, Pa·s	0.49	ASTM D4402 [23]
Soften point (R&B), °C	48.3	ASTM D36-95 [24]
Penetration @ 25 °C, 0.1 mm	56	ASTM D5-97 [25]
PG grade	64–22	ASTM D6373 [26]

Table 2. Gradation design of the three types of asphalt mixtures.

Sieve Hole Size/mm	AC			SMA			OGFC	
	16	**13**	**10**	**16**	**13**	**10**	**13**	**10**
16	91	100	100	91	100	100	100	100
13.2	78	95	100	67	99	100	99	100
9.5	63	77	95	47	72	99	78	99
4.75	38	53	60	21	32	56	28	68
2.36	24	37	44	16	25	31	21	21
1.18	16	27	32	15	23	25	17	17
0.6	11	19	23	13	19	21	14	14
0.3	9	14	16	11	15	17	11	11
0.15	6	10	11	10	14	15	7	7
0.075	5	6	6	8	12	12	5	5
Oil-stone ratio	4.4	4.7	4.9	5.4	6.0	6.2	4.8	4.9

3.2. Data Collection and Processing

The 3D laser scanning equipment (LS-40), Artec 3D, Senningerberg, Luxembourg, used in the study relied on a built-in laser transmitter to emit laser light, and moved the laser beam on the surface of the object [27,28]. The reflected laser beam information was received by the high-resolution camera, and the principle of laser triangulation was used to analyze the laser beam information, which is shown in Figure 2. As shown in Figure 2a, the laser beam generated by the laser transmitter came from a known angle of the instrument, and hit the surface of the object at point B. After one reflection, the reflection intensity depended on the type of surface of the object, and part of the reflected light was received by camera C. There was a phenomenon that it could not accurately measure the surface elevation information of the object [29]. On the one hand, this phenomenon was caused by part of the reflected light not being received by the camera. On the other hand, it was attributed to the elevation of measurement points exceeding the measurement range of the instrument (Figure 2b). The results of the final elevation would be displayed as values much larger than the standard range. Therefore, it was necessary to use NRIA to replace the erroneous results in the original data.

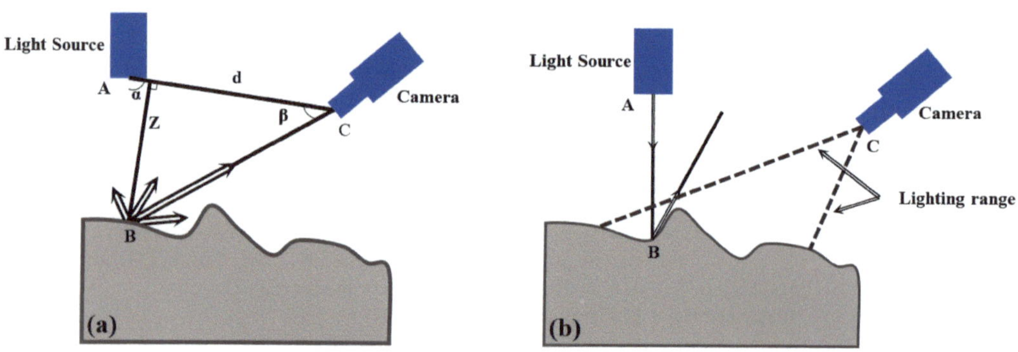

Figure 2. The principle of laser triangulation for the LS-40 equipment (**a**) Laser beam came from a known angle of the instrument, (**b**) Measurement points exceeding the measurement range of the instrument.

Therefore, the elevation data of the three-dimensional discrete point cloud of point B can be calculated as follows [27]:

$$Z = \frac{d}{\cot\alpha + \cot\beta} \quad (1)$$

where d is the distance between two points A and C, α is angle between the incident ray and the AC line, and β is angle between the reflected ray and the AC line. Each experiment was repeated three times and the average value was taken as the experimental value.

The LS-40 equipment collected the linear profile from the target surface. Each measurement scanned a total of 2048 profiles, and each profile line contained 2048 data points, which covered an area of 10.16 cm × 11.43 cm with a horizontal accuracy of 0.05 mm. The LS-40 equipment could measure the elevation texture information within 12.5 mm of the top and bottom of the surface because the instrument was placed on a surface of the specimen with a vertical accuracy of 0.01 mm [28]. The scanning results were displayed in the form of elevation data, which is illustrated in Figure 3.

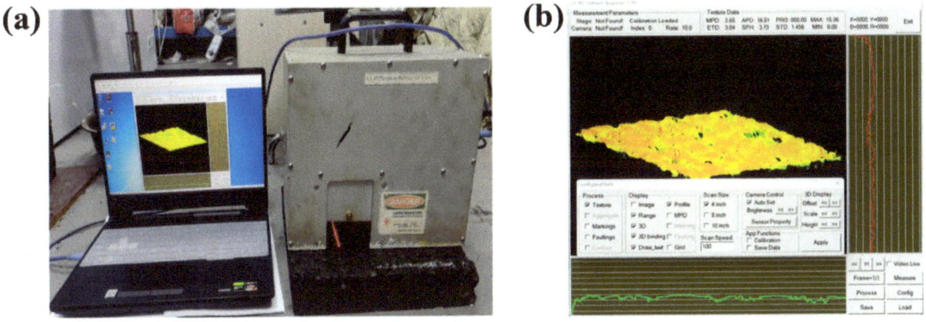

Figure 3. Surface texture elevation data collected by the LS-40 equipment of (**a**) field collection and (**b**) data collection.

The scanned contour of the LS-40 equipment was cut into 1448 × 1448 data points, and the original texture data of the contour scanned by the LS-40 equipment is shown in Figure 4. It can be seen from Figure 5 that there are outliers and signal noise in the original texture data, and there are some outliers and glitches in the 3D and 2D images.

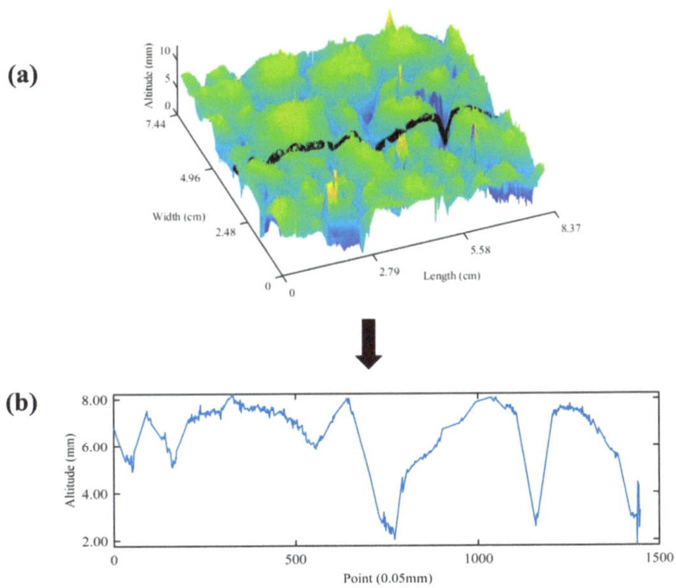

Figure 4. Schematic diagram: (**a**) color map; (**b**) waveform of original texture.

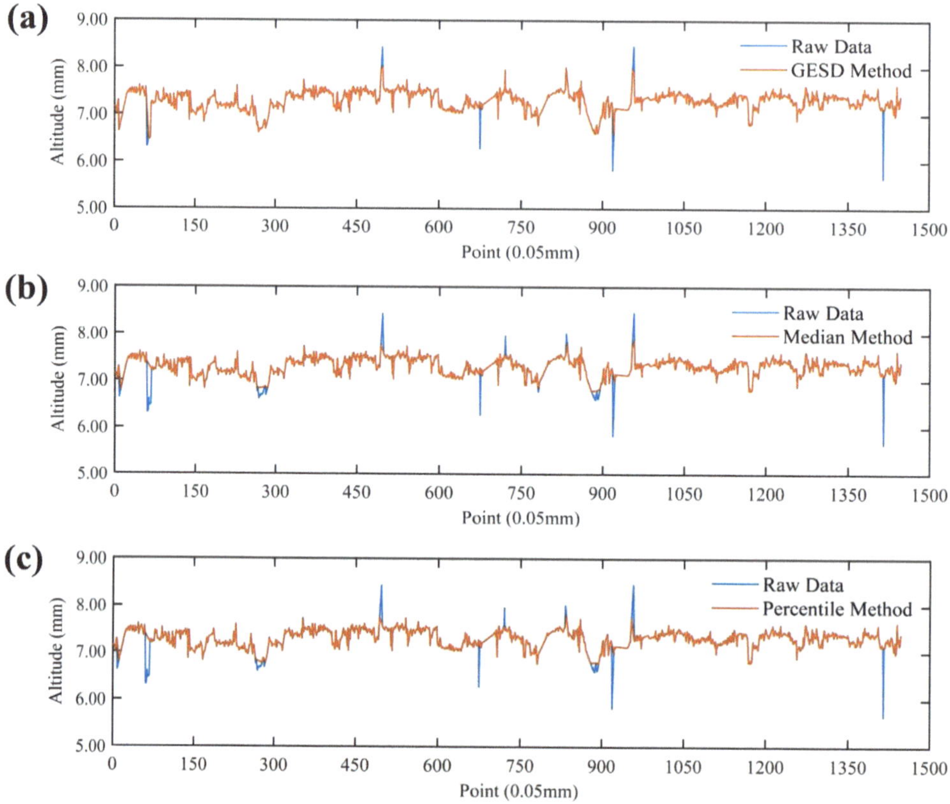

Figure 5. Comparison of outlier identification methods: (**a**) GESD method; (**b**) median method; (**c**) parametric method.

3.2.1. Outlier Processing

Compared with most of the normal data, there were significant differences in the outliers [30]. The elevation values of the pavement textures usually changed continuously, but the measured elevation value deviated from larger values due to the reflective, oily, and occlusion characteristics of the specimen surface. The existence of outliers will lead to the inexact reconstruction and evaluation of topographical characteristics for pavement textures. Therefore, it is necessary to identify and replace the outliers.

Figure 5 demonstrates the comparative effect of each outlier identification method, and it can be seen that the median and parametric methods identify part of the pavement textures contour information as outliers in the process of identifying the pavement texture outliers, while the GESD method can accurately identify the outliers of pavement textures and the elevation profile information of pavement textures is relatively completely preserved compared with the median and parametric methods. Therefore, the outlier recognition effect is better.

Based on this, the outlier identification is performed on the raw texture data in the experiment using the GESD method, and the identified outliers are replaced by the NRIA of the adjacent non-outliers as shown in Equation (2) [31]

$$z_i = z_n + \frac{z_n - z_m}{n - m}(i - m) \tag{2}$$

where i is the identified outlier number, n is the data number of the non-outlier immediately before i, m is the data number of the non-outlier immediately after i, z_i is the replacement value of the outlier at i, z_n is the contour elevation value of number n, and z_m is the contour elevation value of number m.

3.2.2. Filter Denoising

The noise is the interference superimposed on the real value due to a system error and signal transmission, and the actual measured pavement texture data often contains real and noise data at the same time, which also causes the collected original road surface texture to appear frizzy and oscillating [32]. The filtering can remove the noise signal in the original texture information, and both the mean and median filtering belong to the smoothing methods in the spatial domain [33]. The mean filtering is a type of linear filtering, which replaces element values in the center of the template by calculating the average value of elements within the coverage of the template to achieve the purpose of denoising. Therefore, the filtering effect of the mean filter is related to the size of the selected template radius. It was reported that the mean filtering had advantages of a fast running speed, a simple algorithm, and easy implementation, and the calculation formula of the mean filter is as follows [33]:

$$g(x,y) = \frac{1}{(2i+1)^2} \sum_{i=-r}^{r} \sum_{j=-r}^{r} I(x+i, y+i) \quad (3)$$

where $g(x, y)$ is the filtered element value, $I(x, y)$ is the original signal element value, and i is the calculation window template radius. When $i = 1$, the window template is as follows:

1/9	1/9	1/9
1/9	1/9	1/9
1/9	1/9	1/9

As a nonlinear filter, the median filter sorts the element values in a window template, and replaces the element values in the center of the template with a median value of the elements in the window template (Equation (4)), which shows good noise smoothing properties [33]

$$g(x,y) = med\{f(x-k, y-l)\} \quad (4)$$

where $f(x, y)$ is the original signal element value, and k and l are both calculation window template sizes.

The wavelet threshold denoising is performed by setting a threshold in each wavelet decomposition scale. The wavelet coefficient part is judged as a noise signal because the decomposed wavelet coefficient is less than the threshold value. Then, the original signal is reconstructed by a new wavelet coefficient to achieve the purpose of denoising [34].

Figure 6 compares the filtering effects of mean and median filtering and wavelet threshold denoising. As shown in Figure 7, the wavelet threshold denoising is the closest to the original signal, and retains most of the characteristics of the original signal, which also leads to incomplete filtering of the wavelet threshold denoising such as some glitches. Both mean and median filtering can eliminate the glitch points, and the median filter is in the middle of the wavelet threshold denoising and mean filtering to the extent that it is close to the original signal. The mean filtering is calculated by the average value of all elements in the window. This will discard signal details in the filtering process, and the image is blurred in a larger range. Therefore, the median filtering retains more original details than the mean filtering.

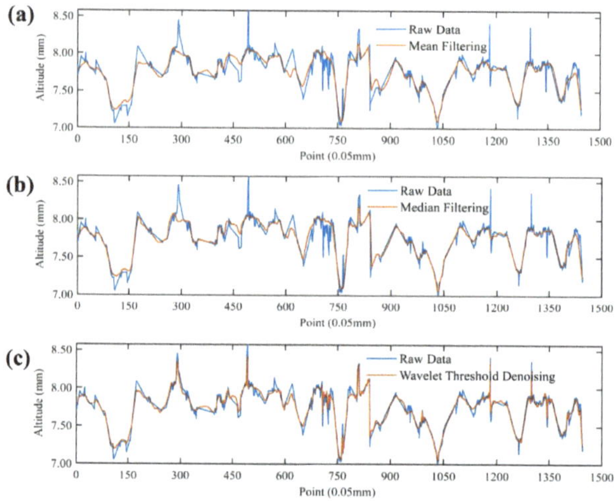

Figure 6. Comparison of the filter denoising methods: (**a**) mean filtering; (**b**) median filtering; (**c**) wavelet threshold denoising.

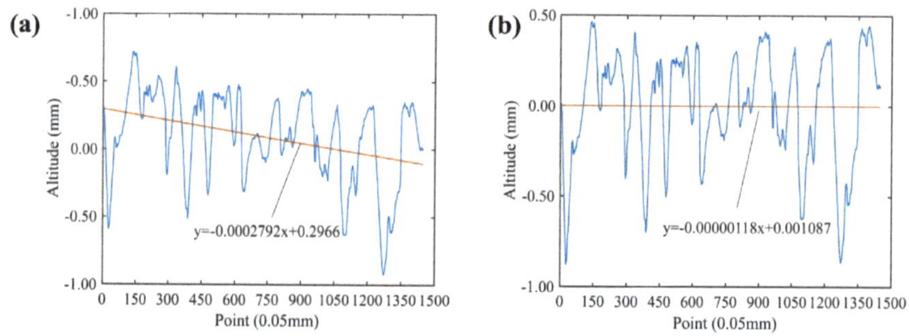

Figure 7. The pavement profile (**a**) before and (**b**) after tilt correction.

3.2.3. Tilt Correction

In addition, the placement of the mixture specimen cannot reach a completely horizontal state, which will cause the reconstructed surface to appear inclined and also affect the calculation of the pavement texture index and the evaluation of the texture topography. Therefore, it is necessary to eliminate the slope and offset errors of the elevation profile data, and the specific implementation formula of the tilt correction is as Equations (5)–(7) [34]

$$H_i = h_i - b_1 i - b_0 \quad i = 0, \ldots, n-1 \tag{5}$$

$$b_1 = \frac{12\sum_{i=0}^{n-1} ih_i - 6(n-1)\sum_{i=0}^{n-1} h_i}{n(n+1)(n-1)} \tag{6}$$

$$b_0 = \frac{1}{n}\sum_{i=0}^{n-1} h_i - b_1 \cdot \frac{n-1}{2} \tag{7}$$

where i is the number of the contour elevation data point; H_i is the slope correction contour elevation corresponding to number i; h_i is the height of the original contour corresponding to number i; N is the total number of collection points for the contour line; b_1 is the elimination coefficient of slope error, which can be calculated as Equation (6); and b_0 is the bias error item calculated by Equation (7). The pavement profile before and after the eliminated slope error and offset error are demonstrated in Figure 7.

The asphalt pavement surface texture is reconstructed after intercepting the collected original texture data, identifying and replacing outliers, filtering to remove noise signals, and inclination correction [35]. Figure 8 illustrates the three-dimensional reconstructed surfaces for various types of specimens, and it can be seen that the height distribution becomes uneven as the types of specimen changes from the AC to OGFC specimen, and the OGFC-13 specimen presents the largest elevation difference in the same reconstructed surface.

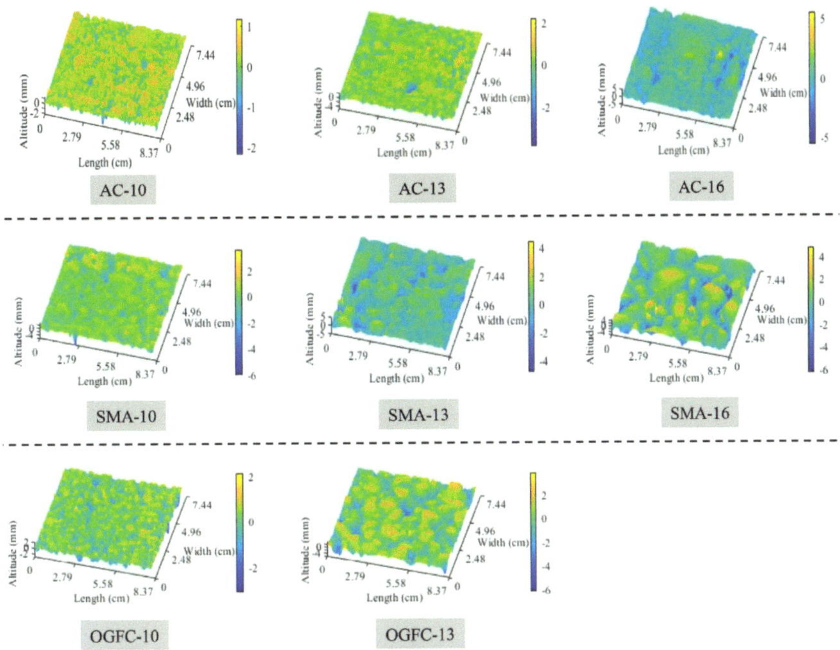

Figure 8. 3D reconstructed surfaces of various types of specimens.

3.3. Separation of Macro- and Micro-Texture

3.3.1. Separation Details

The surface texture of asphalt pavement always contains the texture of various wavelengths [36], and the macro- and micro-texture of pavement have different contributions on the friction mechanism between the tire and the pavement surface. Therefore, to investigate the influence of macro- and micro-texture on the skid resistance of pavement, it is necessary to separate and extract the macro- and micro-texture from the pavement surface textures according to wavelength differences. As a basic operation in signal processing, Fourier transform can realize the conversion from the space to the frequency domain, and it is widely used in signal processing, optics, probability, statistics, and other fields. Therefore, the Fourier-transform method was used to process the texture data of the road surface to convert the texture data from the spatial domain to the frequency domain in this paper, and the Fourier transform is calculated as follows [36]:

$$F(u,v) = \iint f(x,y)e^{-j2\pi(ux+vy)}dxdy \tag{8}$$

In this paper, a sliding window-based finite impulse response (FIR) digital filter was designed to separate and extract the macro- and micro-texture from collected pavement texture data and the design of the FIR digital filter selects the Hamming window. The band-pass filter was used to extract the macroscopic texture from the road surface, and the upper and lower limit frequencies of the band-pass filter are 2 mm^{-1} and 0.2 mm^{-1}, respectively. The high-pass filter is designed to extract the microscopic texture and the corresponding cut-off frequency and filter order of the high-pass filter is 0.02 mm^{-1} and 400, respectively. Based on this, the results of the macro- and micro-texture separated from the 3D reconstructed pavement of AC-13 are demonstrated in Figure 9.

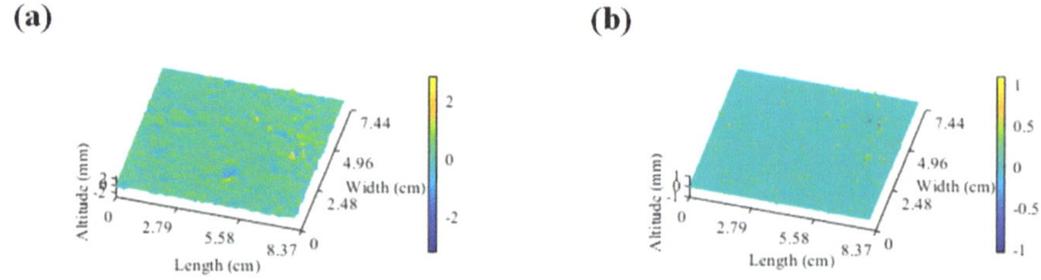

Figure 9. 3D reconstruction of (**a**) macro-texture and (**b**) micro-texture obtained from AC-13.

Figure 10 takes one profile line as an example to illustrate the separation results of the pavement macro- and micro-texture from AC-13. According to Figure 10a, it can be seen the macro-texture of the pavement is basically consistent with the profile shape of the overall profile line, and the macro-texture is smaller than that of the pavement outline. Figure 10b,c show the frequency information of the macro- and micro-texture after frequency domain filtering, respectively. The amplitude of the part with a frequency greater than 2 mm^{-1} is 0 in the frequency domain map of the macroscopic texture, while the amplitude of the part with a frequency less than 2 mm^{-1} is also close to 0 for the frequency domain map of the microscopic texture. The above results show that the pavement texture is converted from the spatial domain to the frequency domain using the fast Fourier transform, and the filter is designed according to the frequency domain information corresponding to the wavelength of the macro- and micro-texture, which can effectively separate the macro- and micro-texture from the pavement textures.

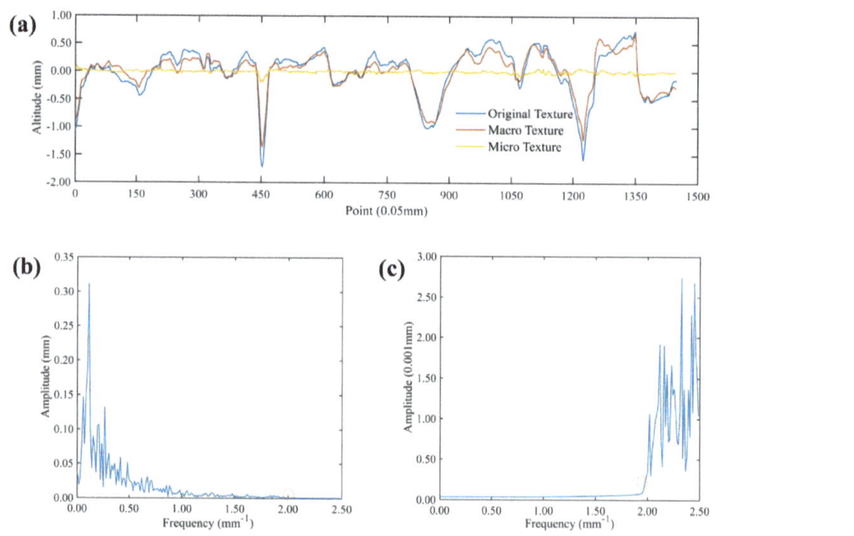

Figure 10. Pavement macro- and micro-texture separation and frequency domain map of (**a**) the contour line, (**b**) the frequency domain of the macro-texture, and (**c**) the frequency domain of the micro-texture from AC-13.

3.3.2. Pavement Texture Indexes

The texture characteristics of asphalt pavement have a great influence on the contact behavior between the tire and the pavement surface. Therefore, it is necessary to comprehensively evaluate pavement texture characteristics from different angles and scales. On the basis of existing research [37,38], a number of parameters are selected from aspects of the pavement topography height, protruding body shape, comprehensive distribution, and fractal characteristics to comprehensively evaluate the macro- and micro-texture of pavement. This can also analyze the correlation between texture parameters and swing value, which aims to investigate the texture characteristics of asphalt pavement and its influence on the skid resistance of asphalt pavement.

The root-mean-square deviation (RMSD) R_q is a parameter describing texture height properties, which characterizes the discreteness of the pavement profile and weights amplitude of the larger absolute value in data. This will further highlight the influence of the larger absolute amplitude on the texture height feature, and the calculation of R_q of the discrete texture data is as follows [37]:

$$R_q = \sqrt{\frac{1}{N}\sum_{i=1}^{n} {z_i}^2} \tag{9}$$

The Monte Carlo expectation method is used to process the macro- and micro-texture data of pavement to obtain the mean texture depth (MTD). This method is used to calculate statistical eigenvalues as numerical solutions of practical problems through continuous sampling and gradual approximation based on the theory of probability and statistics, which has the advantage of a clear program structure and the convergence speed is not affected by the dimension of the problem [39,40]. The process of calculating pi is as follows. The radius of the circle is unit 1, and points are randomly cast into the circumscribed square area of the circle. The total number of cast points is m, and the number of cast points recorded in the circle is a, then the approximate value of pi can be obtained as follows [39]:

$$\pi = \frac{4a}{m} \tag{10}$$

During this process, the parametric points of the overall data are considered as the upper and lower limits of the rectangular frame. It was reported that the sand patch method could also be used to exactly calculate MTD, but the experimental process is not easy to operate manually compared with the Monte Carlo expectation method, and the calculation process could be found in previous literature [41,42]. Therefore, Figure 11 compares the MTD obtained by the sand patch method and the Monte Carlo expectation method with different parametric points. From Figure 12, the MTD calculated by the Monte Carlo expectation method is the closest to that obtained by the sand-laying method because the 98% and 2% parametric points are used as the upper and lower boundaries of the rectangular box, and more original elevation data information are retained at the same time. Therefore, the 98% and 2% parametric points of the overall data are used as the upper and lower limits of the Monte Carlo expectation method in this paper.

The peak half angle (PHA) α can characterize the particle shape of the protruding peak on the surface of the specimen, and the larger PHA indicates a wider and flatter protruding peak, which can be calculated as follows [42]:

$$\alpha = \frac{1}{2}\left[\arctan\left|\frac{x_i - x_{i-1}}{y_i - y_{i-1}}\right| + \arctan\left|\frac{x_{i+1} - x_i}{y_{i+1} - y_i}\right|\right] i = 2,3,\ldots,n \tag{11}$$

where x_i is the abscissa of the i-th extreme point, x_{i+1} and x_{i-1} are the abscissa corresponding to the bottom of the valley, and y_i is the i-th extreme point elevation value as the i-th extreme point is the peak.

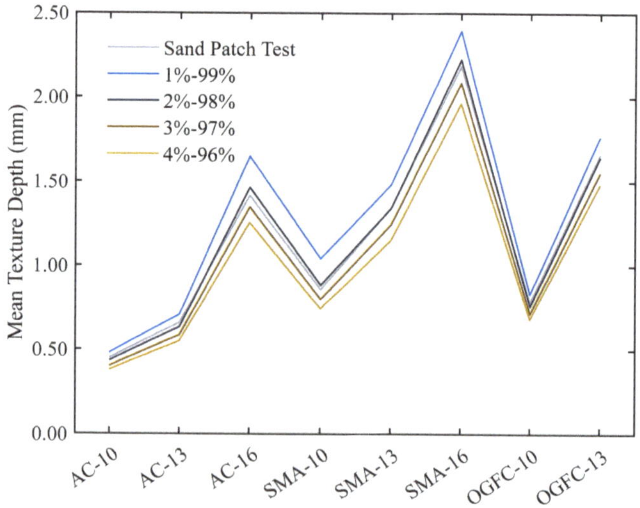

Figure 11. The MTD calculated by the sand patch and the Monte Carlo expectation methods.

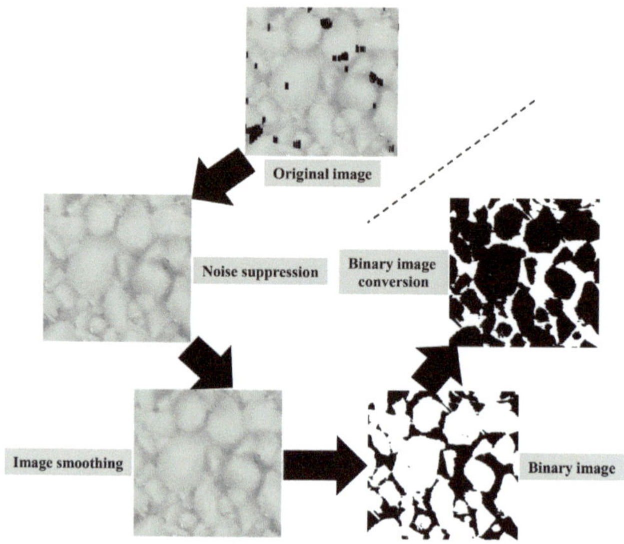

Figure 12. Schematic diagram of image segmentation of the specimen.

The mean peak curvature (MPC) K can reflect the sharpness of the profile peak of the micro-bulge, and the sharpness of micro-texture affects the cutting effect of the micro-bulge. The smaller K value indicates that the point where the road surface contacts rubber is more rounded, and the point where the road surface contacts the rubber is sharper because the K value is large. The LS-40 equipment collects the discrete elevation data points, and Equation (12) for the MPC of discrete vertices is as follows [43]:

$$K_i = \left| \frac{\arctan\left(\frac{y_{i+2}-y_{i+1}}{x_{i+2}-x_{i+1}}\right) - \arctan\left(\frac{y_{i+1}-y_i}{x_{i+1}-x_i}\right)}{\sqrt{(x_{i+1}-x_i)^2 + (y_{i+1}-y_i)^2}} \right| \tag{12}$$

The kurtosis parameter R_{ku} is used to evaluate the steepness of the texture curve, and mainly reflects characteristics of overall distribution for contour elevation data, which is determined as [43]

$$R_{\text{ku}} = \frac{1}{R_q{}^4} \cdot \frac{1}{M \times N} \sum_{i=1}^{N} \sum_{i=1}^{M} z^4(x_i, y_i) \tag{13}$$

where M and N are the number of lateral and longitudinal contours of the pavement, R_q is the RMSD of the contour, and $z\ (x, y)$ is the elevation of the pavement contour.

The height change of the surface profile for the pavement is an approximate random process, which characterizes the self-similarity in a certain level [44]. The fractal analysis is a powerful tool for the investigation of irregular shaped objects and can be used to describe the fractal characteristics of the pavement texture. In this paper, the box counting method is used to calculate the fractal dimension of the pavement surface texture images. Firstly, the grayscale image scanned by the LS-40 equipment is smoothed through the median filter. Then, the segmentation threshold is usually selected using the Otsu method [45]. Finally, the image can be divided into the background and target in the process of using this threshold to convert the grayscale image into a binary image. The image segmentation process is shown in Figure 12.

According to the above method, the MTD, RMSD, PHA, MPC, and kurtosis of the profile peak are calculated for the macro- and micro-texture, respectively. Then, the fractal dimension of the road surface is calculated by using the box counting method according to the analysis of the pavement surface image.

4. Results and Discussion

4.1. Macroscopic Texture Features and Skid Resistance Analysis

4.1.1. Height Characteristic Parameters

The macro-texture of asphalt pavement is related to the gradation type of the mixture and the maximum nominal particle size. Therefore, it is very important to discuss the influence of composition of the asphalt mixture on the macro-texture parameters of asphalt pavement and to analyze the relationship between the macro-texture parameters and the British pendulum number (BPN) of asphalt pavement.

Figure 13 illustrates the height characteristic parameters of the macro-texture for different graded asphalt mixture specimens and their relationship with the BPN. It can be seen from Figure 14 that the MTD of the texture is consistent with the RMSD of the profile, and both show characteristics that are significantly related to the gradation type of the mixture and the maximum nominal particle size. The MTD and RMSD of the macroscopic texture increase with the increase in the maximum nominal particle size for mixtures with the same gradation type. The macro-texture height of the open-graded pavement is greater than that of the dense-graded pavement, which shows this order as OGFC > SMA > AC for the same maximum nominal particle size. At the same time, the BPN of asphalt pavement shows a significant positive correlation with the macro-texture height parameters. As the height of the macro-texture increases, the skid resistance of the pavement improves, which is due to the increased height of the macro-texture of asphalt pavement increasing the unevenness of the pavement surface and the energy loss increases using the elastic hysteresis in friction between the tire and the pavement. The fitting coefficient of the RMSD and the BPN is slightly larger than that of the MTD and the BPN, which indicates that Rq exhibits a more significant correlation with the BPN than the MTD. This can be attributed to the convex part of the pavement surface mainly affecting the elastic hysteresis and the RMSD amplifying the profile offset distance of the convex part.

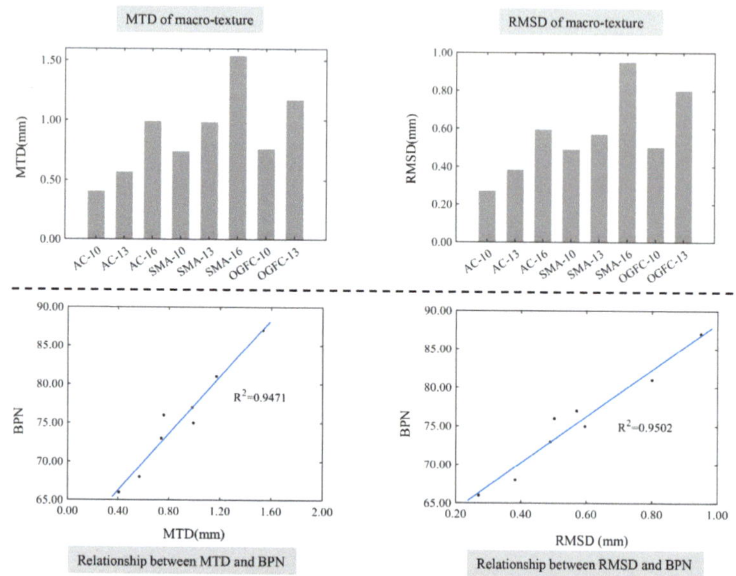

Figure 13. The relationship between height characteristics of the macro-texture and the BPN.

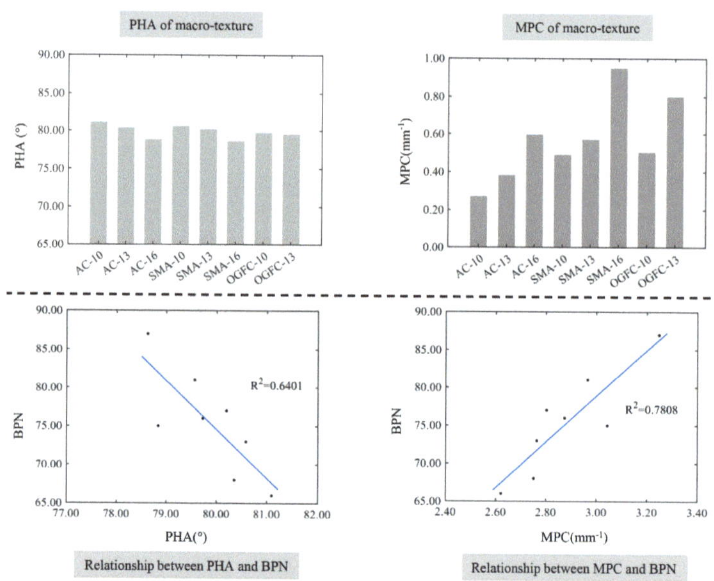

Figure 14. The relationship between the shape characteristics of the macro-texture protrusions and the BPN.

4.1.2. Protrusion Shape Feature Parameters

The relationships between the shape characteristic parameters of macro-texture protrusions and the BPN of different graded asphalt mixture specimens are demonstrated in Figure 14. The PHA and MPC also show a significant correlation with the mixture gradation and maximum nominal particle size. The profile peak of the pavement macroscopic texture is mainly caused by the irregular shape of the aggregate, and asphalt pavement tends to form a sharper convex peak as the proportion of the coarse aggregate or particle size of the aggregate is large, which leads to the larger MPC and smaller PHA. The increased amount of fine aggregate in the mixture leads to the larger PHA and a smaller MPC. This can be explained by the small and dense voids being easier to form on the surface of the pavement, and it is not easy to form a protruding profile peak. Interestingly, there is a negative correlation between the PHA and the BPN, and the fitting coefficient R^2 is 0.6401. The MPC and BPN has a positive relationship, whose fitting coefficient R^2 is 0.7808. On the one hand, the sharper macro-texture profile peak can lead to a greater contact stress and an increased elastic hysteresis generated during the contact between the rubber and the pavement surface. On the other hand, the increased sharpness of the macro-texture profile peak will cause a decrease in the contact area, which in turn reduces the frictional resistance caused by the adhesion.

4.1.3. Comprehensive Distribution Characteristic Parameters

Figure 15 illustrates the relationship between the macro-texture comprehensive distribution characteristics and the BPN. It can be seen from Figure 15 that the elevation distribution of the surface profile for SMA-10 is more concentrated than that of OGFC-13, which is attributed to the fact that the kurtosis of the macro-texture for SMA-10 is greater than that of OGFC-13. The steeper elevation data of the pavement contour indicates that most of the contour elevations are distributed around mean values, and there are fewer elevation data deviated from the mean value. It can also be concluded that the kurtosis of the macroscopic texture is related to the gradation of mixture and the maximum nominal particle size. The kurtosis of the macro-texture for open-graded asphalt pavement is smaller than that of dense-graded asphalt pavement, which indicates that the distribution of data furthest from the mean is more dispersed in the elevation data of the open-graded pavement profile. This is consistent with the performance of larger void depths on open-graded pavement surfaces. The kurtosis of the macro-texture for asphalt pavement decreases slightly as the maximum nominal particle size increases, which can be explained by the increase in the amount of coarse aggregate leading to an increase in the amount of protruding aggregate on the pavement.

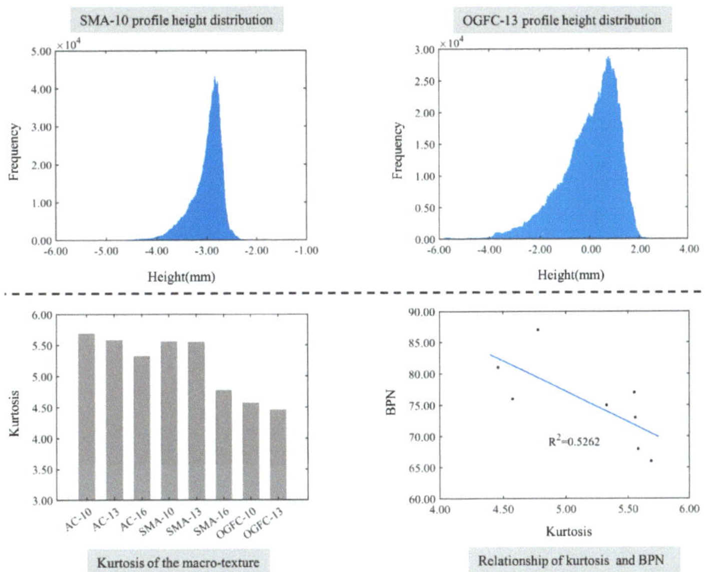

Figure 15. The relationship between the comprehensive distribution characteristics of the macro-texture and the BPN.

Interestingly, the increased kurtosis of the macro-texture will lead to a decrease in the BPN, and the fitting coefficient is R^2 = 0.5262. The degree of correlation is smaller than the height parameter of the macro-texture and the shape of the protrusion. This can be explained by the kurtosis of the macro-texture at the pavement surface being relatively large, and the distribution curve of the profile elevation being sharper than the normal distribution curve [41,42]. Therefore, the kurtosis of the macro-texture only shows a dispersion degree of the profile height distributed near the mean values and has no evident impact on the skid resistance of the pavement surface.

4.2. Microscopic Texture Features and Skid Resistance Analysis

4.2.1. Height Characteristic Parameters

The microscopic texture of asphalt pavement mainly describes the microstructure of the aggregate surface, which is determined by the surface structure of the aggregate and the amount of asphalt. The relationship between the microscopic texture height characteristics and the BPN is demonstrated in Figure 16. It can be seen from Figure 17 that the height parameter of the microscopic texture is generally small, and increases with the addition of the maximum nominal particle size of the asphalt mixture. The height of the aggregate surface is related to the size of the aggregate, and the size of the surface structure of the aggregate with a larger particle size is larger. Therefore, the larger size and the amount of coarse aggregate in the asphalt mixture can lead to a larger height of the microscopic texture.

In addition, there is a positive correlation between the characteristic parameters of the height of the micro-texture and the BPN. This is because the height of the micro-texture mainly affects the micro-cutting effect of pavement micro-bulges on the rubber. The deformation of the microscopic texture is large, and the microscopic texture is enveloped in the rubber as the tire is running on the road. Therefore, the increased height of the microscopic texture can increase the cutting depth of the rubber by the micro-bulge and the friction component caused by the cutting action of the micro-bulge, which can enhance the skid resistance of the pavement surface.

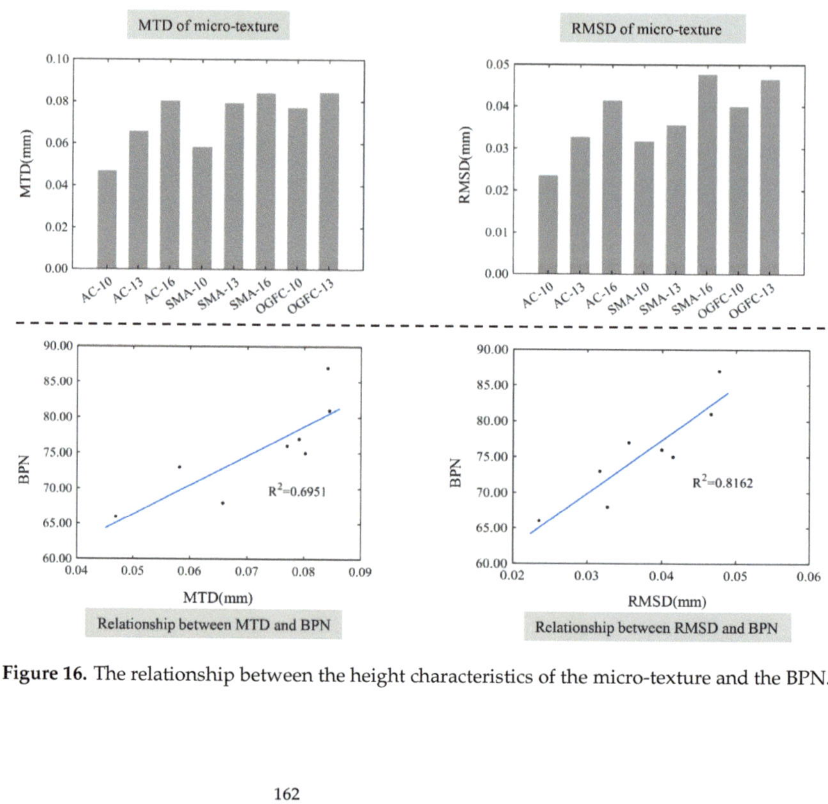

Figure 16. The relationship between the height characteristics of the micro-texture and the BPN.

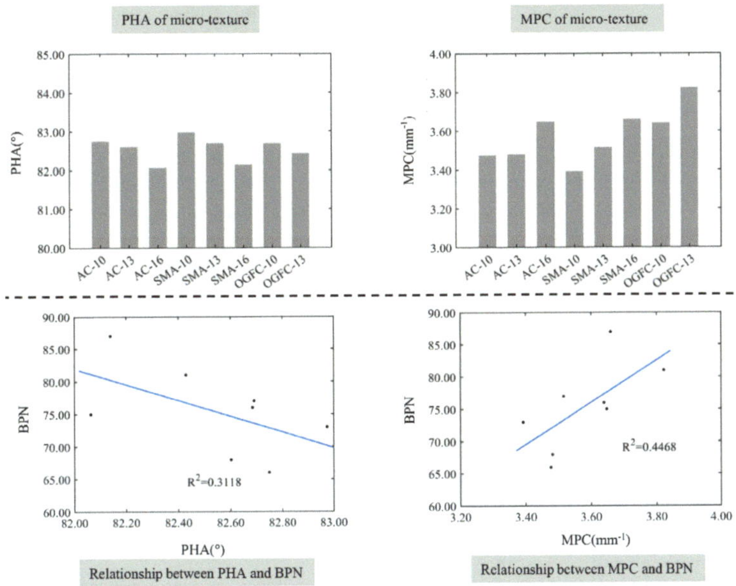

Figure 17. The relationship between the shape characteristics of the micro-texture protrusions and the BPN.

4.2.2. Shape Feature Parameters of Textured Protrusions

Figure 17 demonstrates the relationship between the shape characteristics of micro-textured protrusions and the BPN. The PHA of the microscopic texture for each specimen is relatively close to the MPC, the average PHA is 82.53°, and the difference between the maximum and minimum values is 0.81°. The mean value of MPC for the profile is 3.58 mm^{-1}, and the difference between the maximum and minimum values is 0.43 mm^{-1}. The PHA of the microscopic texture is related to the maximum nominal particle size within a small range of variation and decreases with the increase in the maximum nominal particle size. The addition of the maximum nominal particle size increases the MPC, and PHA on the surface of the SMA type mixture achieves the maximum, but the MPC is the smallest. This is due to the addition of lignocellulose into the SMA reducing the sharpness of the profile peak of its microstructure. In addition, according to Figure 18, the fitting coefficient R^2 between the BPN and the shape parameter of the micro-texture is 0.3118 and the fitting coefficient R^2 between the MPC and the BPN is 0.4468. The smaller fitting coefficient indicates that the sharpness of the profile peak of the micro-texture has little effect on the skid resistance of pavement.

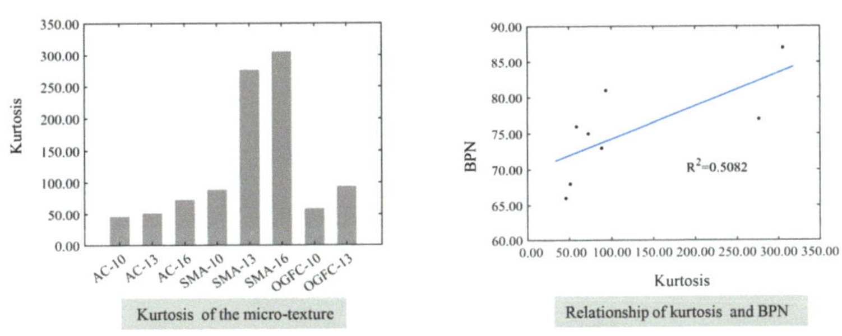

Figure 18. The relationship between the comprehensive distribution characteristics of the micro-texture and the BPN.

4.2.3. Comprehensive Distribution Characteristic Parameters

The relationship between the comprehensive distribution characteristics of the micro-texture and the BPN is shown in Figure 18. The kurtosis of the microscopic texture increases with the increase in the maximum nominal particle size, which is different from the kurtosis of the macroscopic texture. This can be explained by the microscopic texture of the pavement mainly describing the microstructure of the aggregate surface and the structure of the surface for large aggregates is relatively uniform and single compared with the variety of microstructures in the small size of aggregates. Therefore, the elevation distribution of microscopic texture is more concentrated with an increase in the maximum nominal particle size. In addition, the kurtosis of the micro-texture for the SMA-type mixtures is significantly greater than that of the AC and the OGFC due to the covering effect of fine aggregates and asphalt on the surface structure of coarse aggregates. At the same time, the fitting coefficient R^2 between the kurtosis of the micro-texture and the BPN is 0.5082, which indicates that the kurtosis of the micro-texture has little effect on the skid resistance of the pavement surface.

4.3. Fractal Dimension of Asphalt Pavement and Skid Resistance Analysis

The fractal dimension of asphalt pavement reflects the irregularity of the pavement topography and self-similarity on a certain level [46]. Figure 19 illustrates the relationship between the fractal dimension of the pavement surface and the BPN. It can be seen that the fractal dimension of asphalt pavement is related to maximum nominal particle size and gradation type of the asphalt mixture, and the BPN increases as the fractal dimension increases, which indicates that an increased fractal dimension enhances the skid resistance. This can be attributed to the complexity of the pavement surface topography increasing and the texture topography becoming richer when the irregular bump distribution of the pavement increases. Interestingly, the roughness of the pavement increases with an increase in the macroscopic texture because microscopic texture of the pavement is not much different. When the maximum nominal particle size of the mixture increases or the gradation becomes sparser, the textures become coarser and the degree of self-similarity for surface topography increases, which leads to the increase in fractal dimension and an improvement of the skid resistance for asphalt pavement.

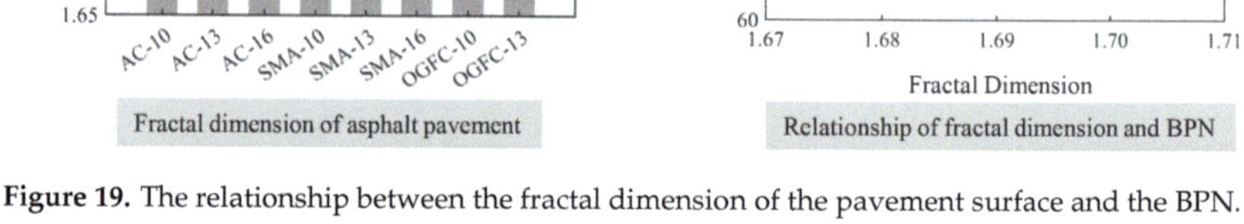

Figure 19. The relationship between the fractal dimension of the pavement surface and the BPN.

5. Conclusions and Recommendations

In this paper, three-dimensional laser image technology was used to extract and separate the high-precision feature information of macro- and micro-texture of asphalt pavement, and the Monte Carlo expectation method was used to calculate the MTD. On this basis, the relationship between pavement texture features and the skid resistance of asphalt pavement was also established through linear fitting. The conclusions can be drawn as follows.

(1) The median and parametric methods will identify part of the pavement texture profile as outliers, and the GESD method can more accurately identify outliers in the pavement texture.
(2) The signal after wavelet threshold denoising can retain most of the original signal features, but there are still some glitches in the denoised signal. The mean filter and median filter can both eliminate glitch points, and the median filter is in the middle of the wavelet threshold denoising and the mean filter methods, which is close to the original signal.
(3) The Monte Carlo expectation method can exactly calculate the MDT of reconstructed pavement, and the 98% and 2% parametric points of overall data as upper and lower limits of the Monte Carlo projection area is performed to obtain the MDT after the selection of the upper and lower limits.
(4) The height of the macro-texture is greater and sharper protruding peaks appear to enhance the skid resistance of the pavement as the maximum nominal particle size increases or the asphalt mixture changes from dense gradation to open gradation.
(5) The larger maximum nominal particle size of the asphalt mixture enriches the aggregate surface structure, reduces the amount of fine aggregate and asphalt, and increases the height and sharpness of the micro-texture of the pavement, which leads to an improved skid resistance of the pavement. The elevation distribution of the micro-texture has little effect on the skid resistance of the pavement.

This paper investigates the relationship between the characteristic of macro- and micro-texture and the skid resistance of asphalt pavement, which can present the interior mechanism and specific relationship between pavement textures and the skid resistance of asphalt pavement. Our future work will further consider more factors with a combination of the macro- and micro-texture to predict the skid resistance of asphalt pavement more comprehensively, use continuous friction measuring equipment (CFME) to determine the skid resistance of asphalt pavement as comparison, perform other materials of the rutting slab and statistical analysis including the response surface method (RSM) to validate the results of the method in this paper, and investigate the effect of aging and wear on roughness.

Author Contributions: Methodology, T.X.; Validation, E.Y. and H.Z.; Investigation, T.X. and Q.C.; Writing—original draft, T.X.; Writing—review & editing, E.Y., Q.C., J.R. and H.Z.; Supervision, E.Y. and Y.Q.; Project administration, E.Y. All authors have read and agreed to the published version of the manuscript.

Funding: This study was supported by the National Natural Science Foundation of China under Grant (52278460, 52178438, 52468057) and the Guizhou Provincial Science Technology Projects (Grant No. Qian Ke He Jichu—ZK [2023] Yiban 080). The authors appreciate Sichuan Provincial Key Laboratory of Road Engineering for providing the macroscopic instruments to carry out the indoor pavement test.

Institutional Review Board Statement: Not applicable.

Informed Consent Statement: Not applicable.

Data Availability Statement: The data presented in this study are available on request from the corresponding author.

Conflicts of Interest: Author Tao Xie was employed by the company Guizhou University Survey and Design Institute Co., Ltd. The remaining authors declare that the research was conducted in the absence of any commercial or financial relationships that could be construed as a potential conflict of interest.

References

1. Wang, Y.; Yang, Z.; Liu, Y.; Sun, L. The characterisation of three-dimensional texture morphology of pavement for describing pavement sliding resistance. *Road Mater. Pavement Des.* **2019**, *20*, 1076–1095. [CrossRef]
2. Guo, W.; Chu, L.; Yang, L.; Fwa, T. Determination of tire rubber-pavement directional coefficient of friction based on contact mechanism considerations. *Tribol. Int.* **2023**, *179*, 108178. [CrossRef]
3. Li, J.; Chen, Z.; Xiao, F.; Amirkhanian, S.N. Surface activation of scrap tire crumb rubber to improve compatibility of rubberized asphalt. *Resour. Conserv. Recycl.* **2021**, *169*, 105518. [CrossRef]
4. Rosenkranz, A.; Costa, H.L.; Baykara, M.Z.; Martini, A. Synergetic effects of surface texturing and solid lubricants to tailor friction and wear—A review. *Tribol. Int.* **2021**, *155*, 106792. [CrossRef]
5. Magnus, C.; Gulenc, I.T.; Rainforth, W. Ambient dry sliding friction and wear behaviour of laser surface textured (lst) ti3sic2 max phase composite against hardened steel and alumina. *Wear* **2022**, *490*, 204184. [CrossRef]
6. Wang, H.-P.; Zhang, M.-Q.; Sun, R.-X.; Cui, S.-J.; Mo, J.-L. Performance improvement strategy of the tbm disc cutter ring material and evaluation of impact-sliding friction and wear performance. *Wear* **2023**, *526*, 204943. [CrossRef]
7. Haritonovs, V.; Tihonovs, J. Use of unconventional aggregates in hot mix asphalt concrete. *Balt. J. Road Bridge Eng.* **2014**, *9*, 276–282. [CrossRef]
8. Xu, Y.; Yang, R.; Chen, P.; Ge, J.; Liu, J.; Xie, H. Experimental study on energy and failure characteristics of rubber-cement composite short-column under cyclic loading. *Case Stud. Constr. Mater.* **2022**, *16*, e00885. [CrossRef]
9. Kogbara, R.B.; Masad, E.A.; Kassem, E.; Scarpas, A.T.; Anupam, K. A state-of-the-art review of parameters influencing measurement and modeling of skid resistance of asphalt pavements. *Constr. Build. Mater.* **2016**, *114*, 602–617. [CrossRef]
10. Yang, G.; Wang, K.C.; Li, J.Q. Multiresolution analysis of three-dimensional (3d) surface texture for asphalt pavement friction estimation. *Int. J. Pavement Eng.* **2021**, *22*, 1882–1891. [CrossRef]
11. Du, Y.; Weng, Z.; Li, F.; Ablat, G.; Wu, D.; Liu, C. A novel approach for pavement texture characterisation using 2d-wavelet decomposition. *Int. J. Pavement Eng.* **2022**, *23*, 1851–1866. [CrossRef]
12. Jahanshahi, M.R.; Jazizadeh, F.; Masri, S.F.; Becerik-Gerber, B. Unsupervised approach for autonomous pavement-defect detection and quantification using an inexpensive depth sensor. *J. Comput. Civ. Eng.* **2013**, *27*, 743–754. [CrossRef]
13. Puzzo, L.; Loprencipe, G.; Tozzo, C.; D'andrea, A. Three-dimensional survey method of pavement texture using photographic equipment. *Measurement* **2017**, *111*, 146–157. [CrossRef]
14. Moldovanu, A.; Huang, L. Development of a portable circular texture meter for road texture depth measurement. *J. Transp. Eng. Part B Pavements* **2018**, *144*, 04018026. [CrossRef]
15. Hoang, N.-D.; Nguyen, Q.-L. A novel method for asphalt pavement crack classification based on image processing and machine learning. *Eng. Comput.* **2019**, *35*, 487–498. [CrossRef]
16. Alhasan, A.; Smadi, O.; Bou-Saab, G.; Hernandez, N.; Cochran, E. Pavement friction modeling using texture measurements and pendulum skid tester. *Transp. Res. Rec.* **2018**, *2672*, 440–451. [CrossRef]
17. Hofko, B.; Kugler, H.; Chankov, G.; Spielhofer, R. A laboratory procedure for predicting skid and polishing resistance of road surfaces. *Int. J. Pavement Eng.* **2019**, *20*, 439–447. [CrossRef]
18. Li, Q.J.; Zhan, Y.; Yang, G.; Wang, K.C. Pavement skid resistance as a function of pavement surface and aggregate texture properties. *Int. J. Pavement Eng.* **2020**, *21*, 1159–1169. [CrossRef]
19. Plati, C.; Pomoni, M.; Georgouli, K. Quantification of skid resistance seasonal variation in asphalt pavements. *J. Traffic Transp. Eng.* **2020**, *7*, 237–248. [CrossRef]
20. Wang, D.; Zhang, Z.; Kollmann, J.; Oeser, M. Development of aggregate micro-texture during polishing and correlation with skid resistance. *Int. J. Pavement Eng.* **2020**, *21*, 629–641. [CrossRef]
21. *ASTM D70*; Standard Test Method for Density of Semi-Solid Asphalt Binder (Pycnometer Method). ASTM: Washington, DC, USA, 2021.
22. *ASTM D113-99*; Standard Test Method for Ductility of Bituminous Materials. ASTM: Washington, DC, USA, 2010.
23. *ASTM D4402*; Standard Test Method for Viscosity Determination of Asphalt at Elevated Temperatures Using a Rotational Viscometer. ASTM: Washington, DC, USA, 2012.
24. *ASTM D36-95*; Standard Test Method for Softening Point of Bitumen (Ring-and-Ball Apparatus). ASTM: Washington, DC, USA, 2017.
25. *ASTM D5-97*; Standard Test Method for Penetration of Bituminous Materials. ASTM: Washington, DC, USA, 2017.
26. *ASTM D6373*; Standard Specification for Performance-Graded Asphalt Binder. ASTM: Washington, DC, USA, 2015.
27. Zhou, B.; Chu, L.; Yin, C.; Fwa, T. Improved laboratory laser scanning setup and test procedure for 3-d pavement texture measurement. *Measurement* **2022**, *196*, 111168. [CrossRef]
28. Zou, Y.; Yang, G.; Cao, M. Neural network-based prediction of sideway force coefficient for asphalt pavement using high-resolution 3d texture data. *Int. J. Pavement Eng.* **2022**, *23*, 3157–3166. [CrossRef]
29. Ding, S.; Yang, E.; Wang, K.C.; Wang, G. Texture measurement based on 3d pavement surface images at sub-mm resolutioned. In Proceedings of the International Conference on Transportation and Development: Airfield and Highway Pavements, Pittsburgh, PA, USA, 15–18 July 2018; pp. 392–400.
30. Katicha, S.W.; Flintsch, G.; Bryce, J.; Ferne, B. Wavelet denoising of tsd deflection slope measurements for improved pavement structural evaluation. *Comput.-Aided Civ. Infrastruct. Eng.* **2014**, *29*, 399–415. [CrossRef]

31. Atef, A.H.; Rashed, M.A. Gpr ringing suppression using lateral outliers' swap filter. *J. Appl. Geophys.* **2023**, *208*, 104873. [CrossRef]
32. Bashar, M.Z.; Torres-Machi, C. Deep learning for estimating pavement roughness using synthetic aperture radar data. *Autom. Constr.* **2022**, *142*, 104504. [CrossRef]
33. Yuwono, B. Image smoothing menggunakan mean filtering, median filtering, modus filtering dan gaussian filtering. *Telemat. J. Inform. Dan Teknol. Inf.* **2015**, *7*. [CrossRef]
34. Katicha, S.W.; Mogrovejo, D.E.; Flintsch, G.W.; De León Izeppi, E.D. Adaptive spike removal method for high-speed pavement macrotexture measurements by controlling the false discovery rate. *Transp. Res. Rec.* **2015**, *2525*, 100–110. [CrossRef]
35. Deng, Q.; Zhan, Y.; Liu, C.; Qiu, Y.; Zhang, A. Multiscale power spectrum analysis of 3d surface texture for prediction of asphalt pavement friction. *Constr. Build. Mater.* **2021**, *293*, 123506. [CrossRef]
36. Serigos, P.A.; De Fortier Smit, A.; Prozzi, J.A. Incorporating surface microtexture in the prediction of skid resistance of flexible pavements. *Transp. Res. Rec.* **2014**, *2457*, 105–113. [CrossRef]
37. Chen, S.; Liu, X.; Luo, H.; Yu, J.; Chen, F.; Zhang, Y.; Ma, T.; Huang, X. A state-of-the-art review of asphalt pavement surface texture and its measurement techniques. *J. Road Eng.* **2022**, *2*, 156–180. [CrossRef]
38. Kheradmandi, N.; Mehranfar, V. A critical review and comparative study on image segmentation-based techniques for pavement crack detection. *Constr. Build. Mater.* **2022**, *321*, 126162. [CrossRef]
39. Barnett, R.; Reynolds, P.; Lester, W., Jr. Monte carlo algorithms for expectation values of coordinate operators. *J. Comput. Phys.* **1991**, *96*, 258–276. [CrossRef]
40. Reiter, D. The monte carlo method, an introduction. *Comput. Many-Part. Phys.* **2008**, *739*, 63–78.
41. China, S.; James, D.E. Comparison of laser-based and sand patch measurements of pavement surface macrotexture. *J. Transp. Eng.* **2012**, *138*, 176–181. [CrossRef]
42. Dan, H.-C.; Bai, G.-W.; Zhu, Z.-H.; Liu, X.; Cao, W. An improved computation method for asphalt pavement texture depth based on multiocular vision 3d reconstruction technology. *Constr. Build. Mater.* **2022**, *321*, 126427. [CrossRef]
43. Zhihong, M.; Guo, C.; Yanzhao, M.; Lee, K. Curvature estimation for meshes based on vertex normal triangles. *Comput.-Aided Des.* **2011**, *43*, 1561–1566. [CrossRef]
44. Chen, L.; He, Z.-Y.; Chen, H.-B. Road performance research on cold recycled pavement base with foamed asphalt on the basis of fractal dimension. *J. Highw. Transp. Res. Dev.* **2016**, *10*, 8–12. [CrossRef]
45. Moghaddam, R.F.; Cheriet, M. Adotsu: An adaptive and parameterless generalization of otsu's method for document image binarization. *Pattern Recognit.* **2012**, *45*, 2419–2431. [CrossRef]
46. Liu, C.; Zhan, Y.; Deng, Q.; Qiu, Y.; Zhang, A. An improved differential box counting method to measure fractal dimensions for pavement surface skid resistance evaluation. *Measurement* **2021**, *178*, 109376. [CrossRef]

materials

MDPI

Article

Effects of Moisture Infiltration on Interfacial Characteristics of Fiber Asphalt Mastic-Aggregate and the Cracking Resistance of Mixture

Keke Lou [1,2], Silin Jia [1], Peng Xiao [1,2,*], Haochen Wu [1] and Yuhao Wu [1]

[1] College of Civil Science and Engineering, Yangzhou University, Yangzhou 225100, China; lkkyzu@163.com (K.L.); 18994185805@163.com (S.J.); 17715878456@163.com (H.W.); wyh19980402@163.com (Y.W.)

[2] Research Center for Basalt Fiber Composite Construction Materials, Yangzhou 225127, China

* Correspondence: pengxiao@yzu.edu.cn; Tel.: +86-0514-8797-8669

Abstract: The interfacial properties of fiber asphalt aggregate and the cracking resistance of asphalt mixture are directly affected by moisture infiltration. In order to investigate the correlation between interfacial properties and immersion stability of asphalt mixture, three different types of fiber, including basalt fiber (BF), glass fiber (GF), and polyester fiber (PF); five types of fiber contents (0.1% to 0.5% by mass of the mixtures); and two types of aggregates (basalt and limestone) were selected. Experimental methods such as the Bond Strength Test (BBS), Disk-Shaped Compact Tension test (DCT), and interfacial image processing were used in order to assess the strength of interfacial interaction and resistance to cracking under both dry and wet conditions. The results showed that the addition of fibers could enhance fiber asphalt mastic-aggregate interfacial strength; under the influence of moisture infiltration, the interfacial strength showed a significant downward trend. In the process of fiber content increasing from 0.1% to 0.5%, the peak load and fracture energy of fiber asphalt mixtures were first increased and then decreased. The interface between asphalt mastic and aggregate is easier to spalling after being subjected to moisture infiltration, resulting in a decrease in cracking resistance. Compared with the dry environment, after moisture infiltration, the correlation index between interfacial strength and fracture energy is much higher than other influencing factors. The interfacial strength is still an important factor affecting the fracture energy. These findings provide valuable insights for the design and application of more durable asphalt pavement.

Keywords: asphalt mixture; moisture infiltration; fiber; interfacial strength; cracking resistance

Academic Editors: Paulo Santos and Jorge Carvalho Pais

Received: 10 November 2024
Revised: 13 December 2024
Accepted: 25 December 2024
Published: 26 December 2024

Citation: Lou, K.; Jia, S.; Xiao, P.; Wu, H.; Wu, Y. Effects of Moisture Infiltration on Interfacial Characteristics of Fiber Asphalt Mastic-Aggregate and the Cracking Resistance of Mixture. *Materials* **2025**, *18*, 53. https://doi.org/10.3390/ma18010053

1. Introduction

Asphalt mixture is widely utilized in road construction because of its advantages, such as driving comfort. Throughout the service life of asphalt pavement, continuous loading can cause various types of damage, while precipitation and prolonged exposure to moisture can accelerate the deterioration of these mixtures [1,2]. Water damage poses a significant threat to the durability and performance of the asphalt mixture, as moisture infiltration into the asphalt matrix leads to a range of detrimental effects. These include the detachment of asphalt binders from aggregates, reduced load-bearing capacity, and premature structural failure [3]. Such issues not only compromise the integrity of roadways but also increase maintenance costs and shorten their lifespan [4], attracting considerable attention from researchers and engineers worldwide. Many researchers have extensively explored the mechanisms of water damage, yielding valuable insights. Researchers [5–8] employed various techniques to study the microscopic adhesion between aggregates and asphalt and to assess the water stability of different asphalt-aggregate combinations. Additionally, Kai Zhang [9] employed low-field nuclear magnetic resonance technology to observe the pore structure parameters in asphalt mixtures under cyclic loading, noting that under the action of cyclic load and dynamic water erosion, some micropores expand and crack and form larger pores.

Fiber-reinforced asphalt mixtures present an effective approach to enhancing the mechanical and overall performance of asphalt materials [10]. Recent studies indicate that the incorporation of fibers into asphalt mixtures can significantly improve tensile strength, reduce permanent deformation, and improve resistance to thermal cracking. Various types of fibers, including synthetic and natural fibers, have been used to improve the viscoelastic properties of asphalt, offering a promising solution to the challenges posed by water damage. Carlos J. Slebi-Acevedo [11] suggests that the addition of fibers can improve fatigue resistance, reduce permanent deformation, and increase stiffness. Fibers reinforce hot mix asphalt through a three-dimensional network, increasing the adhesion within the mixture. L.M.G. Klinsky [12] demonstrated that the use of polypropylene and aramid fibers significantly enhanced the mechanical properties of conventional hot mix asphalt, enhancing its resistance to water damage and improving its ability to withstand rutting and permanent deformation. This results in better fatigue cracking resistance and a greater capacity to resist reflective cracking. Jie Wu [13] identified the asphalt adsorption effect and the three-dimensional spatial network structure as the primary reinforcement mechanisms of natural fibers, which effectively suppress rutting and cracking in asphalt pavements. Akihiro Sato [14] used optical microscopy and electron microscopy to characterize the morphology of the cellulose nanofibers and evaluated the rheological and mechanical properties of the fibers. Additionally, Hayder Kamil Shanbara's [15] research showed that using glass and hemp fibers as reinforcing materials significantly improved the water sensitivity of cold-mix asphalt, thereby enhancing its resistance to surface cracking. Extensive research has revealed that the stability of the bond between aggregates and asphalt is the key to the formation of mixture performance, where poor interactions can lead to issues such as pavement cracking, water damage, and permanent deformation. Molecular dynamics simulations have been employed to understand the interactions between asphalt and aggregates at the nanoscale [16,17]. In terms of aggregate characteristics affecting interface interactions, Jizhe Zhang [18] and colleagues conducted peel tests on four different aggregates and two types of asphalt, exploring how the mineral composition of aggregates influences water absorption. Their findings indicated that the mineral composition affects the aggregate's sensitivity to moisture. Curtis et al. [19,20] found that various components within asphalt can influence the bond strength at the interface. Several well-established methods exist for evaluating the characteristics of the asphalt binder-aggregate interface, including mechanical tensile testing [21,22], asphalt stripping tests [23–25], surface energy methods [26], and adhesion fatigue testing [27].

In recent years, the crack resistance of fiber-reinforced asphalt mixtures has garnered vital attention. Research indicates that the incorporation of fibers not only delays crack formation but also enhances the overall toughness of asphalt mixtures. A comprehensive understanding of how environmental conditions interact with the performance of fiber-reinforced mixtures is essential for developing strategies to improve their cracking performance under various service conditions. Md Nafiur Rahman [28] evaluated the crack resistance of polyethylene aramid and nylon fiber-reinforced asphalt mastic by conducting two crack mouth opening displacement-controlled experiments, revealing that nylon fiber exhibits superior crack resistance. Moreover, the inclusion of sugarcane bagasse fibers can enhance the high-temperature stability and low-temperature crack resistance of asphalt mixtures, achieving performance comparable to lignin fiber-reinforced mixtures [29]. The single-edge notched three-point bending tests demonstrated that polypropylene fibers increase the strength and fracture toughness of asphalt mastic, with crack bridging and fiber pullout identified as the two primary toughening mechanisms [30]. Yao Zhang [31] employed X-ray computed tomography, discrete element simulation techniques, digital image processing, and low- and medium-temperature cracking tests to obtain the optimal fiber length distribution so as to enhance crack resistance. Qin Tang [32] investigated the effects of different fiber types, fiber contents, and aggregate types on interface properties. The results indicated a strong correlation between the interface interaction strength of fiber-reinforced asphalt mastic and the fracture energy and peak crack mouth opening displacement of the asphalt mixtures.

Currently, research on the interfacial and cracking performance of fiber-reinforced asphalt mixtures is limited, particularly regarding the interfacial characteristics and cracking resistance of fiber asphalt mastic-aggregate in a water environment. Therefore, it is crucial to conduct in-depth studies on the influence of moisture infiltration on these properties. The flow chart for this study is shown in Figure 1. By investigating the interaction between moisture infiltration and the properties of the mixture, this study aims to provide valuable insights for the design and application of more durable asphalt pavement.

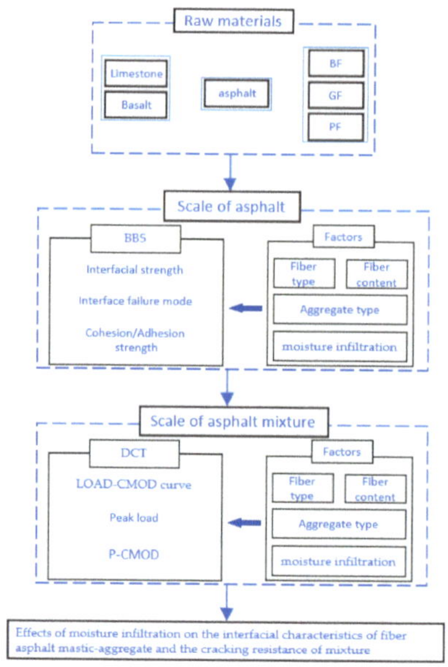

Figure 1. Flowchart of this study.

2. Materials and Methods

2.1. Raw Materials

SBS-modified asphalt was used in this study. The tests were conducted in accordance with the standard [33,34]. The penetration (25 °C, 0.1 mm) is 51.8, the ductility (5 cm/min, 5 °C) is 30 cm, and the softening point is 77 °C, respectively. The coarse aggregate was made of limestone and basalt. Three types of fiber were used in this experiment, including basalt fiber, glass fiber, and polyester fiber, and the surface morphology of the three kinds of fiber is shown in Figure 2, and the technical parameters are shown in Table 1. The basalt fiber is hereinafter referred to as BF, glass fiber as GF, and polyester fiber as PF.

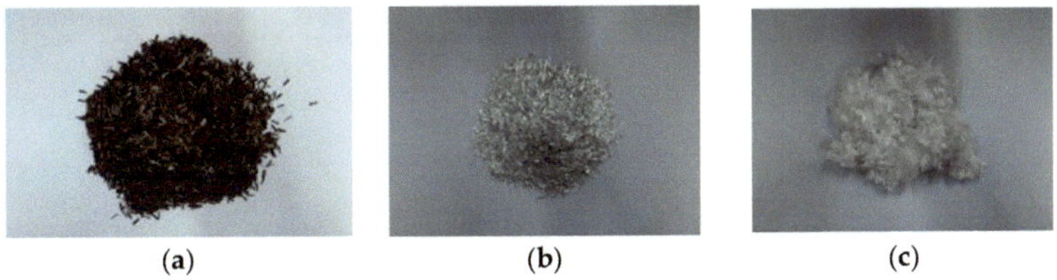

Figure 2. Macroscopic; (**a**) BF; (**b**) GF; and (**c**) PF.

Table 1. Fiber oil absorption rate.

Fiber Type	Basalt Fiber	Glass Fiber	Polyester Fiber
Tensile strength MPa	2561	2318	979
Oil absorption rate/times	1.04	0.5	3.43

2.2. Preparation for Fiber-Reinforced Asphalt Mixture and Asphalt Binder

The dense-graded gradation of AC-13, featuring a nominal maximum aggregate size (NMAS) of 13.2 mm, was created using the Marshall method. The gradation curve is shown in Figure 3. BF, GF, and PF were blended into the asphalt mixture according to 0.1%, 0.2%, 0.3%, 0.4%, and 0.5% of the mass of the mixture to prepare a fiber-reinforced asphalt mixture.

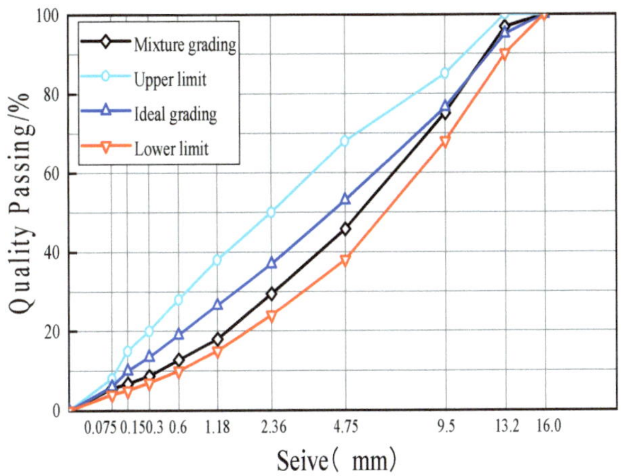

Figure 3. Gradation curve.

Table 2 illustrates the optimum asphalt content of the asphalt mixture under different fiber types and content. The fiber content selected in this paper is from the previous research results [32].

Table 2. Optimum asphalt content of asphalt mixture under different fiber types and content.

Fiber Type	Fiber Content/%				
	0.1	**0.2**	**0.3**	**0.4**	**0.5**
BF	4.9	5.0	5.1	5.2	5.2
GF	4.9	5.0	5.1	5.2	5.2
PF	5.0	5.1	5.2	5.3	5.3

The asphalt mastic powder-to-binder ratio prepared in this study was 1:1, and the fiber content in the asphalt mastic was set as 1%, 2%, 3%, 4%, and 5%. To ensure the uniformity of samples, a precision electric stirrer was used to prepare the fiber asphalt mastic. The process for preparing fiber asphalt mastic began by placing asphalt and mineral powder in an oven set at 175 °C for 2 h. After this, the mineral powder was added, and the mixture was stirred at a rotational speed of 2000 rpm for a duration of 5 min. Finally, the fiber was added to the asphalt mastic three times to decrease the agglomeration and was held for 10 min at 1000 rpm. In the process of sample preparation, three sets of parallel samples were prepared.

2.3. Test Methods

2.3.1. Binder Bond Strength Test (BBS)

BBS was conducted using an interfacial pullout test system. An automatic adhesion tester, a mineralogical stone plate, and a spindle are included in this system as shown in Figure 4. The technical parameters of the instrument are shown in Table 3. The basic principle is to apply the asphalt material to the surface of the aggregate or spindle and apply vertical tension to the spindle for pulling, and the interfacial strength between asphalt and aggregate can be reacted by tensile force or tensile strength.

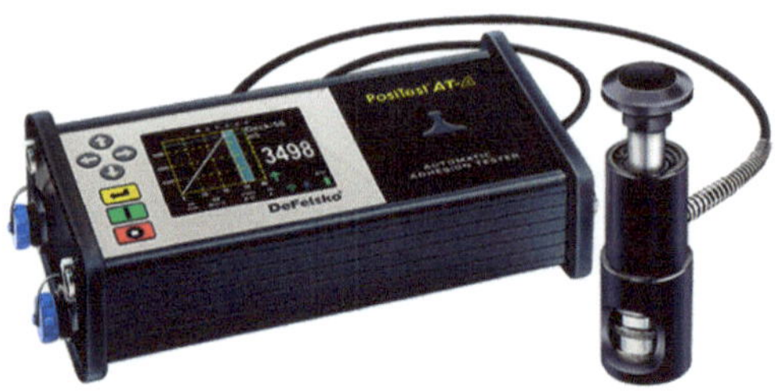

Figure 4. Test system.

Table 3. Technical parameter.

Spindle Diameter/mm	Resolution/MPa	Accuracy	Range/MPa
20	0.01	±1%	0–20

The interfacial strength can be calculated by Equation (1).

$$\sigma = \frac{4F}{\pi D^2} \tag{1}$$

where σ is the interfacial strength (MPa); F is the tensile force (kN); D is the inner diameter of the spindle (mm).

The testing steps are as follows:

(1) Heat the spindle in the oven at 170 °C for 1 h, and the substrate in the oven at 50 °C for 2 h.

(2) Place the silicone ring in the middle of the substrate to prevent the overflow of samples.
(3) Apply pressure to the center of the spindle for 5 min to make it fully bonded.
(4) Clean excess residual asphalt, and carry out the test after 2 h of heat preservation.

2.3.2. Disk-Shaped Compact Tension Test (DCT)

The cracking performance of asphalt mixtures was tested using the DCT test method with a strain control method to ensure that the rate of opening displacement (CMOD) was 1 mm/min, and the test was terminated when the test stress level peaked and then recovered to 0.1 kN. This study investigates the effect of different fiber types (PF, GF, and BF), different fiber content (0%, 0.1%, 0.2%, 0.3%, 0.4%, and 0.5%), and aggregate (limestone and basalt) types on the cracking resistance of fiber-reinforced asphalt mixture.

Fracture energy (G_f) can be used to evaluate the cracking resistance of asphalt mixtures, which can be calculated by Equation (2), and the envelope area of the LOAD–CMOD curve can be calculated by Equation (3).

$$G_f = \frac{Area}{B \cdot (W - a)} \tag{2}$$

where G_f is fracture energy (Jm^2); Area is the area of envelope of LOAD–CMOD curve (mm·kN); B is the thickness of the specimen (m); W-a is the initial ligament length of the specimen (m).

$$Area = \sum_{i=1}^{n} (x_{i+1} - x_i) \cdot (y_i) + 0.5 \cdot (x_{i+1} - x_i) \cdot (y_{i+1} - y_i) \tag{3}$$

where Area is area of envelope of LOAD–CMOD curve (mm·kN); x is the opening displacement of the specimen (mm); y is the load (kN); n is the number of data points when the stress level peaks and then recovers to 0.1 kN.

2.3.3. Interface Image Processing Method

In order to analyze the damage mode of the interface between asphalt mastic and aggregate after the binder bond strength test, a digital camera acquisition system was used to collect the images, as shown in Figure 5; with the help of the image analysis software (Image-Pro Plus V 7.0, IPP), the section images of the BBS-pulled specimen were analyzed, and the area of the asphalt mortar was extracted and bare aggregate area, indicating cohesion damage and adhesion damage, respectively.

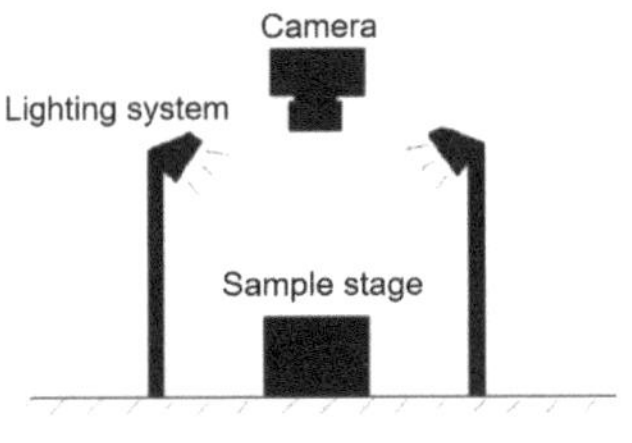

Figure 5. Imaging system.

The interface strength obtained by the test is composed of two parts: bonding strength (σ_a) and cohesion strength (σ_c). The adhesive damage area (S_a) and cohesive damage area (S_c) were obtained by image acquisition and recognition, and the adhesive strength and cohesive strength could be calculated by the formula [32].

2.3.4. Moisture Infiltration Treatment Method

Moisture infiltration treatment of BBS and DIC samples was carried out according to the immersion Marshall test in the water stability test of the asphalt mixture. The specimens were immersed at 60 °C for 7 days. After the immersion, the samples were kept at 25 °C for 4 h, and then the BBS and DIC tests were carried out under the same experimental conditions.

3. Results and Discussion

3.1. Interfacial Strength of Fiber Asphalt Mastic and Aggregate

3.1.1. Effect of Fiber and Aggregate on the Interfacial Strength

Figure 6 shows the influence of two factors of fiber (fiber type and content) on the interfacial strength of fiber asphalt mastic and aggregate at 25 °C. It can be observed from Figure 7 that compared with the neat one, the addition of fibers can significantly enhance the interfacial strength. With the rise in fiber content, the interfacial strength of fiber asphalt mastic and aggregate increased significantly and then decreased slightly. With the fiber content reaching 3–4%, the improvement effect on the interfacial strength is the best, which increases by 30.6–61.3%. Furthermore, BF has the best effect on the enhancement of interfacial strength, followed by GF and PF. This may be due to BF having the best mechanics and being able to play a better bridging role.

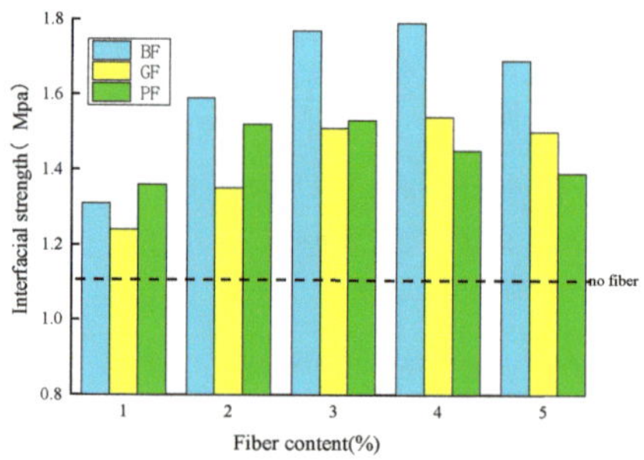

Figure 6. Effect of fiber content and type on the interfacial strength.

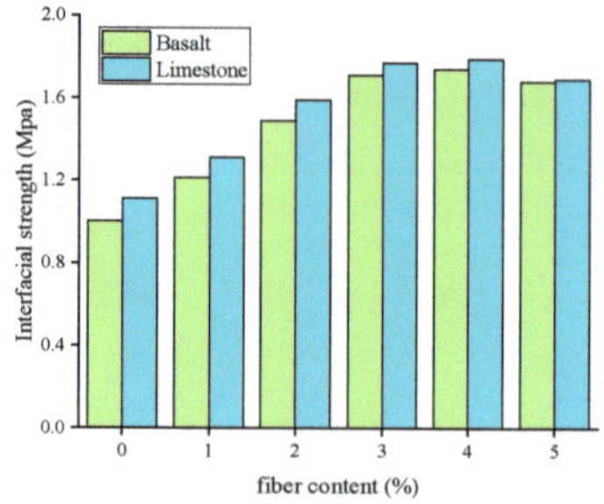

Figure 7. Effect of aggregate type on the interfacial strength.

Figure 7 illustrates the effect of aggregate type on the interfacial strength. The results show that the interfacial strength between asphalt mixture and limestone aggregate (LA) and basalt aggregate (BA) is 1.11 MPa and 1.00 MPa, respectively.

3.1.2. Effect of Moisture Infiltration on Interfacial Strength

Figure 8 illustrates the interfacial strength of asphalt mastic and aggregate under dry environment, and moisture infiltration; the reduction of interfacial strength after moisture infiltration is shown in Figure 9.

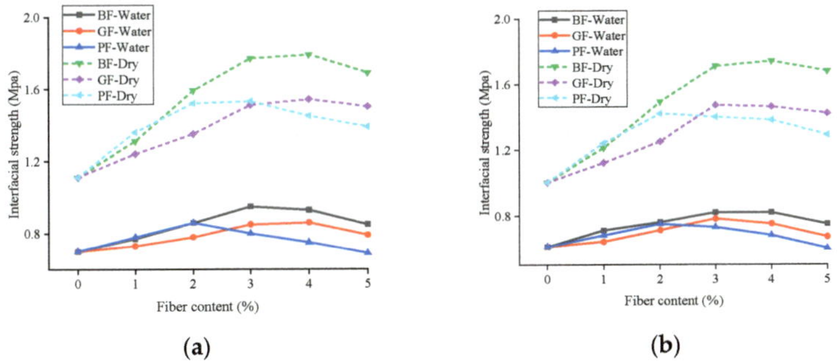

Figure 8. Interfacial strength under different environments: (**a**) LA; (**b**) BA.

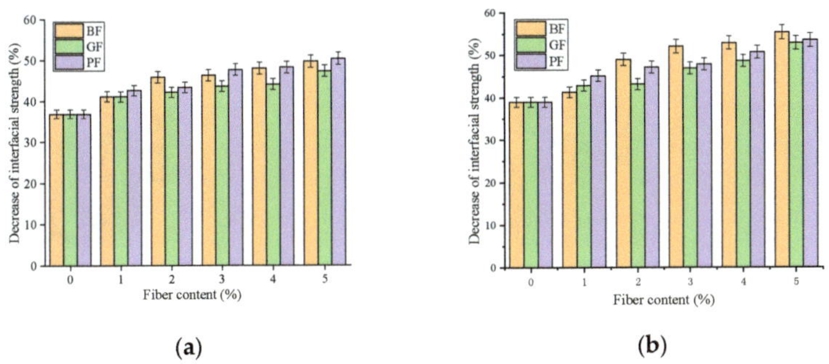

Figure 9. Decrease in interfacial strength after moisture infiltration: (**a**) LA; (**b**) BA.

It can be observed from Figure 8 that in a dry environment, the interface strength is 1.11–1.77 MPa; in a water environment, the interface strength is 0.7–0.95 MPa, and the interfacial strength decreases significantly after the influence of moisture infiltration. Compared with the dry state, with the rise in fiber content, the reduction of interfacial strength after moisture infiltration increased, up to 53.49%. This is due to water penetrating from the edge of the asphalt film to the inside of the aggregate and mastic, damaging the bonded interface and reducing the interfacial strength. Taking the limestone aggregate as an example, the interfacial strength of BF, GF, and PF asphalt mastic and aggregate decreased 41.22–49.70%, 41.13–47.33%, and 42.65–50.36%, respectively, after moisture infiltration. This indicates that the moisture infiltration had the greatest negative effect on the interfacial strength between PF asphalt mastic and aggregate. Furthermore, it can be seen that under the influence of moisture infiltration, the interfacial strength of limestone aggregate and asphalt decreases by 36.94–50.36%, and the interfacial strength of basalt aggregate and asphalt decreases by 39.00–53.49%, with a greater decline. This is mainly because higher SiO_2 content will accelerate the occurrence of water damage in asphalt mixture [33], so moisture infiltration has little effect on the interfacial strength of limestone aggregate and asphalt.

3.2. Effect of Moisture Infiltration on Failure Mode of Fiber Asphalt Mastic and Aggregate Interface

Figure 10 shows the images of interface failure between asphalt mastic and aggregate after moisture infiltration. It can be seen that interface failure is a comprehensive mode, including both adhesive failure and cohesive failure, but the two kinds of aggregate interface failure patterns are different. Specifically, for basalt aggregate, the interface failure is from the edge to the center; on the contrary, limestone aggregate is from the center to the edge. This is mainly due to the difference in the path of moisture infiltration. The red arrow represents the path of moisture infiltration. It can be observed from Figure 10 that there are many weak points at the edge of the interface between asphalt mastic and basalt aggregate; moisture will invade weak points at the edge of the interface and then gradually invade the inside of the interface, replacing the asphalt film on the surface of the aggregate, resulting in extensive spalling of fiber asphalt mastic on the surface of the aggregate. On the contrary, the boundary of the asphalt mastic and limestone aggregate interface has fewer weak points and fewer moisture infiltration points, resulting in more complete edge adhesion.

(a) (b)

Figure 10. Images of interface failure between asphalt mastic and aggregate after moisture infiltration: (**a**) Basalt aggregate; and (**b**) limestone aggregate.

Figure 11 compares the interfacial interaction strength under a dry environment and moisture infiltration conditions. As can be seen from the picture, regardless of adhesive strength or cohesive strength, the interfacial strength in a dry environment is always higher than moisture infiltration. This is due to the gradual invasion of water, replacing the asphalt membrane on the surface of the aggregate so that the interface strength is reduced.

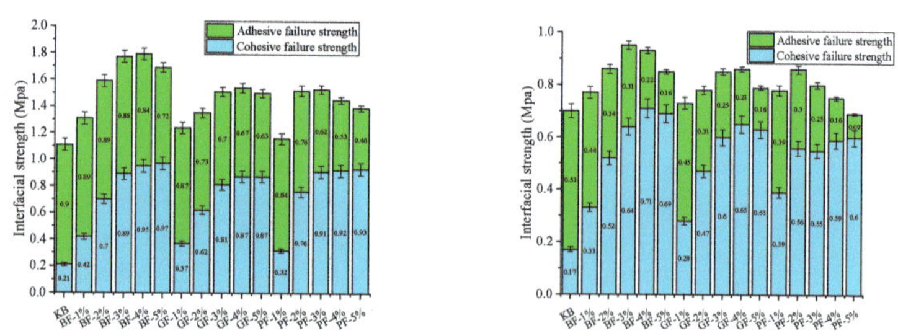

(a) (b)

Figure 11. Interfacial strength:(**a**) Dry environment; (**b**) moisture infiltration.

Figures 12 and 13 illustrate the comparative results of adhesive strength and cohesive strength under different circumstances, respectively. After infiltration, the interfacial cohesive strength decreased from 19.05% to 39.56%, and the interfacial adhesive strength decreased from 41.11% to 80.43%. It can be seen that the main reason for the decrease in interaction strength under the influence of moisture infiltration is the decrease in adhesion strength. This is because the addition of fiber can not only improve the cohesive strength but also slightly reduce the adhesive strength between the interfaces. However, as the water gradually invades and replaces the asphalt membrane on the aggregate surface in the moisture infiltration, the adhesive strength between the interfaces is further weakened and finally leads to a rapid decline in the interaction strength between the fiber asphalt mastic-aggregate interface after infiltration.

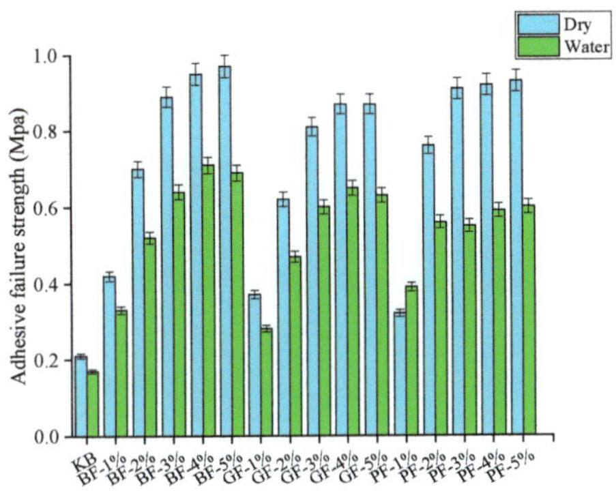

Figure 12. Adhesion strength.

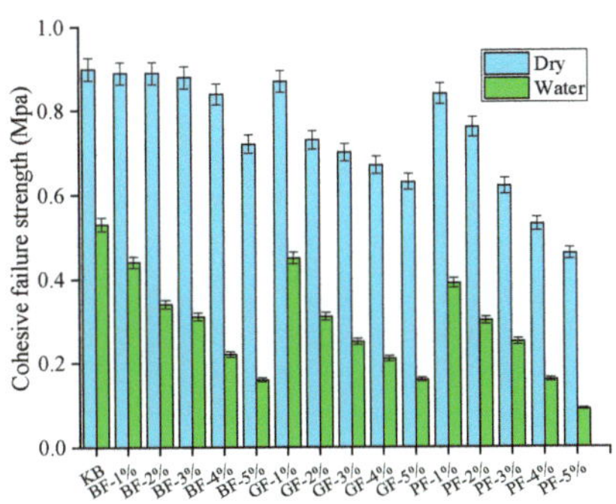

Figure 13. Cohesive strength.

3.3. Cracking Resistance of Fiber-Reinforced Asphalt Mixture

3.3.1. Effect of Fiber and Aggregate on Cracking Resistance

Figure 14 illustrates the effect of fiber content and type on the cracking resistance of fiber-reinforced asphalt mixture. As shown in Figure 14a,b, when the fiber content increased from 0.1% to 0.5%, the peak load of the fiber asphalt mixture first increased and then decreased, and the fracture energy also changed in this trend; the values were the largest, which were 0.506 kN and 1327.5 J/m^2, respectively. Compared with the neat one, the values are increased by 24.02% and 35.32%, respectively, indicating that the medium fiber content has the best effect on the peak load and fracture energy of the fiber asphalt mixture. It can be seen from Figure 14c that the P-CMOD of the mixture increases with the increase of BF and GF fiber content, while the P-CMOD of the PF asphalt mixture increases first and then decreases with the increase of fiber content.

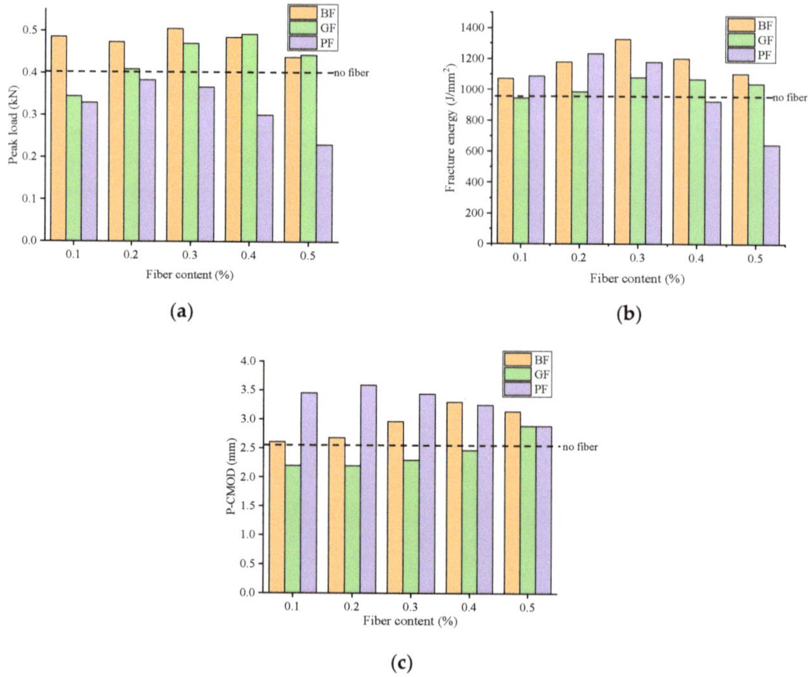

Figure 14. Effect of fiber content and type on cracking resistance: (**a**) Peak load; (**b**) fracture energy; and (**c**) P-CMOD.

Figure 15 illustrates the impact of aggregate type on the cracking resistance of fiber-reinforced asphalt mixture. As can be seen from Figure 15, the peak load of fiber-reinforced asphalt mixture with limestone aggregate and basalt aggregate is 0.408–0.506 kN and 0.43–0.537 kN, respectively, and the fracture energy is 981–1327.5 J/m^2 and 833–1214.5 J/m^2, respectively. P-CMOD is 2.56–3.143 mm and 2.44–2.92 mm, respectively. It can be seen that limestone aggregate asphalt mixture is superior to basalt aggregate in the experimental results.

3.3.2. Effect of Moisture Infiltration on Cracking Resistance

Figures 16 and 17 show the CMOD-load curve, peak load, fracture energy, and P-CMOD results of the DCT test under wet and dry conditions. It can be observed that the peak load, fracture energy, and P-CMOD of asphalt mixture decrease to some extent after moisture infiltration. Furthermore, the decrease range of peak load is 5.15–15.08%, the decrease range of fracture energy is 14.57–24.65%, and the decrease range of P-CMOD is 7.02–15.01%. It can be seen that the decreased range of fracture energy is the largest, indicating that moisture infiltration has the most significant effect on the fracture energy of the asphalt mixture. The explanation could be that the load reduction rate of the asphalt mixture after reaching the peak load is significantly accelerated after moisture infiltration, resulting in a significant decrease in fracture energy. In the process of the initial load gradually rising to the peak load, the coincidence degree of the test curve before and after moisture infiltration is high, so the decreased range of peak load and P-CMOD after moisture infiltration is obviously smaller than the decreased range of fracture energy.

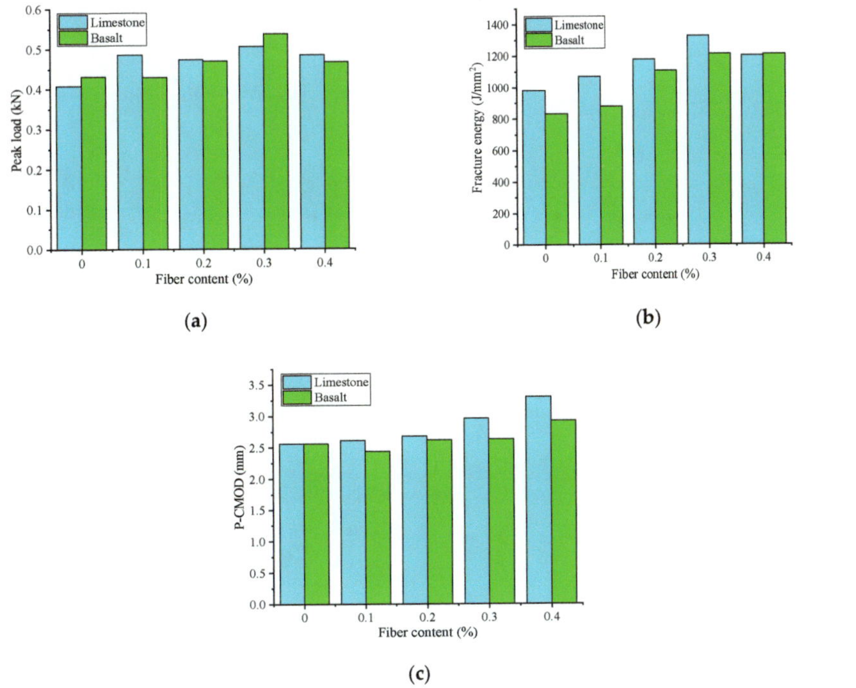

Figure 15. Result of aggregate type on cracking resistance: (**a**) Peak load; (**b**) fracture energy; and (**c**) P-CMOD.

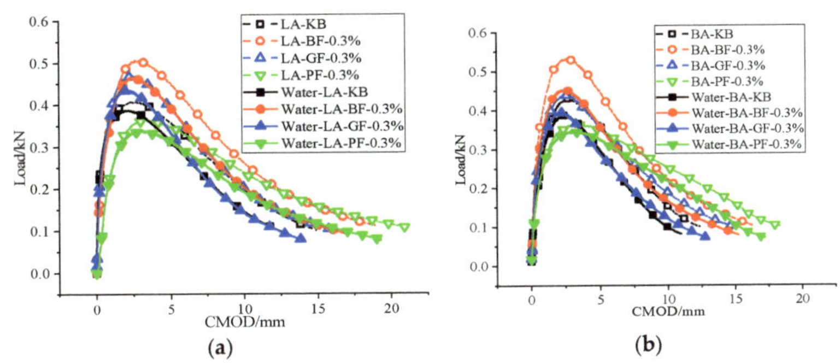

Figure 16. DCT test LOAD–CMOD curve: (**a**) LA; and (**b**) BA.

3.3.3. Effect of Moisture Infiltration on Failure Mode of the Mixture

Figure 18 shows the state of the DCT test specimen after the test. It can be observed that during the test, cracks develop from the tip of the initial precut seam to the other end of the specimen. When the tensile load applied to the specimen is less than 0.1 kN, the test is terminated. At this time, cracks did not completely penetrate the asphalt mixture specimen, as shown in Figure 18b.

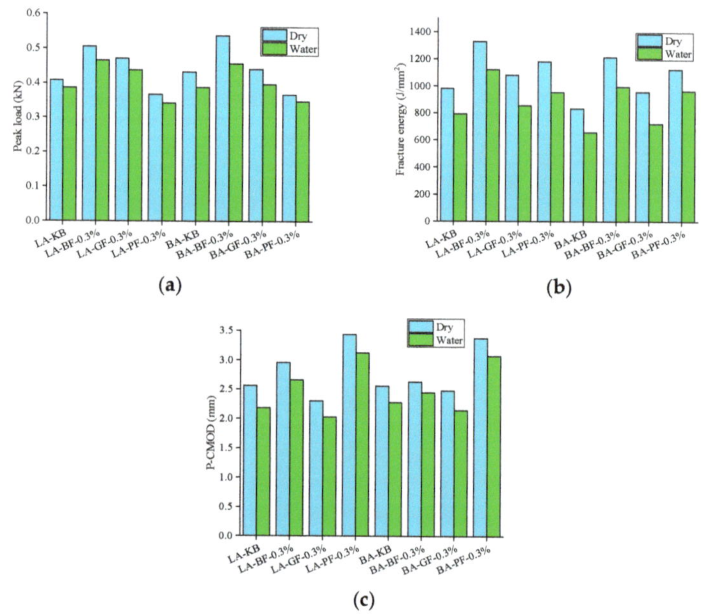

Figure 17. Effect of moisture infiltration on cracking resistance: (a) Peak load; (b) fracture energy; and (c) P-CMOD.

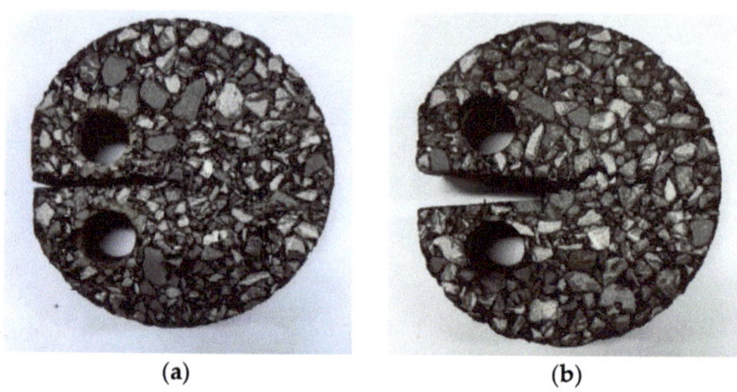

Figure 18. DCT test specimen; (a) Before test; (b) after test.

In order to further observe the fracture morphology of the fiber asphalt mixture specimen after the DCT test, the load was continuously applied until the specimen was completely broken and destroyed, and then the fracture morphology was collected, as shown in Figure 19. It is clear that the exposed aggregate area in the fracture of limestone aggregate specimen without moisture infiltration in Figure 19a is small, while the exposed aggregate area in the fracture of limestone aggregate specimen with moisture infiltration in Figure 19b is significantly increased. This indicates that the moisture infiltration causes the interface failure between the aggregate and the asphalt mastic. Under the action of tension, the phenomenon of stripping occurs between the asphalt mastic and the aggregate, which leads to a decrease in cracking resistance [35].

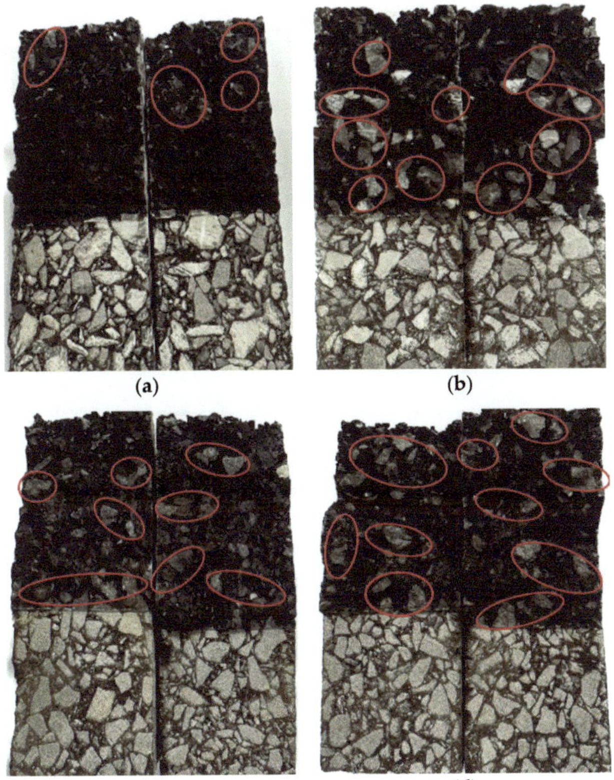

Figure 19. Fracture morphology of the specimen: (**a**) LA; (**b**) LA-Water; (**c**) BA; and (**d**) BA-Water.

Comparing Figure 19a,c, it can be seen that the exposed aggregate of limestone aggregate specimen is significantly less than that of basalt in fracture before moisture infiltration. This shows that the combination of limestone aggregate and asphalt mastic is generally better than that of basalt aggregate. Furthermore, after moisture infiltration, the exposed aggregate in the fracture of the basalt aggregate specimen increased significantly and spread throughout the fracture, which also explains the reason why the cracking resistance of the basalt aggregate asphalt mixture decreased significantly after being subjected to moisture infiltration.

3.3.4. Correlation Analysis of Interfacial Properties and Cracking Resistance of Mixture Under Moisture Infiltration

Previous studies [32] showed that the correlation between the interfacial strength and the cracking resistance indexes is high in the dry state. Especially, the interfacial strength has the greatest influence on the fracture energy and P-CMOD during the cracking of asphalt mixtures. In order to further investigate the influence of the moisture infiltration on the correlation between the interfacial strength and the cracking resistance, the grey correlation index of interfacial strength and peak load, fracture energy, and P-MOD were calculated, respectively.

Six samples with a fiber content of 0.3% were selected; Figure 20 shows the correlation index of multiple factors on peak load, fracture energy, and P-CMOD under different conditions. As can be seen from Figure 20, the correlation index between interfacial strength and peak load after being affected by moisture infiltration is 0.788. Compared with the dry state, the value increased by 7.21%. This indicates that the influence of interface strength on the peak load of asphalt mixture is increased to some extent under the influence of moisture infiltration. After moisture infiltration, the correlation index between interfacial strength and fracture energy is 0.812, which decreases by 4.92% compared with the dry state. However, the correlation between interfacial strength and fracture energy is much higher than that of other influencing factors, whether in the dry state or immersed state, indicating that even under the influence of moisture infiltration, interfacial strength is still one of the decisive factors affecting the fracture energy of asphalt mixture. After moisture infiltration, the correlation index of interfacial strength, surface energy of aggregate, asphalt content, and P-CMOD are 0.771, 0.769, and 0.763, respectively, and the values are very close. This indicates that the interfacial strength, surface energy of aggregate, and asphalt content have a great influence on the speed of the asphalt mixture reaching peak load after moisture infiltration.

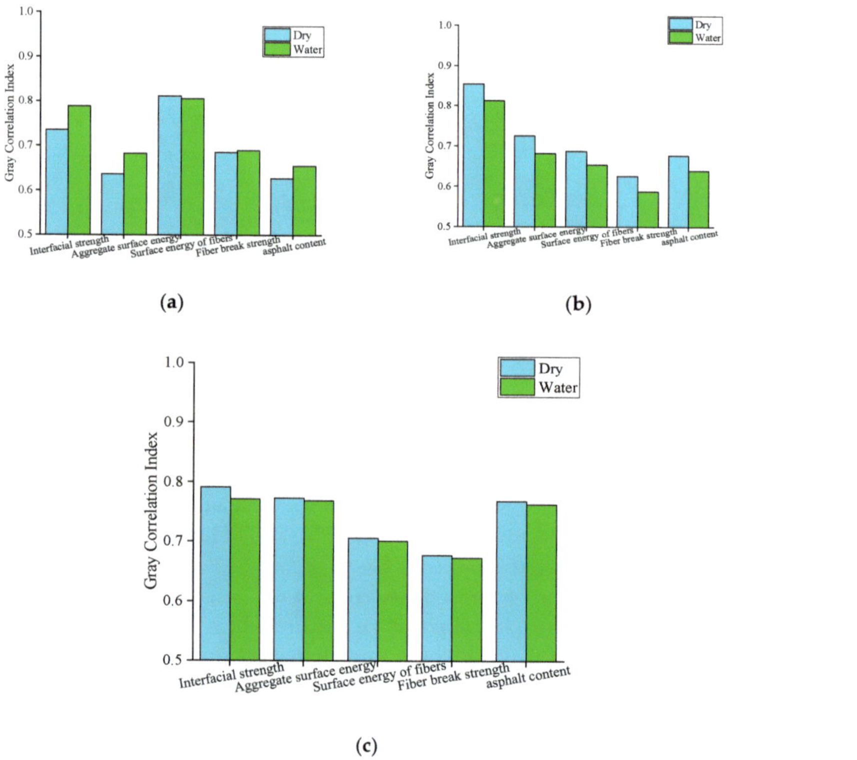

Figure 20. Gray correlation index; (**a**) Peak loads; (**b**) fracture energy; and (**c**) P-CMOD.

4. Conclusions

In this paper, different fiber contents and types were selected to prepare the fiber-reinforced asphalt mastic and asphalt mixture. Bond Strength Test (BBS), Disk-Shaped Compact Tension test (DCT), and interfacial image processing were used to investigate the effects of moisture infiltration on the interfacial characteristics of fiber asphalt mastic-aggregate and the cracking resistance of the mixture. The main conclusions are as follows:

(1) The addition of fibers could enhance fiber asphalt mastic-aggregate interfacial strength. Among the three kinds of fibers, basalt fiber has the most effective effect on interfacial

strength; compared with basalt aggregate, limestone aggregate has a greater impact on interfacial strength.

(2) Under the influence of moisture infiltration, the interfacial strength between fiber asphalt mastic and aggregate shows a significant decrease, especially in the interfacial properties of PF asphalt mastic and aggregate. Compared to basalt aggregate, limestone aggregate is able to resist moisture infiltration more effectively, thus reducing the decline in interfacial strength.

(3) The addition of fibers can effectively improve the cracking resistance of asphalt mixture. BF can effectively increase the peak load and fracture energy, and PF can effectively increase the P-CMOD.

(4) The interface between asphalt mastic and aggregate is more prone to spalling after being subjected to moisture infiltration, resulting in a decrease in cracking resistance. Compared with basalt aggregate, the cracking resistance of limestone aggregate asphalt mixture is not significantly reduced under the influence of moisture infiltration. Limestone aggregates have better resistance to moisture infiltration, so they can be used in rainy regions.

(5) Compared with the dry environment, after moisture infiltration, the correlation index between interfacial strength and fracture energy is much higher than other influencing factors. The interfacial strength is still an important factor affecting the fracture energy of the asphalt mixture even though it is affected by moisture infiltration.

(6) The mechanism of asphalt peeling from the aggregate surface under moisture infiltration can be studied by means of analytical dynamic simulation.

Author Contributions: Conceptualization, P.X. and K.L.; methodology, P.X. and K.L.; software, S.J. and Y.W.; formal analysis, S.J., H.W. and Y.W.; data curation, H.W. and Y.W.; writing—original draft preparation, S.J.; writing—review and editing, K.L.; funding acquisition, K.L. All authors have read and agreed to the published version of the manuscript.

Funding: This research was funded by the National Natural Science Foundation of China, grant number 52408498, and the Yangzhou Natural Science Foundation, grant number YZ2024176. China Postdoctoral Science Foundation, grant number 2024M762740.

Institutional Review Board Statement: Not applicable.

Informed Consent Statement: Not applicable.

Data Availability Statement: The original contributions presented in this study are included in the article. Further inquiries can be directed to the corresponding author.

Conflicts of Interest: The authors declare no conflict of interest.

References

1. Wang, W.; Wang, L.; Xiong, H.; Luo, R. A review and perspective for research on moisture damage in asphalt pavement induced by dynamic pore water pressure. *Constr. Build. Mater.* **2019**, *204*, 631–642. [CrossRef]
2. Saeed, F.; Rahman, M.; Chamberlain, D.; Collins, P. Asphalt surface damage due to combined action of water and dynamic loading. *Constr. Build. Mater.* **2019**, *196*, 530–538. [CrossRef]
3. Hung, M.A.; Goodwin, A.; Fini, H.E. Effects of water exposure on bitumen surface microstructure. *Constr. Build. Mater.* **2017**, *135*, 682–688. [CrossRef]
4. Muhammad, R.K.; Meor, O.H.; Jan, V. A review on moisture damages of hot and warm mix asphalt and related investigations. *J. Clean. Prod.* **2015**, *99*, 39–58.
5. Sun, E.; Zhao, Y.; Cai, R. Characterization of microstructural evolution of asphalt due to water damage using atomic force microscopy. *Constr. Build. Mater.* **2024**, *438*, 137175. [CrossRef]
6. Zhang, Y.; Yao, K.; Yi, Z.; Li, Z.; Li, H.; Zang, W.; Zhang, Y. Evaluation for water stability of bitumen mixture with multi-scale method. *Constr. Build. Mater.* **2024**, *412*, 134836. [CrossRef]

7. Guo, M.J.; Neng, T.F.; Ling, P.; Kovalskiy, V.P. Exploring the short-term water damage characteristics of asphalt mixtures: The combined effect of salt erosion and dynamic water scouring. *Constr. Build. Mater.* **2024**, *411*, 134310. [CrossRef]
8. Wu, H.; Ji, X.; Song, W.; Wei, J.; Sun, Y.; Zhan, Y.; Zou, X. Mechanical behaviors and meso analysis of asphalt mixtures under moisture damage induced by hydrodynamic impacts. *Constr. Build. Mater.* **2024**, *438*, 137048. [CrossRef]
9. Zhang, K.; Lu, H.; Zhu, R.; Weng, L.; Zhou, X.; Kong, X.; Liu, C.; Jiang, Y. Study on pore structure evolution and water damage of asphalt mixture under cyclic loading. *Constr. Build. Mater.* **2024**, *440*, 137461. [CrossRef]
10. Hui, Y.; Men, G.; Xiao, P.; Tang, Q.; Han, F.; Kang, A.; Wu, Z. Recent Advances in Basalt Fiber Reinforced Asphalt Mixture for Pavement Applications. *Materials* **2022**, *15*, 6826. [CrossRef]
11. Slebi-Acevedo, J.C.; Lastra-González, P.; Pascual-Muñoz, P.; Castro-Fresno, D. Mechanical performance of fibers in hot mix asphalt: A review. *Constr. Build. Mater.* **2019**, *200*, 756–769. [CrossRef]
12. Klinsky, L.; Kaloush, K.; Faria, V.; Bardini, V. Performance characteristics of fiber modified hot mix asphalt. *Constr. Build. Mater.* **2018**, *176*, 747–752. [CrossRef]
13. Wu, J.; Zhao, Z.; Jiang, C.; Yang, Y.; Sun, Z.; Yuang, J.; Xiao, F. Recent development and application of natural fiber in asphalt pavement. *J. Clean. Prod.* **2024**, *449*, 141832. [CrossRef]
14. Sato, A.; Yoshimura, T.; Kabusaki, D.; Okumura, H.; Homma, Y.; Sano, H.; Nakatsubo, F.; Kuboki, T.; Yano, H. Influences of dispersion media for chemically modified cellulose nanofibers on rheological and mechanical properties of cellulose nanofiber reinforced high-density polyethylene. *Cellulose* **2021**, *28*, 4719–4728. [CrossRef]
15. Shanbara, K.H.; Ruddock, F.; Atherton, W. A laboratory study of high-performance cold mix asphalt mixtures reinforced with natural and synthetic fibres. *Constr. Build. Mater.* **2018**, *172*, 166–175. [CrossRef]
16. Xu, J.Y.; Ma, B.; Mao, W.J.; Si, W.; Wang, X. Review of interfacial adhesion between asphalt and aggregate based on molecular dynamics. *Constr. Build. Mater.* **2023**, *362*, 129642. [CrossRef]
17. Jiao, B.; Pan, B.; Liu, F.; Yin, P.; Li, Z. Evaluating the interfacial properties between crumb rubber modified asphalt and aggregates using molecular dynamics simulation methods. *Constr. Build. Mater.* **2023**, *400*, 132809. [CrossRef]
18. Zhang, J.Z.; Apeagyei, A.K.; Airey, G.D.; Grenfell, J.R. Influence of aggregate mineralogical composition on water resistance of aggregate–bitumen adhesion. *Int. J. Adhes. Adhes.* **2015**, *62*, 45–54. [CrossRef]
19. Curtis, C.W. Investigation of Asphalt-Aggregate Interactions in Asphalt Pavements. *Am. Chem. Society.* **1992**, *37*, 1292–1297.
20. Huang, W.; Zhou, L. Evaluation of adhesion properties of modified asphalt binders with use of binder bond strength test. *Transp. Res. Rec. J. Transp. Res. Board* **2017**, *2632*, 88–98. [CrossRef]
21. Cala, A.; Caro, S.; Lleras, M.; Rojas-Agramonte, Y. Impact of the chemical composition of aggregates on the adhesion quality and durability of asphalt-aggregate systems. *Constr. Build. Mater.* **2019**, *216*, 661–672. [CrossRef]
22. Aguiarmoya, J.P.; Salazardelgado, J.; Baldisevilla, A.; Leiva-Villacorta, F.; Loria-Salazar, L. Effect of Aging on Adhesion Properties of Asphalt Mixtures with the Use of Bitumen Bond Strength and Surface Energy Measurement Tests. *Transp. Res. Rec.* **2015**, *2505*, 57–65. [CrossRef]
23. Xu, Y.; Zhang, C.F.; Feng, Y.P.; Guo, X. Low-temperature cohesive and adhesive strength testing of contact surface between bitumen and mineral aggregates by image analysis. *Constr. Build. Mater.* **2018**, *183*, 95–101. [CrossRef]
24. Guo, M.; Tan, Y.; Zhou, S. Multiscale test research on interfacial adhesion property of cold mix asphalt. *Constr. Build. Mater.* **2014**, *68*, 769–776. [CrossRef]
25. Tan, Y.; Guo, M. Using surface free energy method to study the cohesion and adhesion of asphalt mastic. *Constr. Build. Mater.* **2013**, *47*, 254–260. [CrossRef]
26. Rossi, C.O.; Caputo, P.; Baldino, N.; Szerb, E.I.; Teltayev, B. Quantitative evaluation of organosilane-based adhesion promoter effect on bitumen-aggregate bond by contact angle test. *Int. J. Adhes. Adhes.* **2016**, *72*, 117–122. [CrossRef]
27. Safaei, F.; Castorena, C.; Kim, Y.R. Linking asphalt binder fatigue to asphalt mixture fatigue performance using viscoelastic continuum damage modeling. *Mech. Time-Depend. Mater.* **2016**, *20*, 299–323. [CrossRef]
28. Rahman, M.N.; Castro, S.; Noorvand, H.; Kaloush, K.; Mamlouk, M.; Ozer, H. Effect of low dosage fiber reinforcement on cracking performance of asphalt concrete. *Constr. Build. Mater.* **2024**, *411*, 134380. [CrossRef]
29. Li, Z.; Zhang, X.; Fa, C.; Zhang, Y.; Xiong, J.; Chen, H. Investigation on characteristics and properties of bagasse fibers: Performances of asphalt mixtures with bagasse fibers. *Constr. Build. Mater.* **2020**, *248*, 118648. [CrossRef]
30. Wang, S.; Mallick, B.R.; Rahbar, N. Toughening mechanisms in polypropylene fiber-reinforced asphalt mastic at low temperature. *Constr. Build. Mater.* **2020**, *248*, 118690. [CrossRef]
31. Zhang, Y.; Sang, T.; Kang, A.; Wang, B.; Wu, X.; Zhao, Y. Investigation of fiber length distribution on crack resistance mechanism of basalt fiber asphalt mixture based on digital image processing technology. *Constr. Build. Mater.* **2023**, *401*, 132753. [CrossRef]
32. Tang, Q.; Xiao, P.; Lou, K.; Wu, Y. Interfacial characteristics of fiber asphalt mastic and aggregates. Impact on mixture crack resistance performance. *Constr. Build. Mater.* **2024**, *414*, 134866. [CrossRef]
33. Yin, Y.; Chen, H.; Kuang, D.; Song, L.; Wang, L. Effect of chemical composition of aggregate on interfacial adhesion property between aggregate and asphalt. *Constr. Build. Mater.* **2017**, *146*, 231–237. [CrossRef]

34. *JTG E20-2011*; Test Rules for Asphalt and Asphalt Mixtures for Highway Engineering. Sheben: Benjing, China, 2011.
35. Pipintakos, G.; Sreeram, A.; Mirwald, J.; Bhasin, A. Engineering bitumen for future asphalt pavements: A review of chemistry, structure and rheology. *Mater. Des.* **2024**, *244*, 113157. [CrossRef]

materials

Article

Evaluation of SMA-13 Asphalt Mixture Reinforced by Different Types of Fiber Additives

Haochen Wu [1], Peng Xiao [1,*], Ziyun Fei [1], Aihong Kang [1] and Xing Wu [2]

1 College of Architectural Science and Engineering, Yangzhou University, Yangzhou 225100, China; 17715878456@163.com (H.W.); 13961740804@163.com (Z.F.); ahkang@yzu.edu.cn (A.K.)
2 Department of Architecture Built Environment and Construction Engineering, Politecnico di Milano, Piazza Leonardo da Vinci, 32, 20133 Milan, Italy; xing.wu@polimi.it
* Correspondence: pengxiao@yzu.edu.cn

Abstract: This research aims at systematically evaluating the properties of SMA-13 asphalt mixture reinforced by several fiber additives including flocculent lignin fiber (FLF), granular lignin fiber (GLF), chopped basalt fiber (CBF), and flocculent basalt fiber (FBF). Firstly, the thermal stability, moisture absorption, and oil absorption property of these fiber additives were analyzed. Secondly, the property of SMA-13 reinforced using four types of single fibers and two kinds of composite fibers (FLF + CBF and FLF + FBF) was comprehensively analyzed. Specifically, the high-temperature performance was evaluated using the uniaxial penetration test and the rutting test, the medium-temperature anticracking property was evaluated using the IDEAL-CT test, the low-temperature property was analyzed using the beam bending test, and the water stability was studied by the freeze–thaw splitting test. Thirdly, the dynamic mechanical response of different-fibers-modified SMA-13 was evaluated using the uniaxial compression dynamic modulus test. Finally, correlation analysis between the results of dynamic modulus and the high-, medium-, and low-temperature mechanical performance was carried out. The research results reveal that the stability of CBF and FBF under thermal action is better than that of GLF and FLF, and FBF shows the best thermal stability. The oil absorption property of FLF is better than that of GLF, followed by FBF and CBF. The comprehensive mechanical properties of CBF- and FBF-reinforced SMA-13 are better than those of FLF- and GLF-modified SMA-13. CBF can better reinforce the mechanical property of SMA-13 under low and medium temperature, while FBF can better reinforce the performance of SMA-13 at high temperature. FLF/CBF- and FLF/FBF-composite-modified SMA-13 show better high-temperature mechanical performance than that of the single-fiber-reinforced mixture, and FLF has some negative impact on the properties of FLF/FBF-composite-modified SMA-13 at low temperature. Fibers have no significant influence on the water stability of the mixtures. Meanwhile, the linear correlation between the mechanical performance of all the fiber-reinforced SMA-13 and the dynamic modulus result is good.

Keywords: fiber additives; SMA-13 mixture; mechanical property; dynamic mechanical response; basalt fiber; lignin fiber

Citation: Wu, H.; Xiao, P.; Fei, Z.; Kang, A.; Wu, X. Evaluation of SMA-13 Asphalt Mixture Reinforced by Different Types of Fiber Additives. *Materials* **2024**, *17*, 5468. https://doi.org/10.3390/ma17225468

Academic Editor: Jānis Andersons

Received: 14 October 2024
Revised: 1 November 2024
Accepted: 4 November 2024
Published: 8 November 2024

1. Introduction

Asphalt pavement structure nowadays suffers from various road distresses under the comprehensive effects of human behavior and severe environmental conditions [1–3]. Rutting, cracks, etc., have presented road surfaces with a lot of challenges, which also increases the cost for constructing and repairing the road surfaces. Therefore, researchers around the world have focused on finding different modifiers to reinforce the properties of asphalt pavement materials [4,5].

Some researchers adopted polymer modifiers to reinforce the properties of asphalt mixtures [6–8]. For instance, B. Wang et al. studied the properties of asphalt mixture reinforced by antirut agents and developed a permeative antirutting agent using the Buton rock asphalt (BRA) and epoxy-based resin [9]. G. Huang et al. and S. Lv et al. evaluated the rutting and fatigue resistance of asphalt mixture reinforced by high-modulus agents [10,11]. Other researchers used fiber additives such as lignin fiber (LF), basalt fiber (BF), etc., to enhance the asphalt mixture property [12–15]. For example, Y. Hui et al. analyzed the cracking resistance of asphalt mixture modified using basalt fiber (BF), and the results indicated that BF can greatly improve the cracking resistance [16]. D. Ren et al. adopted some surface sizing agents to modify BF and studied the performance of an asphalt mixture prepared using the modified BFs [17,18]. Tamrin et al. studied the property of an asphalt mixture reinforced using the composite modifier of lignin fiber and an organic modifier [19].

Among all the additives, fiber additives are more and more popular because they are eco-friendly and the raw materials are easily found on Earth [20–22]. References indicate that LF (lignin fiber) and BF (basalt fiber) are two types of plant and mineral fiber being used in asphalt mixture [23,24]. Therefore, this paper chooses these to analyze their impact on asphalt mixtures. Regarding the mixture type, this paper focuses on the stone matrix asphalt with the nominal maximum aggregate size of 13 mm (SMA-13), for the reason that it is one of the most widely used mixture types for the upper surface [25].

However, apart from the traditional flocculent LF (lignin fiber) and short-cut BF (basalt fiber), two new forms of LF and BF, namely, granular LF (GLF) and flocculent BF (FBF), have emerged. The differences between FBF/FLF and BF/LF are the morphology. The research related to the effect of the morphology of fiber on the fiber-reinforced asphalt concrete or mixture material is limited. Thus, it is worth analyzing its effect on the fiber-reinforced mixture. Furthermore, the hybrid use of fiber additives with different morphology is also worthy of exploration.

Therefore, this paper focuses on analyzing the effect of flocculent lignin fiber (FLF), granular lignin fiber (GLF), chopped basalt fiber (CBF), and flocculent basalt fiber (FBF) on the high-temperature antirutting performance, low- and medium-temperature crack resistance, water stability, and dynamic mechanical response of SMA-13. Meanwhile, since the flocculent lignin fiber (FLF) is suggested for use in SMA-13 to absorb the oil in the literature, the combined use of FLF and CBF or FBF in SMA-13 mixture is analyzed. This paper could help to guide the scientific selection of fiber additives in the design and application of SMA-13 mixtures.

2. Materials and Methods

2.1. Materials

2.1.1. Raw Materials

The coarse aggregate used in the current study was grid from basalt rock, and the limestone fine aggregate was selected. Filler powder was adopted as limestone powder, with the properties shown in Table 1. SBS-modified asphalt with the grade of PG76-22 was used, and the properties are shown in Table 2. Four type of fibers, flocculent basalt fiber (FBF), chopped basalt fiber (CBF), flocculent lignin fiber (FLF), and granular lignin fiber (GLF), were selected, and the fiber properties are shown in Table 3. Macroscopic morphologies of the different fiber modifiers are shown in Figure 1.

Table 1. Indexes of mineral powder properties.

Index		Result
Moisture content/%		0.8
Relative density		2.702
Hydrophilic coefficient		0.8
Particle size range	<0.6 mm	100
	<0.15 mm	99.0
	<0.075 mm	88.5

Table 2. Indexes of asphalt performance.

Index		Result
Needle penetration (25 °C)/0.1 mm		56
Softening point/°C		81
Ductility (5 cm/min, 5 °C)/cm		48
Elastic recovery (25 °C)/%		96
Residue after RTFOT	Mass change/%	−0.1
	Penetration ratio/%	86
	5 °C Residual ductility/cm	37

Table 3. Indexes of different types of fibers.

Index	FLF	GLF	CBF	FBF
Diameter/μm	≈13	4.1	16	5
Length/mm	0.8	3~6	6	3~6
Density/g·cm^{-3}	0.910	1.112	2.712	2.711
Tensile strength/MPa	<300	-	≥2000	-
Oil absorption rate	4.3	6.5	1.2	2.3
Hygroscopicity/%	12.7	30.3	2.7	5.8

Figure 1. Morphologies of fiber additives. (**a**) Flocculent basalt fiber; (**b**) chopped basalt fiber; (**c**) flocculent lignin fiber; (**d**) granular lignin fiber.

2.1.2. Gradation Design

The gradation type of the mixture is determined as SMA-13 (stone matrix asphalt whose nominal maximum aggregate size is 13.2 mm). The gradation curve is shown in Figure 2. The upper and lower limits of the gradation curve are obtained from the specification of JTG F40 2004 [26]. The final adopted gradation curve of the SMA-13 asphalt mixture in this study is called the synthetic gradation curve, because it is obtained by the combination of several different kinds of stones with different size ranges. The adopted fiber addition design, the corresponding optimum asphalt content, and the volumetric parameters are shown in Table 4. The Cantabro test result and the Schellenberg binder drainage test results are shown in Table 5. Generally, there are six types of fiber addition schemes, including four types of single fiber and two types of composite fiber. Specifically, the SMA-13 with the flocculent lignin fiber (FLF) is regarded as the control mixture, because FLF is required to be added by the specification of JTG F40 2004 and many references [27].

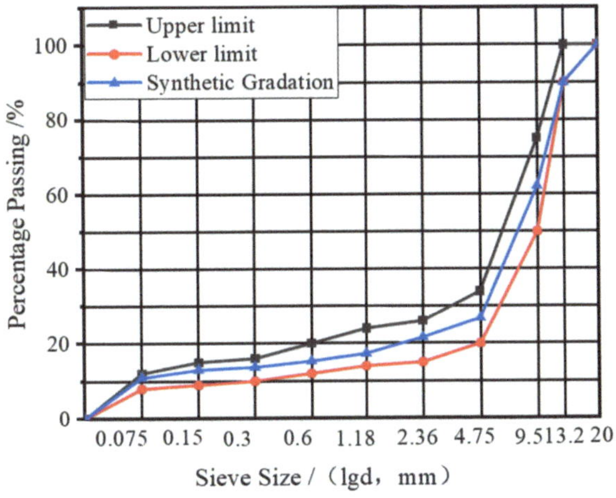

Figure 2. Gradation design curve of SMA-13 mixture.

Table 4. Fiber blending scheme and the mixture gradation design parameters.

Mixture Gradation	Fiber Type	Fiber Content/%	Optimum Asphalt Content/%	Void Ratio VV/%	Voids of Mineral Aggregate VMA/%	Voids Filled with Asphalt VFA/%
SMA-13	GLF	0.4	5.46	3.8	17.1	78.0
	FLF	0.3	5.64	3.5	17.3	79.5
	FBF	0.4	5.46	3.6	16.5	78.2
	CBF	0.4	5.37	3.6	16.6	77.2
	FBF + FLF	0.3 + 0.1	5.55	3.2	17.2	80.3
	CBF + FLF	0.3 + 0.1	5.46	3.2	17.4	81.0

Table 5. Cantabro scattering and Schellenberg binder drainage test results.

Type	Asphalt Scattering Loss ΔS/%	Asphalt Drainage Loss Δm/%
GLF	5.4	0.07
FLF	5.3	0.08
FBF	5.2	0.11
CBF	5.1	0.12
FBF + FLF	5.4	0.09
CBF + FLF	5.5	0.07

2.2. *Experiments*

2.2.1. High-Temperature Property Test Method

The high-temperature rutting test was adopted for high-temperature performance analysis according to JTG E20-2011 [28]. As is indicated in the regulation, the specimen is a cuboid with the size of 300 × 300 × 50 mm (length × width × height). The temperature during the experiment was controlled at 60 °C, while 0.7 MPa was set as the test wheel pressure. The mathematical calculating method of dynamic stability is illustrated in Equation (1).

$$DS = \frac{(t_2 - t_1) \times N}{d_2 - d_1} \times C_1 \times C_2 \quad (1)$$

where DS is index of dynamic stability of the specimen; d_1 represents the rutting deformation at forty-five minutes; d_2 corresponds to the sixty-minutes deformation; C_1 and C_2 are the testing parameters and their value is set as 1; N is the rolling speed of the testing wheel and is normally determined as 42 times per minute.

2.2.2. Low-Temperature Property Test

The low-temperature performance test was conducted by trabecular test following JTG E20-2011, in which trabecular specimens are prismatic beam with length × width × height of 250 mm × 30 mm × 35 mm, respectively. The experiment was conducted using a UTM-25 testing machine, with the temperature maintained at −10 °C and the pressure velocity of the specimens maintained at 50 mm/min. The UTM-25 testing machine is made by IPC Global Co., Ltd., Melbourne, Australia. The calculation formulas for flexural tensile strength and maximum flexural strain are shown in Equations (2) and (3).

$$R_B = \frac{3LP_B}{2bh^2} \tag{2}$$

$$\varepsilon_B = \frac{6hd}{L^2} \tag{3}$$

where b and h are the width and height of the specimen; L is the test span of the sample in the test; P_B is the peak load when the specimen breaks; d is the deflection at mid-span when the sample reaches failure.

2.2.3. Medium-Temperature Performance Test

The medium-temperature property was evaluated by the IDEAL-CT test [29–31]. The cylinder-shaped specimens were prepared by using the rotatory-compaction machine, and the diameter and thickness of the test sample were 150 mm and 62 mm. The experiment was carried out by UTM-25 testing machine, and the displacement loading rate and experiment temperature was set as 50 mm/min and 25 °C. The evaluation indexes are listed in Equations (4) and (5).

$$CT_{\text{Index}} = \frac{G_f}{|m_{75}|} \times \left(\frac{l_{75}}{D}\right) \tag{4}$$

$$|m_{75}| = |(p_{85} - p_{65})/(l_{85} - l_{65})| \tag{5}$$

where CT_{Index} represents the cracking propagation property; G_f is the cracking energy during the experimentation process; $|m_{75}|$ is the slope value of the post peak curve when the force is 75% of the maximum force, and l_{75} is the corresponding displacement at this post peak curve point; D is 150 mm and refers to the diameter of the test sample.

2.2.4. Water Stability Test

The water stability was tested by the freeze–thaw splitting test (FTST) following the instruction listed in JTG E20-2011. The calculation formulas of the test indexes are shown in Equations (6)–(8).

$$R_{T1} = 0.006287 P_{T1} \times h_1 \tag{6}$$

$$R_{T2} = 0.006287 P_{T2} \times h_2 \tag{7}$$

$$TSR = \frac{R_{T2}}{R_{T1}} \times 100 \tag{8}$$

where R_{T1} represents the splitting tensile strength (STS) of the test samples which are not under the freeze–thaw actions, and this is regarded as the test group 1; R_{T2} represents the STS of the second test groups, in which the samples have been through the freeze–thaw process. P_{T1} and P_{T2} are the tensile force of the first and second test group. h_1 and h_2 are the specimen height. TSR is the ratio between the average STS value of the second (R_{T2}) and the first (R_{T1}) specimen group.

2.2.5. Dynamic Modulus Test Method

The compression dynamic modulus test (CDM) was adopted to evaluate the dynamic response property of the mixture, and standard test specimens with height of 150 mm and diameter of 100 mm were used. A multifunctional instrument was selected to apply sine wave load to the top of the specimen. The test temperature was selected as −10, 5 °C, 20 °C, 35 °C, and 50 °C, and six loading frequencies were selected for each temperature (25 Hz, 10 Hz, 5 Hz, 1 Hz, 0.5 Hz, and 0.1 Hz). The experiment started from low to high temperature, and was conducted from high to low frequency at the same temperature. Then, a mathematical model was adopted to fit the master curve of dynamic modulus. The calculation formulas are shown in Equations (9)–(11).

$$\log(|E^*|) = \delta + \frac{\alpha-\delta}{e^{\beta+\gamma\log t_r}} \tag{9}$$

$$t_r = \frac{t}{\alpha_T} \tag{10}$$

$$\log(t_r) = \log(t) - \log(\alpha_T) \tag{11}$$

where $|E^*|$ is the dynamic modulus in the main master curve; t_r is the master curve frequency at the reference temperature; δ and α are the parameters of the model shown in Equation (11); β, γ are the descriptions parameters for the shape of the master curve; t is the action frequency of the load at the converted temperature; α_T is the shift factor.

3. Results and Discussion

3.1. High-Temperature Rutting Test Results

The dynamic stability (DS) results of the rutting test are illustrated in Figure 3. It can be seen from Figure 3 that the DS values of the mixtures with BF are better than those of mixtures with LF and the improvement range is between 12.3% to 32.7%, indicating that BFs have better reinforcing effect on the high-temperature dynamic stability than LFs. In addition, the dynamic stability of SMA-13 with flocculent BF (FBF) is better than the mixture with chopped BF (CBF), and the DS of the mixture with FBF increases by 11.3% compared to the mixture with granular lignin fiber (GLF). Similar differences can be found between the mixtures with flocculent lignin fiber (FLF) and GLF. This indicates that fiber morphologies also have an influence on the anti-high-temperature property, and flocculent fibers have better reinforcing effect on the high-temperature dynamic stability. Furthermore, DS values of the mixture with composite fibers are much higher than those of mixtures with single fiber, despite the fiber type, and the composite of FBF + FLF-reinforced mixture presents the best high-temperature rutting resistance.

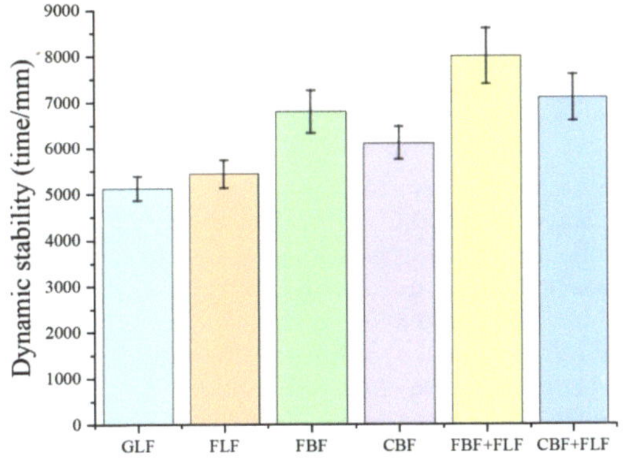

Figure 3. The results of rutting test at high temperature.

3.2. Low-Temperature Trabecular Bending Test Results

The maximum flexural strain and flexural tensile strength results of the low-temperature test are shown in Figure 4. It can be observed from Figure 4 that maximum flexural strain of SMA-13 mixtures with BFs are over 3000 με, which can be used for the extreme cold regions with temperature as low as −37 °C. Meanwhile, the maximum flexural strain of mixtures with LFs are around 2500 με, which can be used for cold regions with temperatures as low as −37 °C to −21.5 °C, according to the JTG-F40-2004. This means that BF has a better reinforcing effect on the low-temperature resistance than LF. These findings are consistent with the results from reference [32]. In addition, the maximum flexural strain of the asphalt mixture with CBF is 10.5% higher than that of asphalt mixture with FBF, while the flexural strain of mixtures modified by LFs with different morphologies is almost the same. Furthermore, it is found that CBF has a better reinforcing effect on the maximum flexural strain of FLF-modified SMA-13 than FBF. Meanwhile, FLF could decrease the maximum flexural strain of the FBF-modified mixture and increase the maximum flexural strain of the CBF-modified mixture. This phenomenon is mainly because CBF could form a solid three-dimensional network structure in the mixtures, and FLF could absorb the redundant oil in the mixture [33,34]. These two fibers could have synergic effects on the properties of the mixture in the low-temperature cracking failure process of the mixture. However, when FLF and FBF are used together, the network structure formed by these two flocculent fibers might not be good as the FLF/CBF-modified mixture, because the two types of flocculent fibers might agglomerate with each other, resulting in some stress concentration at low temperatures [35].

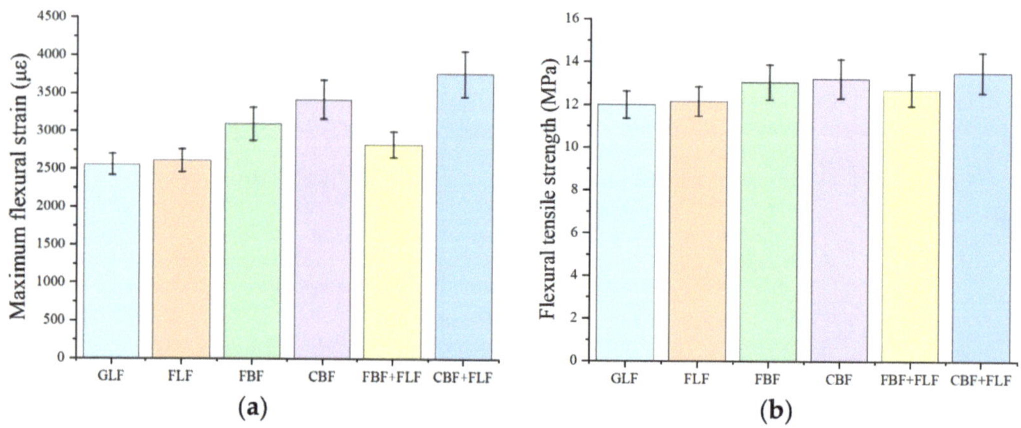

Figure 4. Low-temperature trabecular bending test results: (**a**) Maximum bending tensile strain. (**b**) Flexural tensile strength.

3.3. Medium-Temperature IDEAL-CT Test Results

The cracking energy and CT_{Index} results of the IDEAL-CT test are illustrated in Figure 5. The cracking energy and CT_{Index} calculated from the IDEAL-CT load–displacement curve indicate that the cracking resistance and the anticracking propagation ability of SMA-13 with BFs is better than those with LFs. The CT_{Index} of the mixture with BFs improves by 9.2% to 24.2% compared to the mixture with LFs. The cracking resistance and anticracking propagation ability of the samples with composite fiber are better than those with single fibers. The modification effect of flocculent LF (FLF) is better than granular lignin fiber (GLF), and the reinforcing degree of flocculent BF (FBF) is better than the chopped BF (CBF). The CT_{Index} of mixture with FBF increases by 14.2% compared to the mixture with CBF. This pattern is also observed in the composite-fiber-modified SMA-13. Flocculent BF (FBF) can better increase the anticracking ability of the traditional SMA-13 with flocculent LF (FLF) than the chopped BF (CBF).

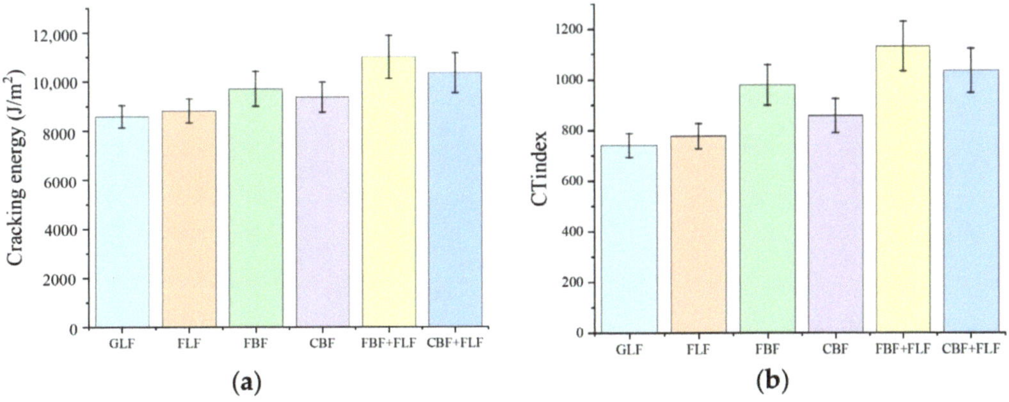

Figure 5. The results of the IDEAL-CT test: (**a**) Cracking energy. (**b**) Crack resistance index.

3.4. Freeze–Thaw Splitting Test Results

The TSR results of the freeze–thaw splitting test are illustrated in Figure 6. It can be seen from Figure 6 that the TSR values of all the six type of mixtures are within the range of 85–89%, which indicates that the six types of mixtures all meet the minimum limitation value of 80% according to JTG E20-2011. This means that these fibers have no significant influence on the water stability of the mixtures.

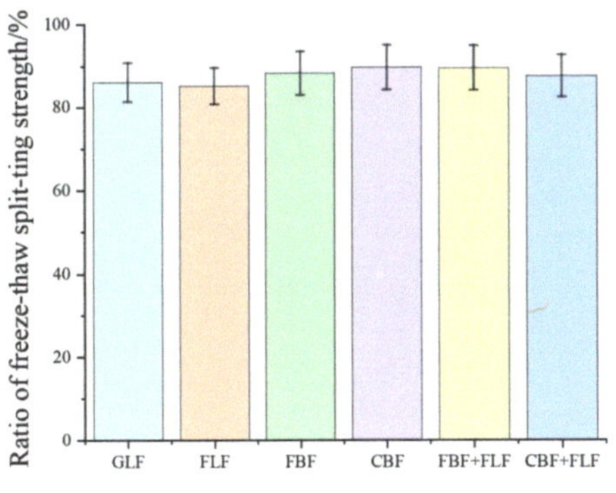

Figure 6. The results of the water stability test.

3.5. Dynamic Modulus Test Results

The asphalt mixture is mainly elastic at low temperature and viscous at high temperature, which is primarily due to the complexity of the asphalt [36]. Figure 7 shows the variation of dynamic modulus of different mixtures with temperature. As shown in Figure 7, the DM values of the mixture with all types of fibers decrease with the increase in temperature. Meanwhile, the changing trend is similar and consistent for all test samples. When the temperature is below 20 °C, the dynamic modulus decreases sharply with the increase in temperature; at 20 °C, there is a significant inflection point. When the temperature exceeds 20 °C, the decreasing rate of the dynamic modulus tends to flatten out; when the temperature rises to 50 °C, the dynamic modulus at each frequency tends to be close to each other. These findings indicate that the DM values vary more obviously at low temperatures.

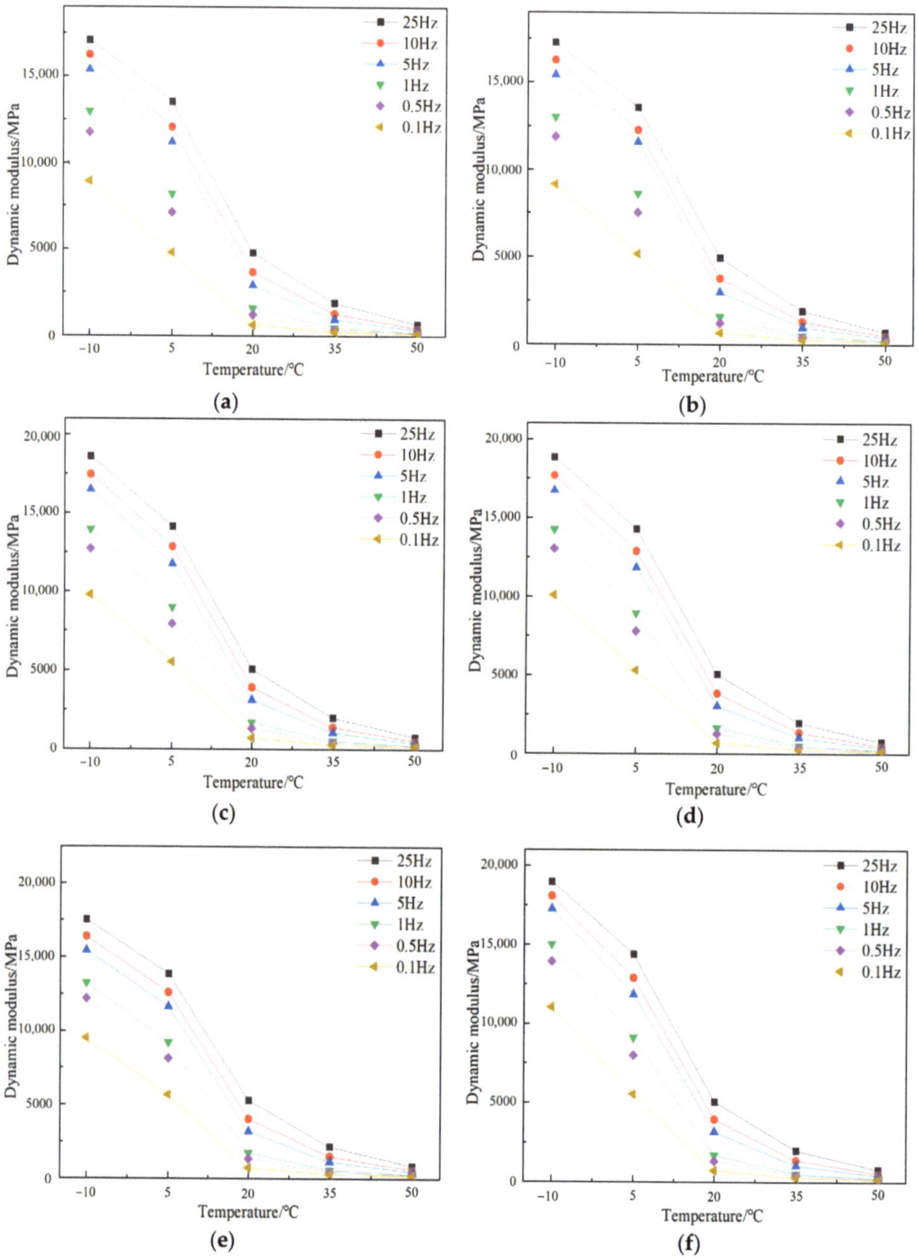

Figure 7. Dynamic modulus of different-fiber-modified SMA-13 under various temperatures: (**a**) GLF, (**b**) FLF, (**c**) FBF, (**d**) CBF, (**e**) FBF + FLF, and (**f**) CBF + FLF.

Figure 8 shows the variation of dynamic modulus with frequency. From Figure 8, it can be seen that DM increases with the frequency at the same temperature. Meanwhile, the DM value at low temperature varies more greatly than the DM value at high temperature when the frequency varies from 0.1 to 25 Hz. The changing degree of DM value is the smallest when the frequency is small and the temperature is high.

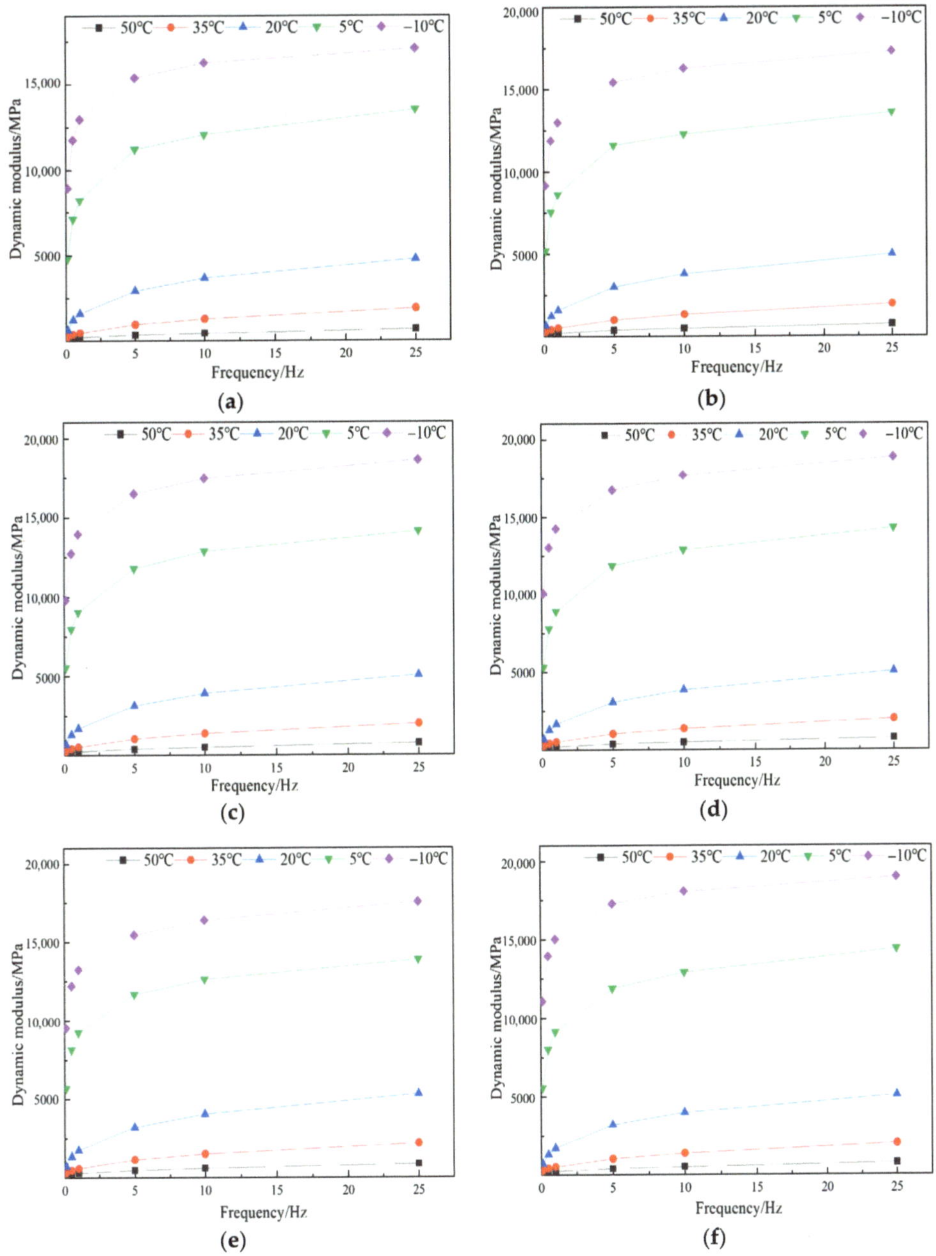

Figure 8. Dynamic modulus of different-fiber-modified SMA-13 under different frequencies: (**a**) GLF, (**b**) FLF, (**c**) FBF, (**d**) CBF, (**e**) FBF + FLF, and (**f**) CBF + FLF.

From Figure 9a–f, it can be seen that there is a functional relationship between shift factor and temperature, and the variation range of shift factor can reflect the impact degree of temperature on the dynamic mechanical behavior of the mixtures. It reveals that the shift factor decreases with the increase in temperature. The master curve can extend the test frequency of 0.01~25 Hz to 10^{-6}~10^{6} Hz. It breaks through the limitation of test equipment and has a wider frequency domain and a higher DM value. The DM value of the fiber-modified mixture is positively correlated with the loading frequency, in which the increasing trend is gentle at the high frequency; an S-shaped curve is observed for all the samples.

However, Figure 9a–f cannot well show the influence of various fiber additives on the dynamic modulus value. To further analyze the specific effect of fiber types on the dynamic mechanical response of the mixture samples, all the master curves of the test specimens at 20 °C are illustrated in Figure 10.

From Figure 10 it can be seen that in the low-frequency (corresponding to the high-temperature) area, the difference of DM values of different fiber SMA-13s is obvious, indicating that the fiber type and fiber morphology had a greater influence under these conditions. In this area, the DM value of SMA-13 with composite fibers is greater than the mixture with single fibers. BFs had better enhancing effect on the DM value than LFs. The FBF + FLF mixture had the largest DM value at low frequencies. This indicates that it has the highest temperature resistance performance, which is consistent with the results of the rutting experiment.

Meanwhile, the effect of different fibers on the DM values of different fiber SMA-13 shows different trends in the high-frequency (corresponding to low-temperature) area. The CBF + FLF-modified mixture has the biggest DM value at high frequencies, and the FBF + FLF mixture has a smaller DM value than the BF-modified mixture. This is also in accordance with the low-temperature maximum bending tensile strain and flexural tensile strength results introduced before.

In order to analyze the correlation between the dynamic mechanical behavior and the various road performances of the SMA-13s studied before, the measured dynamic moduli at 50 °C, 20 °C, and −10 °C under 10 Hz were adopted to conduct a correlation analysis between the high-temperature dynamic stability, medium-temperature anticrack ability, and low-temperature property of the asphalt mixtures. As shown in Figure 11a–c, the correlation coefficients reach 0.90, 0.93, and 0.93, respectively, indicating good consistency between the dynamic modulus and road performance of fiber asphalt mixtures.

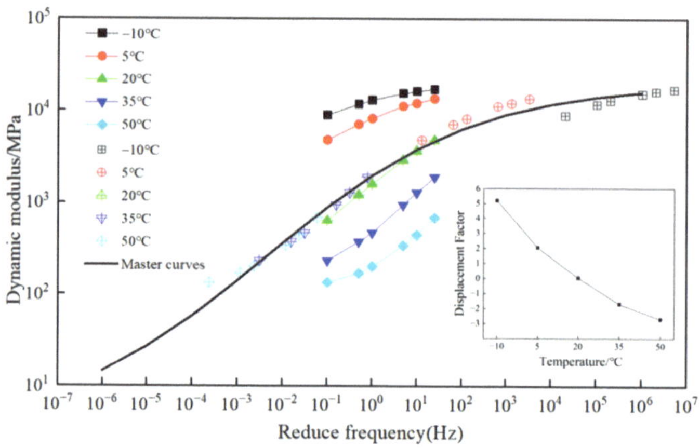

(a)

Figure 9. *Cont.*

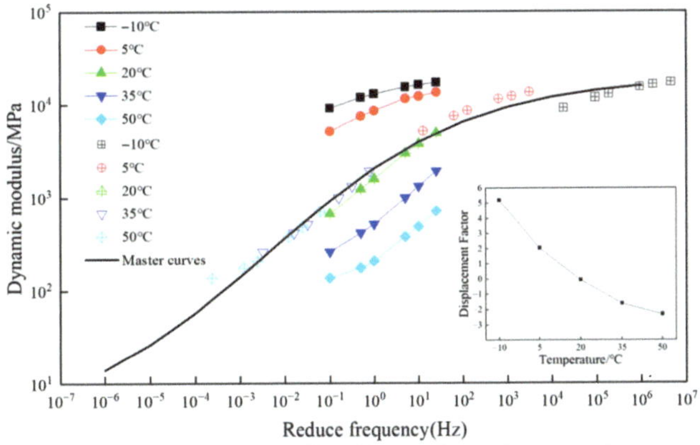

(b)

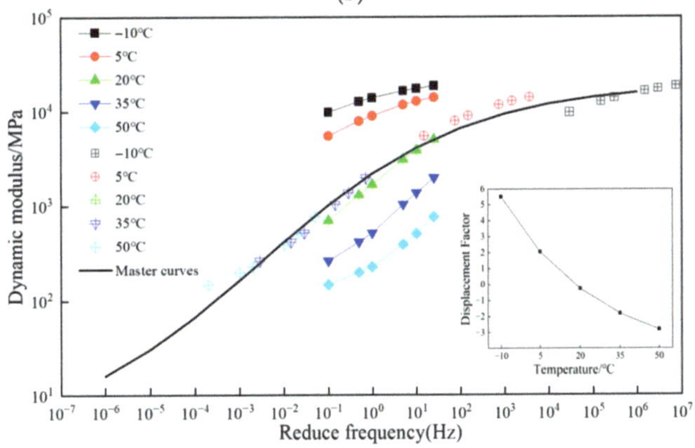

(c)

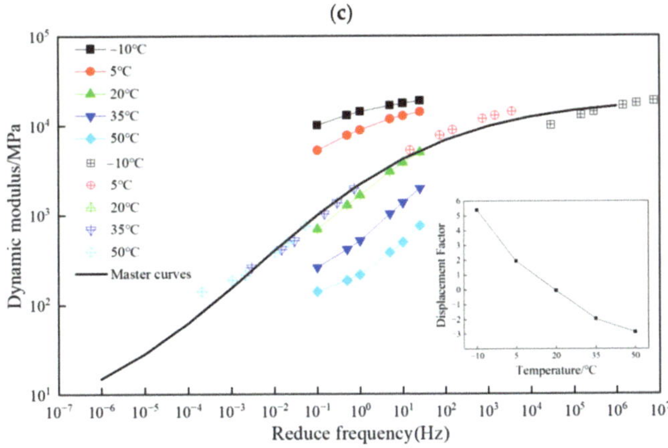

(d)

Figure 9. *Cont.*

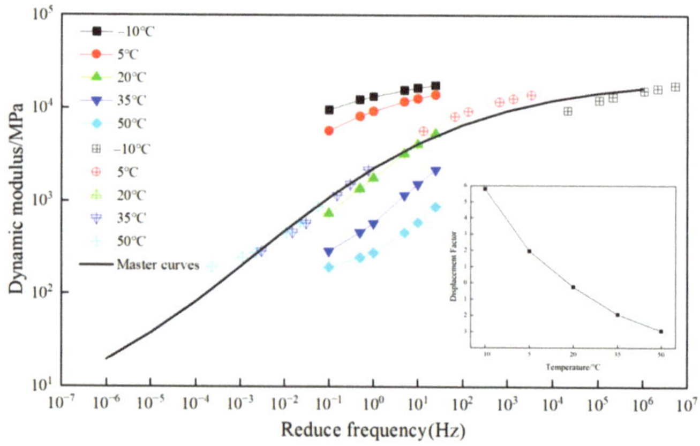

(**e**)

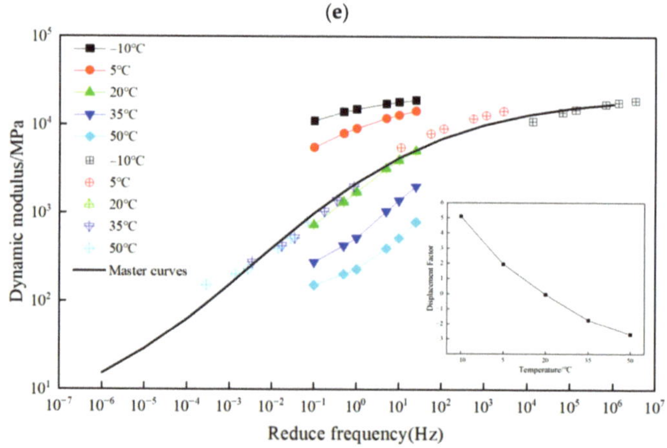

(**f**)

Figure 9. Master curve of different types of fiber asphalt mixture: (**a**) GLF, (**b**) FLF, (**c**) FBF, (**d**) CBF, (**e**) FBF + FLF, and (**f**) CBF + FLF.

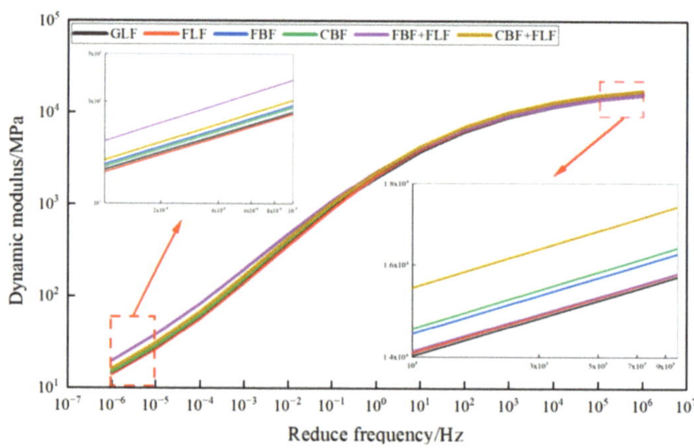

Figure 10. Total master curve.

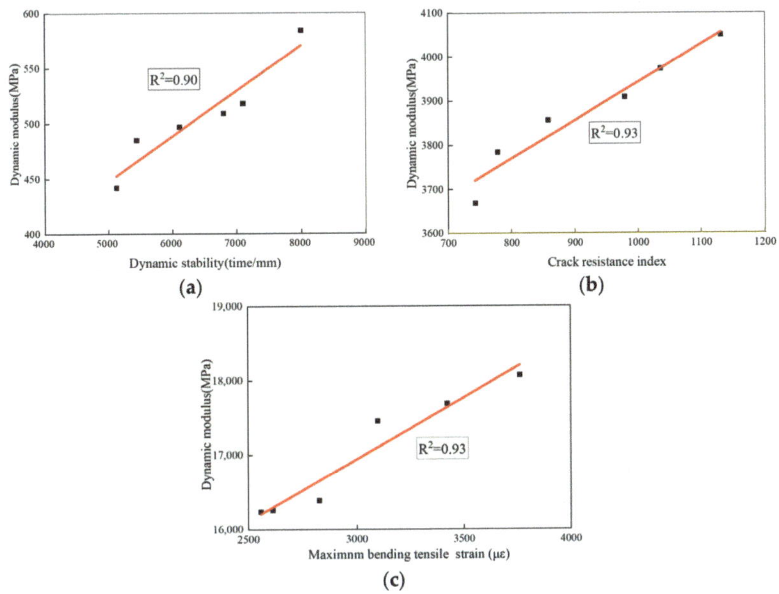

Figure 11. Correlation analysis: (**a**) Correlation analysis between dynamic stability and dynamic modulus; (**b**) correlation analysis between crack resistance index and dynamic modulus; (**c**) correlation analysis between maximum bending strain and dynamic modulus.

4. Conclusions

In this study, different types of fibers (GLF, FLF, CBF, FBF, FBF + FLF, CBF + FLF) were selected to prepare SMA asphalt mixtures. The influence of various fibers on the high-temperature antirutting performance, low- and medium-temperature crack resistance, water stability, dynamic mechanical response, etc., were analyzed. The following conclusions are drawn:

(1) BFs have a better reinforcing effect on the high-temperature dynamic stability and medium-temperature cracking resistance than LFs. Flocculent fibers have a better reinforcing effect on the high-temperature dynamic stability and medium-temperature cracking resistance than the fibers with other morphologies. The FBF + FLF-reinforced mixture presents the best high-temperature rutting resistance and medium-temperature cracking resistance.

(2) BF has a better reinforcing effect on the low-temperature resistance than LF. The maximum flexural strain of the asphalt mixture with CBF is 10.5% higher than that of the asphalt mixture with FBF, while the flexural strain of mixtures modified by LFs with different morphologies is almost the same. CBF has better reinforcing effect on the low-temperature maximum flexural strain of FLF-modified SMA-13 than FBF. FLF could decrease the maximum flexural strain of the FBF-modified mixture and increase the maximum flexural strain of the CBF-modified mixture.

(3) The TSR values of all the six type of mixtures meet the minimum limitation value of 80% according to the specification. This means that these fibers have no significant influence on the water stability of the mixtures.

(4) The DM value is negatively correlated with the test temperature and positively correlated with frequency. The changing degree of DM value is the smallest when the frequency is small and temperature is high. Shift factor decreases with the increase in temperature.

(5) According to the master curve of DM of each fiber SMA-13, it can be inferred that the FBF + FLF mixture has the largest DM value at low frequencies and the CBF + FLF

mixture has the biggest modulus at high frequencies. The differences between the DM value of different SMA-13 with various fibers is smaller when the frequency increases.

(6) The correlation between the dynamic modulus and the high-temperature dynamic stability, medium-temperature cracking resistance, and low-temperature property of fiber asphalt mixtures is good, and the R^2 values are all over 0.9. This indicates that the dynamic modulus values can be used to reveal the road performance of mixture to some extent.

Overall, the scope of this study aimed at analyzing the effect of fiber type, fiber morphology, fiber combination, etc., on the performance of the asphalt mixture. The findings could be regarded as references for the design of fiber-reinforced asphalt mixtures.

Author Contributions: Conceptualization, P.X. and A.K.; data curation, P.X. and A.K.; formal analysis, H.W.; funding acquisition, P.X.; investigation, H.W. and Z.F.; methodology, H.W., P.X., X.W. and A.K.; project administration, P.X., A.K. and H.W.; supervision, P.X.; visualization, A.K., H.W. and Z.F.; writing—original draft, H.W.; writing—review and editing, P.X., X.W. and A.K. All authors have read and agreed to the published version of the manuscript.

Funding: This study was funded by the National Natural Science Foundation of China (grant number 52178439) and the Postgraduate Research & Practice Innovation Program of Jiangsu Province (Yangzhou University) (grant number KYCX23_3537).

Institutional Review Board Statement: Not applicable.

Informed Consent Statement: Not applicable.

Data Availability Statement: The data presented in this study are available on request from the corresponding author.

Acknowledgments: The authors would like to thank Yangzhou University Test Center for providing experimental support for this research.

Conflicts of Interest: The authors declare no conflicts of interest.

References

1. Zhang, Q.; Yang, S.; Chen, G. Regional variations of climate change impacts on asphalt pavement rutting distress. *Transp. Res. D Transp. Environ.* **2024**, *126*, 103968. [CrossRef]
2. Li, Y.; Zou, Z.; Zhang, J.; He, Y. Study on the evolution of airport asphalt pavement integrated distress based on association rule mining. *Constr. Build. Mater.* **2023**, *369*, 130565. [CrossRef]
3. Liu, B.; He, J.; Huang, K.; Feng, S.; Wang, Z.; Wu, X.; Huang, B. Organic expanded vermiculite as an alternative to filler for improving aging resistance of asphalt mixture. *Case Stud. Constr. Mater.* **2024**, *20*, e03307. [CrossRef]
4. Ullah, S.; Qabur, A.; Ullah, A.; Aati, K.; Abdelgiom, M.A. Enhancing High-Temperature Performance of Flexible Pavement with Plastic-Modified Asphalt. *Polymers* **2024**, *16*, 2399. [CrossRef]
5. Liu, W.; Wu, S.; Liu, Q.; Wang, J.; Wan, P.; Xu, H.; Jiang, Q. Research on the Mechanical, Thermal and Induction Healing Properties of Asphalt Wearing Course with Steel Fibers. *Materials* **2024**, *17*, 2040. [CrossRef]
6. Zhao, L.-C.; Xu, L.; Yazdi, A. Laboratory evaluation of the effect of waste materials on mechanical properties of asphalt binder and mixture containing combined natural binder and waste polymer. *Constr. Build. Mater.* **2023**, *403*, 132995. [CrossRef]
7. Al-Kafaji, M.; Al-Busaltan, S.; Kadhim, M.A.; Dulaimi, A.; Saghafi, B.; Al Hawesah, H. Investigating the Impact of Polymer and Portland Cement on the Crack Resistance of Half-Warm Bituminous Emulsion Mixtures. *Sustainability* **2023**, *15*, 15256. [CrossRef]
8. Queiroz, R.F.R.; Rodrigues, J.K.; Patricio, J.D.; Silva, P.D.D.; Carvalho, J.R.; Neto, O.D.M.M.; Rodrigues, L.G.; Lima, R.K.B.D. Linear viscoelastic properties and fatigue S-VECD based evaluation of polymer-modified asphalt mixtures. *J. Build.* **2023**, *75*, 106916. [CrossRef]
9. Wang, B.; Hu, X.; Cao, C.; Wan, J.; Gan, W.; Chen, Z.; Cai, C. Development and characterization of permeative anti-rutting agent for asphalt mixture enhancement. *Constr. Build. Mater.* **2023**, *364*, 129937. [CrossRef]
10. Huang, G.; Zhang, J.; Hui, B.; Zhang, H.; Guan, Y.; Guo, F.; Li, Y.; He, Y.; Wang, D. Analysis of Modulus Properties of High-Modulus Asphalt Mixture and Its New Evaluation Index of Rutting Resistance. *Sustainability* **2023**, *15*, 7574. [CrossRef]
11. Lv, S.; Liu, Y.; Xia, C.; Liu, B.; Zhao, S.; Liu, T.; Liu, Y. Unified characterizing fatigue performance of high modulus asphalt concretes under diverse stress state. *Constr. Build. Mater.* **2022**, *326*, 126805. [CrossRef]
12. Chen, L.; Li, W.; Chen, M.; Qian, Z.; Chen, X.; Zheng, Z. Bleeding mechanism and mitigation technique of basalt fiber-reinforced asphalt mixture. *Case Stud. Constr. Mater.* **2023**, *19*, e02442. [CrossRef]
13. Guo, Q.; Chen, Z.; Liu, P.; Li, Y.; Hu, J.; Gao, Y.; Li, X. Influence of basalt fiber on mode I and II fracture properties of asphalt mixture at medium and low temperatures. *Theor. Appl. Fract. Mech.* **2021**, *112*, 102884. [CrossRef]

14. Xia, W.; Wang, J.; Xu, T.; Jiang, N. Effects of Filler-Asphalt Ratio on the Properties of Lignin and Polyester Fiber Reinforced SMPU/SBS Modified Asphalt Mortar. *J. Renew. Mater.* **2023**, *11*, 3387–3402. [CrossRef]
15. Tanzadeh, R.; Tanzadeh, J.; Tahami, S.A. Experimental study on the effect of basalt and glass fibers on behavior of open-graded friction course asphalt modified with nano-silica. *Constr. Build. Mater.* **2019**, *212*, 467–475. [CrossRef]
16. Hui, Y.; Men, G.; Xiao, P.; Tang, Q.; Han, F.; Kang, A.; Wu, Z. Recent Advances in Basalt Fiber Reinforced Asphalt Mixture for Pavement Applications. *Materials* **2022**, *15*, 6826. [CrossRef]
17. Ren, D.; Luo, W.; Su, S.; Wang, Z.; Kong, L.; Ai, C. Study on crack resistance of basalt fiber reinforced asphalt mixture modified by titanate coupling agent based on digital image correlation. *Constr. Build. Mater.* **2024**, *437*, 136934. [CrossRef]
18. Cheng, X.; Liu, J.; Han, C.; Zhang, X.; Wu, Z. Silane coupling agent impact on surface features of modification of basalt fibers and the rheological properties of basalt fiber reinforced asphalt. *Constr. Build. Mater.* **2023**, *366*, 130182. [CrossRef]
19. Tamrin; Ritonga, A.H.; Aritonang, B. The enhancement of mechanical, physical, and thermal properties of modified asphalt mixture with lignin from mahogany sawdust in the presence of diphenylmethane 4,4′-diisocyanate. *Int. J. Adhes. Adhes.* **2023**, *125*, 103431. [CrossRef]
20. Wu, J.; Zhao, Z.; Jiang, C.; Yang, Y.; Sun, Z.; Yuang, J.; Xiao, F. Recent development and application of natural fiber in asphalt pavement. *J. Clean. Prod.* **2024**, *449*, 141832. [CrossRef]
21. Sun, Z.; Ma, Y.; Liu, S.; Li, Y.; Qiu, X.; Luo, Z. Evaluation of engineering properties of Fiber-reinforced Usual-temperature Synthetic Pitch (USP) modified cold mix patching asphalt. *Case Stud. Constr. Mater.* **2022**, *16*, e00997. [CrossRef]
22. Cheng, X.; Zhong, B.; Wu, Z. Material-specific coefficients C1 and C2 in Williams-Landel-Ferry equations and phase transition behaviors of asphalt binder reinforced by various basalt fibers. *Case Stud. Constr. Mater.* **2024**, *21*, e03619. [CrossRef]
23. Long, A.; Sun, X.; Zhang, G.; Yu, Z.; Zhang, B.; Huang, P.; Wang, J.; Wen, L. Study on fracture performance and failure mechanism of hydraulic basalt fiber asphalt concrete by considering temperature effect. *Theor. Appl. Fract. Mech.* **2023**, *125*, 103895. [CrossRef]
24. Wu, X.; Easa, S.; Kang, A.; Xiao, P.; Fan, Z.; Zheng, X. Performance evaluation of lignin-fibre reinforced asphalt mixture modified by anti-rutting agent. *Constr. Build. Mater.* **2022**, *346*, 128152. [CrossRef]
25. Wu, X.; Tang, X.; Liu, L.; He, Z.; He, S. The Friction–Lubrication Effect and Compaction Characteristics of an SMA Asphalt Mixture under Variable Temperature Conditions. *Materials* **2024**, *17*, 1694. [CrossRef]
26. *JTG F40 (2004)*; Technical Specification for Construction of Highway Asphalt Pavements. China Communications Press: Beijing, China, 2004.
27. *JTG E20 (2011)*; Standard Test Methods of Bitumen and Bituminous Mixtures for Highway Engineering. China Communications Press: Beijing, China, 2011.
28. Wu, X.; Kang, A.; Wu, B.; Lou, K.; Fan, Z. Prediction of crack resistance of LFSMA-13 with and without anti-rut agent using parameters of FTIR spectrum under different aging degrees. *Materials* **2021**, *14*, 3209. [CrossRef]
29. Yan, C.; Zhang, Y.; Bahia, H.U. Comparison between SCB-IFIT, un-notched SCB-IFIT and IDEAL-CT for measuring cracking resistance of asphalt mixtures. *Constr. Build. Mater.* **2020**, *252*, 119060. [CrossRef]
30. Chen, H.; Zhang, Y.; Bahia, H.U. The role of binders in mixture cracking resistance measured by ideal-CT test. *Int. J. Fatigue* **2021**, *142*, 105947. [CrossRef]
31. Jin, D.; Ge, D.; Wang, J.; Malburg, L.; You, Z. Reconstruction of Asphalt Pavements with Crumb Rubber Modified Asphalt Mixture in Cold Region: Material Characterization, Construction, and Performance. *Materials* **2023**, *16*, 1874. [CrossRef]
32. Zhang, Y.; Zhang, Y.; Li, B.; Kang, A.; Wang, Y. Evaluation of Cracking Resistance of SMA-13 Hot Recycling Asphalt Mixtures Reinforced by Basalt Fiber. *Materials* **2024**, *17*, 1762. [CrossRef]
33. Pei, Z.; Lou, K.; Kong, H.; Wu, B.; Wu, X.; Xiao, P.; Qi, Y. Effects of Fiber Diameter on Crack Resistance of Asphalt Mixtures Reinforced by Basalt Fibers Based on Digital Image Correlation Technology. *Materials* **2021**, *14*, 7426. [CrossRef] [PubMed]
34. Mishra, V.; Singh, D.; Habal, A. Investigating the condition number approach to select probe liquids for evaluating surface free energy of bitumen. *Int. J. Pavement Res. Technol.* **2020**, *13*, 10–19. [CrossRef]
35. Wang, H.; Liu, L.P.; Sun, L.J. Characterization of OGFC Mixtures Containing Lignin Fibers. *Appl. Mech. Mater* **2012**, *174*, 775–781. [CrossRef]
36. Büchner, J.; Wistuba, P.M.; Remmler, T.; Wang, D. On low temperature binder testing using DSR 4 mm geometry. *Mater. Struct.* **2019**, *52*, 113. [CrossRef]

materials

Article

Biochars from Cotton Seed, Camelia Seed Shell, and Coffee Ground in Modification of Asphalt: Fundamental Properties, Rheological Performance, and Inhibition of VOC Emissions

Xiao Zhang, Yi Zhou and Yongjie Xue *

State Key Laboratory of Silicate Materials for Architectures, Wuhan University of Technology, Wuhan 430070, China; fans-cooler@outlook.com (X.Z.); joejoe-98@live.com (Y.Z.)
* Correspondence: xyjskl@whut.edu.cn; Tel.: +86-027-8716-2595

Abstract: With the increasing requirement for asphalt modification, a new environmentally friendly asphalt modifier is needed. In this study, three varieties of biomass, cotton seed (CO), camelia seed shell (CA), and coffee ground (CG), were chosen for biochar preparation and asphalt modification to find an environmentally friendly asphalt modifier. A segregation test was applied to evaluate the storage stability of the modified asphalt. A dynamic shear rheometer (DSR) temperature sweep and frequency sweep were used to characterize the high-temperature performance. The low-temperature performance was evaluated by the bending beam rheometer (BBR) test. The DSR results indicate that the rutting factor increase for modified asphalt at high temperatures is CO $\approx$ CG > CA, and a high temperature could reflect the biochar's properties better in modified asphalt. Furthermore, the low-temperature deterioration is well controlled in CO and CA biochar-modified asphalt. Finally, the volatile organic compound (VOC) emission behavior was evaluated using gas chromatography–mass spectrometry (GC-MS).

Keywords: asphalt modification; biochar; rheology

Academic Editor: Giovanni Polacco

Received: 14 February 2025
Revised: 21 March 2025
Accepted: 25 March 2025
Published: 27 March 2025

Citation: Zhang, X.; Zhou, Y.; Xue, Y. Biochars from Cotton Seed, Camelia Seed Shell, and Coffee Ground in Modification of Asphalt: Fundamental Properties, Rheological Performance, and Inhibition of VOC Emissions. *Materials* **2025**, *18*, 1504. https://doi.org/10.3390/ma18071504

1. Introduction

Asphalt, a byproduct of the petroleum industry, is an organic viscoelastic material with excellent physical properties. Asphalt binder is widely used in the construction of arterial highways and urban roads for its advantages, including low passing noise, low vibration, comfortable driving experience, and simple maintenance. However, asphalt binder faces various challenges, such as heavy loads and extreme climates, posing difficult challenges for virgin asphalt. Adding a modifier to asphalt binders and preparing a modified asphalt binder are effective solutions to these problems. Currently, styrene–butadiene–styrene block copolymer (SBS), styrene–butadiene rubber (SBR), polyethylene (PE), etc., [1–7], are the most conventional modifiers considering both economy and performance. SBS could improve the high-temperature performance and reduce the temperature sensitivity of asphalt to enhance the anti-deformation ability of asphalt binders [1–3,8]. SBR is the main material of tires; thus, recycled tires are fantastic feedstock for SBR asphalt modifiers. A proper content of SBR in asphalt is an obvious benefit for ductility at low temperatures, elastic recovery rate, and viscosity [4,9,10].

The production of asphalt and modifiers as well as storage is a crucial part of the life cycle of asphalt pavements [11]. As China aims to hit peak carbon dioxide emissions in 2030 and achieve carbon neutrality in 2060, the energy consumption and carbon emissions from the production of pavement materials account for a significant portion of the carbon emissions life cycle assessment of highway facilities. Not only does the production of modifiers consume energy, but the energy required to produce the raw materials also contributes substantially. For rubber asphalt, the value is 42,700 MJ/t and 34,900 MJ/t, respectively. And the comprehensive energy consumption of plastic rubber-modified asphalt and SBS-modified asphalt mixture is 678.8 MJ/t and 743.8 MJ/t, within which the consumption for modifiers is 36 MJ and 98.8 MJ, respectively [12]. These substantial energy consumptions indirectly lead to considerable carbon emissions. In response to carbon reduction and the construction of a circular economy, new low-carbon asphalt modifiers are urgently needed.

Biochar, a product of biomass pyrolysis, emerges as a promising low-carbon modifier alternative for asphalt modification [13]. The energy consumption of biochar production largely depends on the pyrolysis temperature, a parameter that affects both pyrolysis time and yield. However, the overall energy and carbon emissions are still lower than those of the traditional modifiers mentioned above. The carbon in biochar primarily originates from lignin [14], hemicellulose, and cellulose [15] in biomass, which means biomass with high contents of these components is suitable for biochar production. The requirements for feedstock for biomass pyrolysis are not strict, as it can utilize agricultural waste [16,17] without dedicated cultivation requirements, providing a sustainable way to handle agricultural wastes.

The porous structure and surface functional groups of biochar allow it to interact with asphalt, thus enhancing asphalt's performance. Kumar et al. [17] investigated biochar-modified asphalt with biochar contents ranging from 0% to 20% and found that biochar significantly increases asphalt's viscosity and rutting resistance. However, a high biochar content deteriorates fatigue and aging resistance. Zhang et al. [18] studied the influence of biochar particle size on asphalt performance, revealing that smaller biochar particles outperformed larger ones. Furthermore, compared with graphene, the surface of biochar is much tougher, with an abundant porous structure. Ma et al. [19] pointed out that the degradation in the low-temperature performance of biochar-modified asphalt can be managed by adjusting the biochar dosage. Hu et al. [20] prepared corn stalk biochar through the hydrothermal method and explored the molecular weight distribution of biochar-modified asphalt and its relation to biochar contents, finding that a 6 wt% dosage allows for uniform distribution in asphalt with minimal clustering. Zhao et al. [21] demonstrated that biochar enhances the moisture resistance of hot-mixed biochar-modified asphalt mixtures, with a lower impact on crack resistance compared to carbon black and carbon fiber.

Except for modification, biochar also removes the volatile organic compounds (VOCs) in asphalt, which are hazardous and harmful to the environment and human health. Zhou et al. [22] find that biochar could improve the adsorption energy of asphalt, and the adsorption energy would increase with temperatures. Li et al. [23] simulated three representative scenarios with various VOC emission concentrations. Alkenes ($n < 4$), alkanes ($n > 6$), aldehydes, and alkylbenzenes were the main ingredients for the asphalt VOC emission. Zhou et al. [24] compared differential scanning calorimetry (DSC) curves of virgin and biochar-modified asphalt, indicating the chemical adsorption mechanism in VOC inhibition.

While the existing studies on biochar-modified asphalt primarily focus on one single biochar or biomass feedstock, few studies compare the effects of multiple biochar types. The problem of whether different biochars have distinct effects on asphalt performance remains unresolved. Moreover, the relationship between biochar properties and modified asphalt performance is not well understood. Considering both economic and environmental factors, we selected coffee grounds, camellia seed shells, and cotton seed, three types of agricultural and agroindustry wastes, to prepare biochar asphalt modifiers. As coffee is the most popular beverage, the use of its residue, coffee grounds, holds promise. Cotton seed is a byproduct of the cotton industry, and camellia seed shells are the shells of a commonly grown oilseed in the Yangtze River basin. These biomasses, rich in lignin, cellulose and hemicellulose [15,25–27], are ideal raw materials for biochar production. This work aims to find the best biochar asphalt modifier among the 3 types of biochar and investigate the performance improvement of biochar-modified asphalt. We prepared biochar from these three biomasses, performed basic characterizations, and prepared biochar-modified asphalt with dosages of 1%, 6%, and 9% by the weight of virgin asphalt. Penetration, softening point, and ductility tests were conducted to determine the optimal biochar content. The modified asphalt with selected content was further evaluated through DSR and BBR tests, comparing the performance of three types of biochar-modified asphalts. This study provides insights into how different biochars affect asphalt performance, potentially contributing to agricultural waste management and green economy construction. In the end, the biochar-modified asphalt with the best comprehensive physical properties was detected to evaluate the VOC emission behavior through thermal desorption–gas chromatography–mass spectrometry (TD-GC-MS).

2. Materials and Methods

2.1. Materials

Three varieties of biomass: cotton seed (CO), camelia seed shell (CA) and coffee ground (CG), were selected as biochar feedstocks, among which CO and CA were collected from local farms while CG was bought from Starbucks (Wuhan, Hubei Province, China). Asphalt 70# was applied in the preparation of modified asphalt binder, with related parameters presented in Table 1.

Table 1. Conventional properties of virgin asphalt.

Technical Indexes	Units	Results	Requirements	Standards
Penetration (25 °C)	0.1 mm	69	60–80	ASTM D5 [28]
Softening Point	°C	48	≥46	ASTM D36 [29]
Ductility (15 °C)	cm	>100	>100	ASTM D113-17 [30]

2.2. Asphalt Modification

The biomass was first dried in an oven at 80 °C for 12 h. Biochar was then produced via slow pyrolysis, heating the biomass to 600 °C at a rate of 10 °C/min, with a residence time of 90 min under a nitrogen atmosphere at 80 mL/min. Then, the produced biochar was ground with a planet grinding machine for 40 min, as the particle size may influence the biochar dispersion and modification effect. No further treatment was applied. After grinding, the biochar powder was blended and agitated with virgin asphalt using a high-speed shearing mixer for 40 min. Biochar content was set at 1%, 6%, and 9% by the weight of virgin asphalt, with a shearing rate of 3000 rpm at 135 °C [31]; the asphalt was heated with an oil bath.

2.3. Characterization of Biochar

2.3.1. Elemental Analysis

Elemental analysis for the 3 varieties of biochar was conducted, respectively, in order to measure the contents of carbon, hydrogen (H), oxygen (O), nitrogen (N) and sulfur (S). The C, H, N and S contents were measured by a Vario EL Cube elementar (Elementar, Hanau, Germany), while O was measured using a Vario EL Marco Cube elementar (Elementar, Hanau, Germany).

2.3.2. BET

The Brunauer–Emmett–Teller (BET) specified surface area and pore size distribution of the biochar samples were measured using a Micromeritics ASAP 2420 Surface Area & Porosimetry System (Micromeritics, Norcross, GA, USA) with nitrogen adsorbent at a temperature of 77 K.

2.3.3. FTIR

A Fourier transform infrared (FTIR) spectrometer (Bruker Alpha, Rheinstetten, Germany) was utilized to obtain the infrared absorption spectra of the biochar samples from the wave number of 4000 cm^{-1}–400 cm^{-1} with 2 cm^{-1} resolution, employing the potassium bromide (KBr) pellet method for biochar sample preparation.

2.4. Evaluation of Biochar-Modified Asphalt Properties

2.4.1. Conventional Properties

The penetration, softening point and ductility tests were conducted according to ASTM D5, D36 and ASTM D113-17, respectively. The penetration test was conducted at 15 °C, 25 °C and 30 °C, by which the penetration indexes were evaluated using Equations (1) and (2) to evaluate the temperature sensitivity of biochar-modified asphalt.

$$lgP = K + A_{lgPen}T \tag{1}$$

$$PI = \frac{20 - 500A_{lgPen}}{1 + 50A_{lgPen}} \tag{2}$$

2.4.2. Segregation Test

For modified asphalt, the storage stability is critical in the evaluation of long-term properties [4,32–34], which is considered through the compatibility of asphalt with modifiers. A delayed segregation test was applied to evaluate the storage stability of biochar-modified asphalt and the effect of biochar variety. The test was conducted according to ASTM D7173-14 [35] with a 48 h soaking, and ASTM D36 was applied to measure the softening point difference.

2.4.3. Temperature Sweep

In order to assess the rheological properties of biochar-modified asphalt across medium to high temperature, dynamic shear tests were carried out in this research. Temperature sweep and frequency sweep tests were conducted on the virgin asphalt and 3 varieties of modified asphalt in 12% strain-controlled mode using an Anton Paar (Graz, Austria) SmartPave 102e rheometer. A PP25 plate was used in this test, where the sample thickness was 1 mm.

Temperature sweep is a common method to evaluate the temperature sensitivity and high-temperature performance of asphalt binders. The test temperature was from 30 °C to 80 °C with a loading frequency of 10 rad/s; 3 data points were equidistantly taken every 5 °C at the test temperature range. Through the test, basic rheological properties, complex shear modulus G^* and phase angel δ were obtained, from which the rutting factor $G^*/sin\ \delta$ and critical temperature were calculated.

2.4.4. Frequency Sweep

The modulus of typical amorphous polymers increases with loading frequency but decreases when the temperature increases. And curves of the instantaneous modulus as a function of time do not change their shape as the temperature changes, only shifting left or right. It could determine temperature-dependent mechanical properties of linear viscoelastic materials from known properties at a reference temperature. It offers a method to predict curves at various frequencies and temperatures using a shift factor based on the master curve in a given test condition. And this rule is called the time–temperature superposition (TTS) principle.

The TTS principle avoids the inefficiency of measuring polymer behavior over long periods of time, which usually costs a lot of time. According to the introduction above, frequency sweep tests were conducted from 0.01 Hz to 10 Hz at temperatures between 30 °C and 60 °C with an increment of 10 °C. Serval methods were applied to characterize the viscoelastic behavior of biochar-modified asphalt.

The master curve is an effective method in the research of asphalt TTS behavior. It shows the dependance of G* and δ for loading frequency. As a popular method in the research of polymer TTS behaviors, it is also effective for deducing the change of microscopic structure in modified asphalt. For asphalt materials, the master curve could be constructed by the Williams–Landel–Ferry (WLF) model, as shown in Equation (3).

$$log\alpha_T = -\frac{C_1(T - T_0)}{C_2 + (T - T_0)} \tag{3}$$

where C_1 and C_2 are positive constants depending on the material and reference temperature; α_T refers to the shift factor and T_0 is reference temperature. Therefore, through the TTS principle, we can obtain complex dynamic moduli like asphalt complex shear modulus at a required frequency using Equations (4) and (5).

$$G'(\omega, T) = G'(\alpha_T\omega, T_0) \tag{4}$$

$$G''(\omega, T) = G''(\alpha_T\omega, T_0) \tag{5}$$

Han et al. [36,37] discovered that the order–disorder transition of block copolymers could be presented by the double logarithmic diagram of G' versus G''. Though it does not show the relationship between modulus and frequency directly, the higher the modulus, the lower the frequency. Therefore, this method is now widely applied in the research of asphalt. In the Han curve, the graph of G' and G'' is linear-like, following Equation (6) [37].

$$logG' = AlogG'' + B \tag{6}$$

where A and B are constants, in relation with the density, experimental temperature and molecular weight. In the low-frequency terminal zone of the curve, according to previous investigations [38], the plots are expected to have a slope of 2 in the term region for all homogeneous polymeric liquids. This phenomenon is an embodiment of the micro phase change of asphalt at low frequency at the macro scale.

2.4.5. Bending Beam Rheometer Test

Cracking is a common problem for asphalt pavement at low temperature. Since the aggregate properties would not be affected by temperature, the performance of the asphalt binder plays a critical role in cracking resistance. The cracking resistance at low temperature of biochar-modified asphalt was evaluated by bending beam rheometer (BBR) test at −12 °C, −18 °C and −24 °C according to ASTM D6648-16 [39]. The heated modified asphalt was poured into a mold and shaped into a beam specimen (127 mm × 6.35 mm × 12.7 mm). Then, the flexural creep stiffness *S(t)* and creep rate (m value) were measured at each test temperature. According to Superpave, the *S(t)* of asphalt binder should be less than 300 MPa, while the m value must be more than 0.3 [40].

2.5. Asphalt VOC Emission Evaluation

VOCs emitted by the samples were collected by an adsorption tube consisting of the Combination 1 adsorption tube containing Tenax GR and Carbopack B, as stipulated in HJ 734-2014 [41]. Then, the VOCs were detected by a TD-GC-MS (TD: PerkinElmer (Waltham, MA, USA) Turbo matrix ATD 350, GC-MS: Agilent Atomx P&T-Agilent 7890B-5977B, Santa Clara, CA, USA). Qualitative analysis was conducted through NIST mass spectroscopy library. Gas chromatography was applied to record the chromatograms of modified asphalt. Through peak searching, the area of the peaks is calculated and normalized. Considering the complexity and diversity, some original and representative components in asphalt VOCs were detected as the references whose emissions in virgin and modified asphalt are compared, respectively, through which the VOC inhibition of the biochar modifier is evaluated.

3. Results and Discussion

3.1. Biochar Characterization

The porosity and surface reactivity of biochar surpass traditional carbon-based fillers. The elemental analysis (Figure 1) indicates carbon contents reaching approximately 76% for all three biochars, confirming complete pyrolysis of lignin, cellulose, and hemicellulose. Differences in H, O, and N content were observed. The H content was highest in CG (0.624%), followed by CO (0.53%) and CA (0.47%). CG has the highest oxygen content at 13.86%, followed closely by CA at 13.21% and CO at 12.61%. Notably, CG contains 3.58% nitrogen, while CO has 2.74%, with CA showing nearly no nitrogen content.

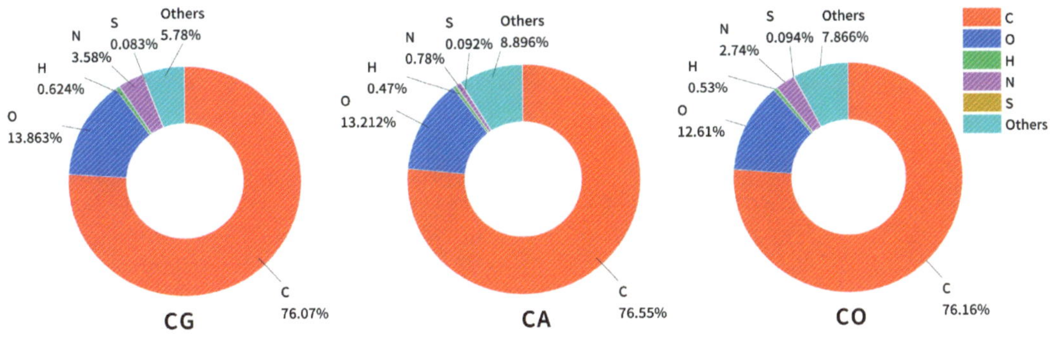

Figure 1. The element contents for biochar.

The surface functional groups could influence physical and chemical properties of biochar [42,43]. FTIR spectra (Figure 2) show the surface functional groups for the three biochars. Peaks between 1868 and 1560 cm^{-1} correspond to the stretching vibrations of C=O, C=C, and potentially some C=N [44] and N-O bonds, with CO and CG displaying a stronger absorption than CA. The absorption peak at 1380 cm^{-1}, mainly from aromatic rings and C-H bonds in methyl and methylene groups [45], is stronger in CA and CG, but weaker in CO. The double peaks around 1100–1000 cm^{-1} indicate C-O and C-N [46] stretching vibrations.

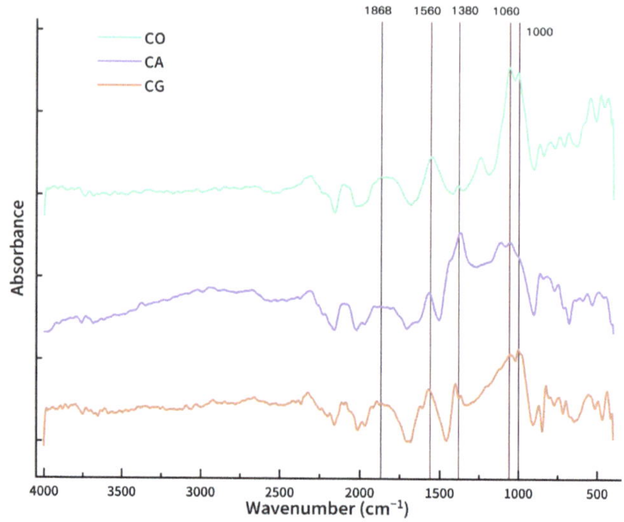

Figure 2. Infrared adsorption spectra for biochar.

Analyzing the spectra from 4000 to 2400 cm^{-1}, combined with the elemental contents, it is obvious that CG reaches the highest absorptivity at this range. It is speculated that a considerable part of it is brought by N-H [47]. The double peaks at 1060 cm^{-1} for CO are noticeably the strongest, and there should also be more C-N bonds in addition to C-O bonds.

Both surface area and porosity are crucial physical properties for biochar [48,49]. Table 2 lists the specific surface area and pore volume for the 3 varieties of biochar through BET, BJH and t-Plot methods. It reveals that CO has the highest specific surface area of 66.42 m^2/g and pore volume of 0.04 cm^3/g among the biochar. Figure 3 shows the isothermal adsorption/desorption plots of the three varieties of biochar. The line within higher saturation accounts for the adsorption process and the other for the desorption process. The isotherms do not close, indicating adsorption hysteresis [50], characteristic of a type IV isotherm [51] with an H2 hysteresis loop [52,53]. In the low relative pressure region, the adsorption of CO increases rapidly and saturates after reaching a certain relative pressure as the adsorption rate slows down. It highlights a relatively high micropore content, consistent with the high specific surface area from the t-Plot method. And then the shape of the 3 adsorption isotherms tends to be consistent, indicating a comparable mesoporous and microporous distribution among the 3 varieties of biochar.

Table 2. Pore area and volume for biochar.

Variety	CO	CG	CA
Surface area (m^2/g)	66.4172	11.3879	15.7413
Pore volume (cm^3/g)	0.0421	0.0137	0.0150
t-Plot micropore area (m^2/g)	41.1906	9.4560	14.7117
t-Plot micropore volume (cm^3/g)	0.0176	0.0037	0.0060
BJH mesopore area (m^2/g)	27.2051	1.0192	1.2887
BJH mesopore volume (cm^3/g)	0.0256	0.0098	0.0092

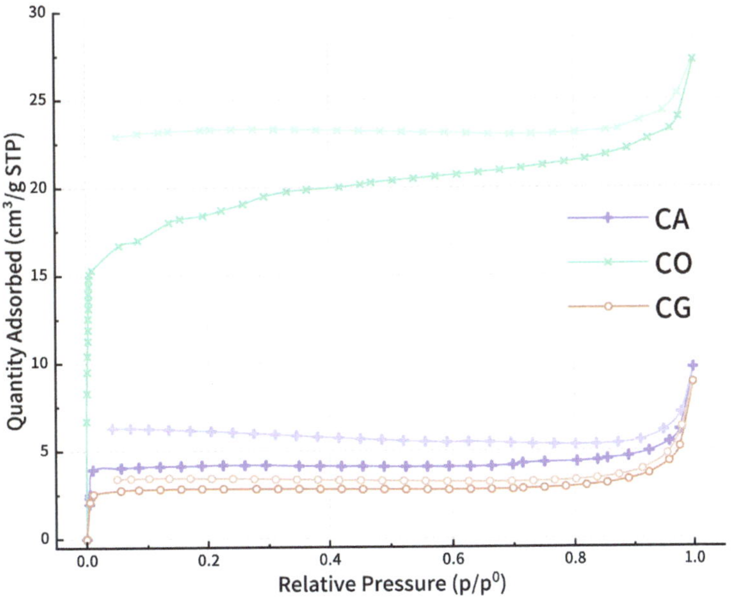

Figure 3. Isotherm adsorption and desorption plot.

3.2. Performance of Modified Asphalt

3.2.1. Selection of Biochar Content Based on Conventional Properties

Table 3 presents the basic properties of biochar-modified asphalt at different biochar contents (1%, 6%, and 9%). The penetration of 9 varieties of modified asphalt were measured at 15 °C, 25 °C, and 30 °C and penetration indexes were calculated. It shows a decrease in penetration values with increasing biochar content, with biochar's modification effect on asphalt increasing with higher test temperatures. It is obvious that the penetration of modified asphalt decreases with the contents of biochar, which is also in line with the discovery from other related studies about biochar-modified asphalt. Furthermore, biochar modification could lower the penetration of asphalt; for virgin asphalt, this value is −0.33. While the modification effect for 1% is negligible, when biochar content reaches 6%, the penetration index is −0.80, which is much lower than virgin asphalt. Biochar-modified asphalt's softening point also rises with biochar content, though the increase was modest. The modified asphalt with the highest softening point is CA-9%, only 2.64 °C ahead of virgin asphalt. However, excessive biochar content (9%) reduces low-temperature ductility significantly, dropping below the threshold of 20 cm at 15 °C, which is unsuitable for engineering applications [54]. Considering both high- and low-temperature performance, a 6% biochar content was selected for further evaluations. The fluorescence microscopic image for modified asphalt at 6% content can be found in the Supplementary Materials.

Table 3. Conventional properties for biochar-modified asphalt.

Sample	Content (%)	Penetration (dmm)			Penetration Index	Softening Point (°C)	Ductility (cm)
		15 °C	25 °C	30 °C			
Virgin	0	29.17	77.62	124.27	−0.32726	49.42	>100
CG	1	28.31	73.30	115.92	−0.14617	49.78	75
	6	23.44	69.82	110.00	−0.78907	51.11	30
	9	18.68	57.23	100.18	−1.25747	51.98	<20

Table 3. *Cont.*

Sample	Content (%)	Penetration (dmm)			Penetration Index	Softening Point (°C)	Ductility (cm)
		15 °C	25 °C	30 °C			
CA	1	25.81	72.11	120.54	−0.71495	50.41	78
	6	23.00	64.27	107.43	−0.71639	51.45	25
	9	22.68	63.97	108.43	−0.80601	52.06	<20
CO	1	28.83	76.64	124.94	−0.39385	50.01	82
	6	24.22	67.87	113.61	−0.73359	50.55	27
	9	23.11	62.69	103.25	−0.52731	51.72	<20

3.2.2. Storage Stability

For asphalt modified with powdered materials, it is essential that the modifier can remain uniformly dispersed in the asphalt matrix to achieve an optimal performance. Storage stability is mainly influenced by two factors: the density of the modifier [55] and surface properties [56] of powder. At 163 °C, asphalt is in a viscous flow state; powders within mismatched densities with asphalt continuous phase may float or sink, causing segregation. Additionally, if the powder surface can interact with asphalt, it increases the resistance to segregation. Figure 4 presents the softening point differences between the upper and lower layers after a 48 h segregation test for the three biochar-modified asphalts. "D" in the figure refers to the difference of the upper and lower layers. After 48 h, the differences for 3 varieties of biochar-modified asphalt are as follows: CA 0.82 °C, CO −0.11 °C, and CG 0.66 °C, respectively, all of which are lower than ±1 °C. No phase separation was observed in the experiment. CO exhibited a smaller difference; it may result from the porosity of the biochar, which is different from others. Generally, a modifier with a difference below 2.5 °C is considered to have good compatibility, as all the modified asphalt fits the requirement, indicating excellent storage stability in this study with 6% biochar content.

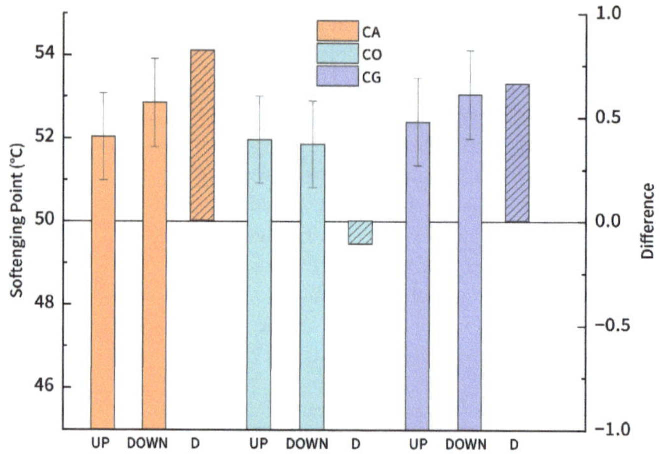

Figure 4. Segregation test of biochar-modified asphalt.

3.2.3. Temperature Sweep Analysis

The temperature sweep results (Figures 5 and 6) show that the complex shear modulus (G^*) and rutting factor $G^*/\sin\delta$ of the three biochar-modified asphalts decrease with increasing temperature, with the rate of decrease slowing as temperature rises. The G^*-T curve fits a Boltzmann function, showing a gradual reduction in modulus as the test temperature increases, which is consistent with our expectation. Around 30 °C, the performance enhancement of modified asphalt compared to virgin asphalt is minimal, with CO-modified asphalt showing an 8% increase in modulus over virgin asphalt, while the performance differences between the biochar-modified asphalts are slight (CO > CA > CG). As the temperature rises, the enhancement for modified asphalts G* increases more significantly, and differences between the biochar-modified asphalts become more pronounced, peaking around 57 °C. At this temperature, CO- and CG-modified asphalts reach an approximate 20% increase in G*, while CA-modified asphalt shows a lower improvement (14%). At temperatures above 70 °C, asphalt begins to transition to a viscous flow state [57,58], and the modulus enhancement effect of biochar begins to decline, with a reduction to around 10% improvement at 80 °C (CO and CG). In this range, the performance of CA-modified asphalt is notably inferior to the other two groups.

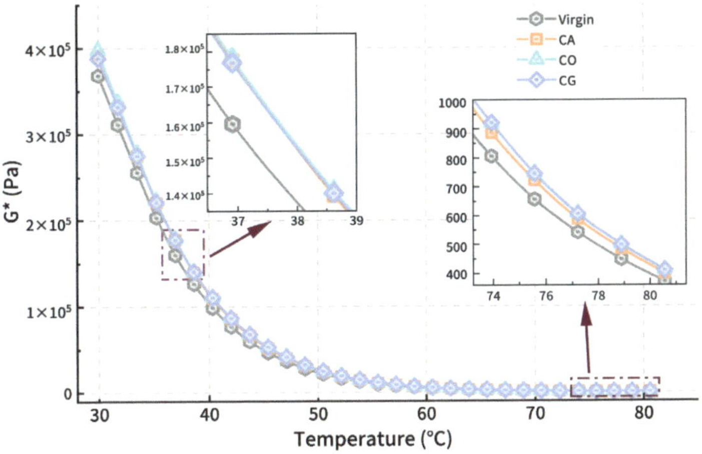

Figure 5. *G**-*T* curve for modified asphalt.

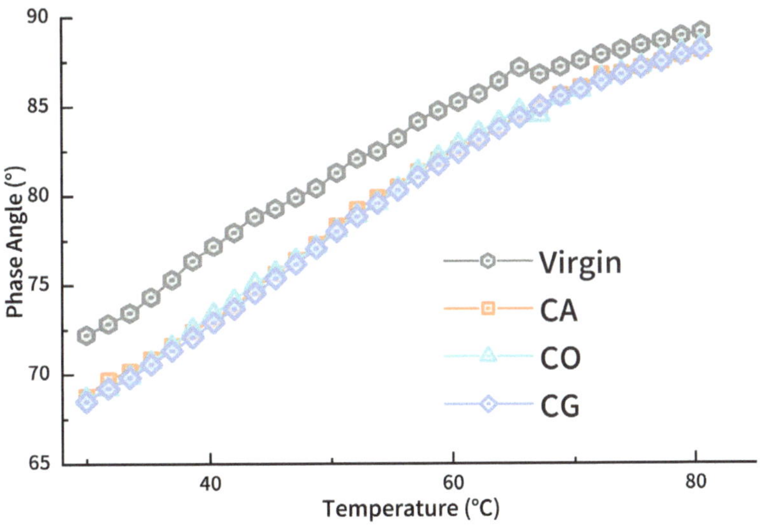

Figure 6. δ-*T* curve for modified asphalt.

The phase angle of asphalt increases linearly as temperature rises from 30 °C to 70 °C, and the difference between modified and virgin asphalt remains stable. After 70 °C, the phase angle of virgin and modified asphalts approaches the maximum of 90°, with minimal differences among the modified asphalts.

Since the phase angle remains above 60° throughout the test temperature range, the rutting factor ($G^*/sin\ \delta$) for asphalt is mainly influenced by G^*. Figure 7 shows the $G^*/sin\ \delta$-T curve, with the rutting factor enhancement for each modified asphalt peaking around 57 °C (CA: 15%, CO: 19%, CG: 20%). At 60°, the δ for CG-modified asphalt is larger than the other 2 modified asphalts; the enhancement for CG is the most obvious. And CA accounts for the lowest enhancing effect during the whole test, while the higher the testing temperature is, the larger the difference is.

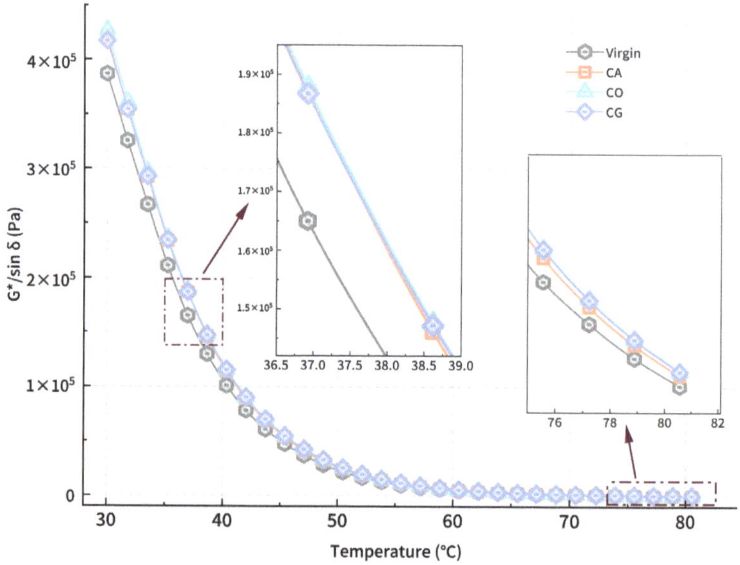

Figure 7. $G^*/sin\ \delta$-T curve for modified asphalt.

SHRP invented the critical temperature to standardize asphalt high-temperature performance evaluation, which is defined as the temperature when the rutting factor is 1 kPa. Figure 8 presents the critical temperatures for both virgin and modified asphalts, with CO- and CG-modified asphalts exceeding 73.1 °C, slightly outperforming CA-modified asphalt, at 72.9 °C. All of the 3 modified asphalt critical temperatures are higher than virgin asphalt at 72.3 °C. With the same contents, the enhancement effects for CO and CG are better than that for CA. The results suggest that the performance differences may relate to the nitrogen and associated functional groups in biochar, as CO and CG biochars contain higher nitrogen contents than CA. The differences may due to the contents and structure of the cellulose, hemicellulose and lignin in the biomass.

The biochar-modified asphalts exhibit a higher G^* and lower δ compared to virgin asphalt, indicating a greater improvement in G' than in G''. This suggests a significant enhancement in the modified asphalt's permanent deformation resistance and anti-rutting performance. The relatively smaller increase in G'' may result from biochar's adsorption of light components in asphalt [59], potentially disrupting the original colloidal structure. Among the three biochars, CO-modified asphalt shows the greatest increase in G^* and the smallest reduction in δ, narrowing the $G^*/sin\ \delta$ difference with other modified asphalts, likely due to the mesoporous structure.

As temperature increases, the performance difference among the three modified asphalts widens. This may be due to the fact that with the increase in testing temperature, the viscosity of asphalt decreases and the molecular motion becomes more intense, which leads to a full adsorption of light components by biochar. The adsorption capacity depends more on the property of biochar itself than on other factors, and the property difference among the 3 biochars could be fully reflected.

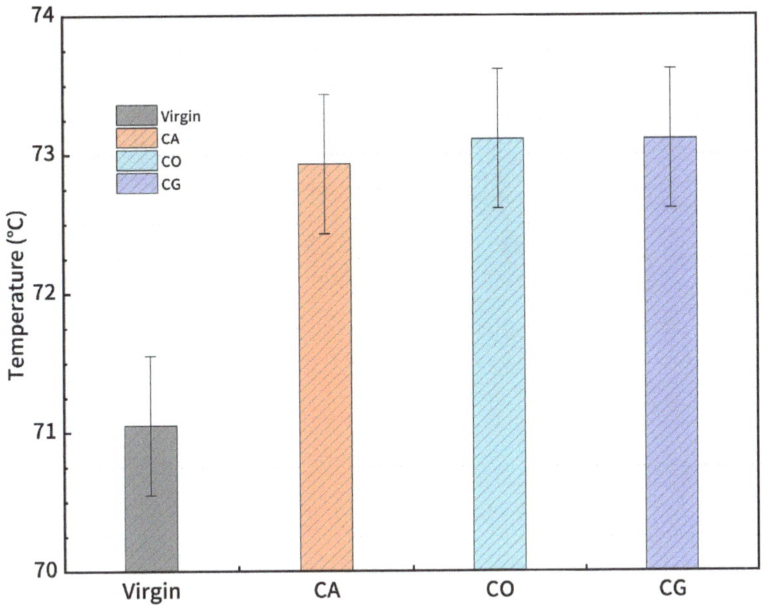

Figure 8. Critical temperatures of modified asphalt.

3.2.4. Master Curve

Based on the DSR frequency sweep, master curves (Figures 9 and 10) were constructed using the TTS principle at a reference temperature of 40 °C to study the rheological performance of biochar-modified asphalts over a wide frequency range. The corresponding α_T values were calculated using a Python program (https://www.python.org/). Linear fitting for the G^*-f double logarithmic curves and R^2 shows that the modified asphalts are in a viscoelastic state where the TTS principle is validated. Master curves exhibit an increase for modified asphalts compared to virgin asphalt in complex shear modulus, especially in the high-frequency region, indicating that the modification successfully enhances the stiffness through a series of physical changes.

According to the TTS principle, the master curve in the low-frequency region represents the rheological properties for polymer materials at high temperature, while high frequency represents the properties at low temperature. In this study, the reference temperature is 40 °C; therefore, the G^* master curve in the mid to low-frequency region could better reflect the rutting resistance and anti-permanent deformation ability for modified asphalt.

In the low-frequency range of the master curve, the G^* of CA-modified asphalt is the highest, followed by CG, with CO exhibiting the lowest permanent deformation resistance. When f = 0.1 Hz, the complex shear modulus for virgin asphalt is 5784 Pa, and for modified asphalts, the values are CA 7148 Pa, CG 6874 Pa, and CO 6743 Pa. The enhancement of biochar for asphalt is evident in this region to distinguish each type of biochar; biochar properties could hardly influence the enhancement effect.

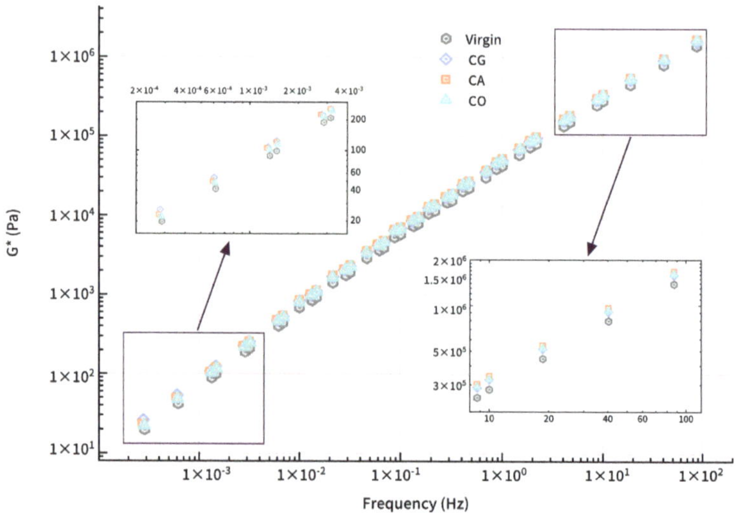

Figure 9. G* master curves for modified asphalt.

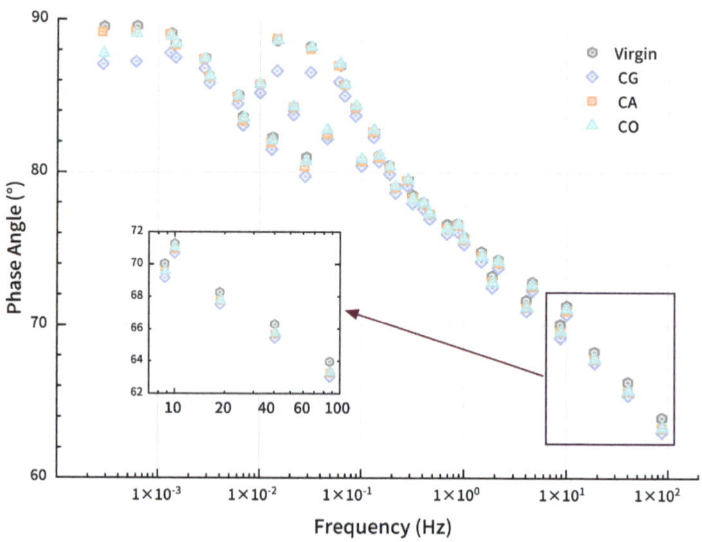

Figure 10. Phase angle master curves for modified asphalt.

In the low-frequency region, the complex shear modulus for modified asphalts is low and increases as frequency increases, reaching the maximum at the fastest frequency, embodying the differences among biochars. In the high-frequency range, asphalt behavior approaches that of a glassy elastic material, where the complex shear modulus could reflect the asphalt's low-temperature cracking resistance. A higher complex shear modulus at high frequency indicates a stiffer, more brittle material prone to cracking. For instance, at 87 Hz, CA-modified asphalt exhibits the highest modulus (1.68 MPa), approximately 20% higher than virgin asphalt (1.39 MPa), followed by CO (1.61 MPa) and CG (1.58 MPa).

The δ-f master curve shows the phase angle of modified asphalt initially increasing linearly with decreasing frequency, then the increasing rate gradually slows. The phase angle reaches a maximum at around 1×10^{-3} Hz before declining. CG-modified asphalt shows a phase angle curve below that of virgin asphalt, while CA and CO phase angle reduction is evident only at medium to high frequency. In the extreme low-frequency range, the phase angle approaches 90°, where G' is almost 0, indicating that asphalt has entered a viscous flow state, with biochar enhancing high-temperature viscosity.

3.2.5. Han Curve

The Han curve (Figure 11) for biochar-modified asphalt displays a linear relation between G' and G'', validating the availability of the TTS principle under test conditions. The Han curve for modified asphalt lies below the reference line (slope of 1, intercept of 0), indicating the materials are in a viscous region. Linear fitting of the Han curve shows high correlation ($R^2 > 0.99$), with deviations observed only at the low-modulus, equivalent low-frequency/high-temperature end, suggesting that the modified asphalt is more elastic than virgin asphalt. Then, as frequency increases, Han curves and the corresponding fitting line gradually approach the reference line, with asphalt entering a viscoelastic state. Contrasting the Han curve for virgin and modified asphalt, the curve for virgin asphalt is at the lower left side of the modified asphalt, and both G' and G'' for modified asphalts are greater than virgin asphalt, indicating the enhancement of biochar for both sides.

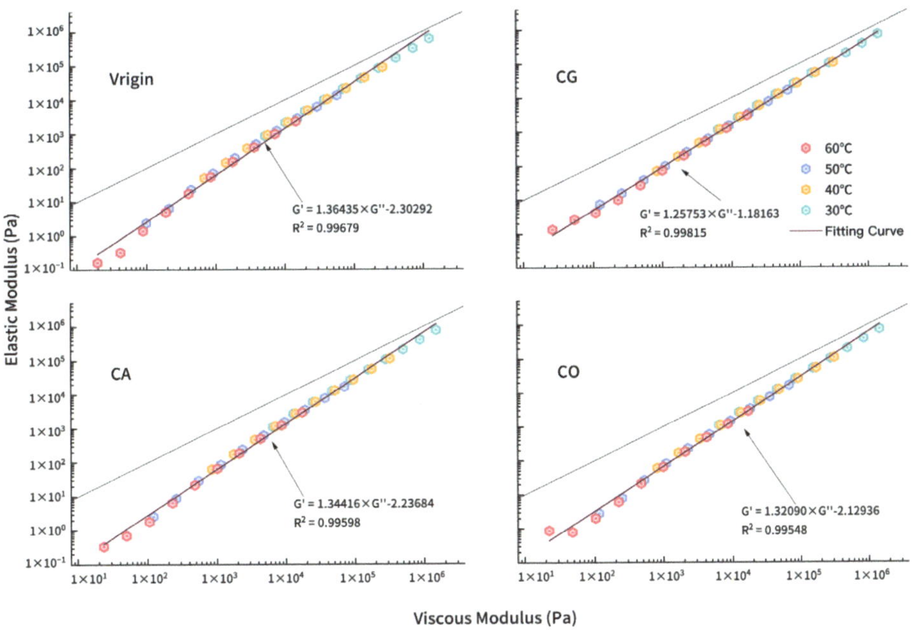

Figure 11. Han curves for modified asphalt.

For the several points with the lowest modulus, G'' of modified asphalt is much higher than G' while the modified asphalt is in a viscous state. At this time, biochar obviously enhances the G'' of asphalt and the enhancement for G' is slight. Then, as frequency increases, the asphalts enter a viscoelastic state, with G' improving rapidly, embodying the enhancement.

For polymer materials, the slope of the Han curve in the low-frequency region would be larger than the entire fitting curve. The slope for homogeneous polymer materials would reach about 2 due to the phase transition at high temperature or low frequency. In Figure 10, only virgin asphalt has a visible slope change to 1.79. The slope for 3 varieties of modified asphalt deviates slightly. There are two main reasons for this phenomenon [37]. One is the test condition; the test temperature is not high enough and the loading frequency is not low enough to produce the above changes in the curve. In this study, the biochar modification treatment increases the complex shear modulus of asphalt, reduces the temperature sensitivity of modified asphalt, and makes the phase transition of asphalt difficult. The second is the polydispersity of asphalt [38]. As a mixture composed of asphaltenes, saturated components, aromatic components and colloids, asphalt has a large difference in molecular weight of different components and strong polydispersity, which can explain the phenomenon that the slope at the end region of the Han curve of virgin asphalt is less than 2. The introduction of biochar and its swelling aggravated the phenomenon of uneven molecular weight distribution in modified asphalt and inhibited the slope change of the end region of the curve.

In conclusion, biochar-modified asphalt exhibits better storage stability than SBS- and rubber-modified asphalt, while the enhancement of rheological properties is lower at high temperature.

3.2.6. Low-Temperature Properties

Figures 12 and 13 show the BBR results of biochar-modified asphalt, including flexural creep stiffness $S(t)$ and m value. ASTM D6816 stipulates that the flexural creep stiffness of asphalt should be less than 300 MPa and the m value should be greater than 0.3; otherwise, the material is considered to fail. It can be discovered that the virgin asphalt and modified asphalt pass the test at −12 °C. However, the flexural creep stiffness of CG-modified asphalt at −18 °C is 360 MPa, which means the material has failed, and the other asphalts are efficient. The failure of CG at −18 °C may be caused by the nitrogen content of biochar. All asphalts fail at −24 °C. Comparing the $S(t)$ and m values of virgin asphalt and modified asphalt, the low-temperature performance of modified asphalt has a certain degree of deterioration, especially the $S(t)$ of CG; thus, the deterioration is the most serious. However, it is still much better than SBS-modified asphalt at the same content. The $S(t)$ and m values of CA- and CO-modified asphalt are not much different, and the difference with virgin asphalt is also relatively little, indicating that the deterioration of low-temperature performance is easy to control depending on biochar types.

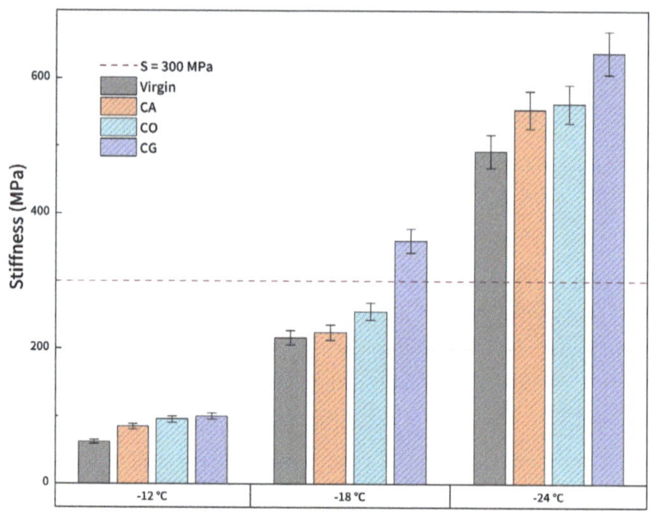

Figure 12. $S(t)$ for modified asphalt.

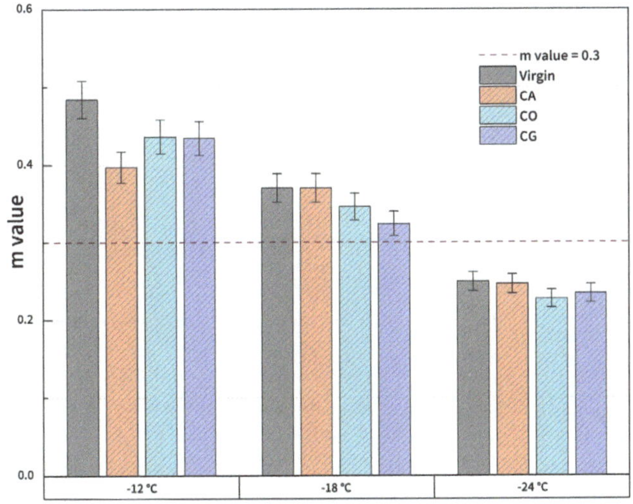

Figure 13. *m* value for modified asphalt.

3.3. Asphalt VOC Emission Behavior Analysis

Based on the results of DSR and BBR tests, CO-modified asphalt demonstrates the best performance among the 3 varieties of modified asphalt. Thus, we evaluated the VOC emission behavior of biochar-modified asphalt. The GC of virgin and CO biochar-modified asphalt is shown in Figure 14. The VOC emission behavior for virgin and modified asphalt are different in the context of residence time. The emission of modified asphalt is more active than virgin asphalt in the first half of the residence time, while virgin asphalt produces more in the second half.

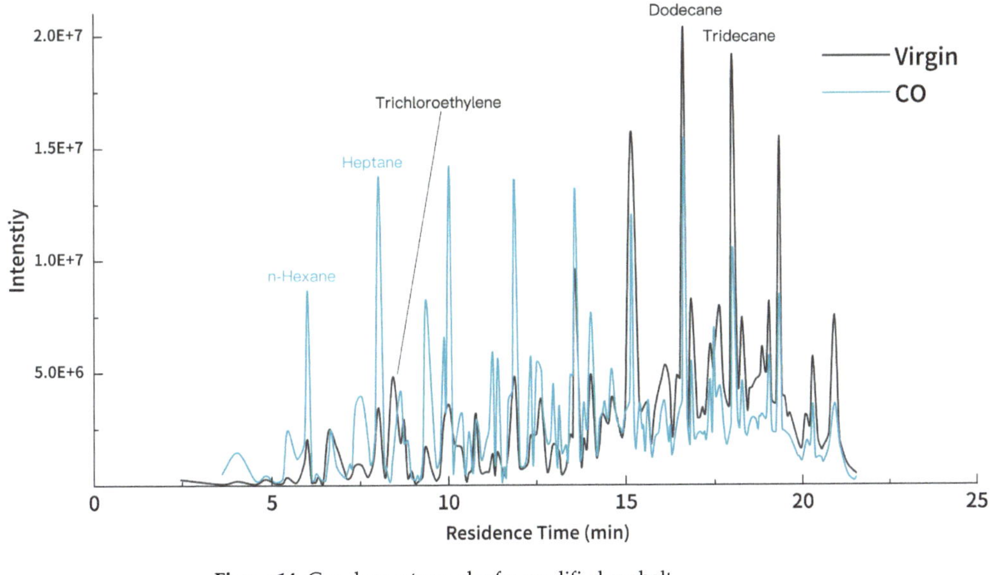

Figure 14. Gas chromatography for modified asphalt.

Asphalt VOCs are composed of alkanes, aromatic hydrocarbons, alkenes and a small number of alcohols and aldehydes, of which alkanes and aromatic hydrocarbons such as benzene and toluene account for the majority. The biochar modifier shows an outstanding inhibition on alkanes (n > 10), aldehydes and naphthalene, similar to prior research [22]. Trichloroethylene accounts for the strongest inhibition, reaching 70%, followed by naphthalene, 2,6-dimethyl- with 65%, showing the adsorption to polycyclic aromatic hydrocarbons [60–62]. The emissions for alkanes, the component with the highest content in asphalt VOCs, are also effectively controlled [63]. Nearly 60% of the octane, 2,6-dimethyl- emission is inhibited in the asphalt as well as tridecane and pentadecane, at more than 40%. Aldehyde VOCs are also inhibited; 44% of butanal and 34% of pentanal emissions are removed. Though there are increases in the emissions of some VOCs, the entire emissions of asphalt are inhibited by biochar modifier.

4. Conclusions

This study prepared biochar-modified asphalt using 3 varieties of biomasses and biochar. Then, we tested the rheological properties at both high and low temperatures for modified asphalt with 6% biochar content. Biochar could significantly enhance the G^* and $G^*/\sin\delta$ at mid to high temperature while reducing phase angle δ; therefore, the permanent deformation ability and rutting resistance of modified asphalt are improved. Among the 3 varieties of biochar-modified asphalt, performance differences increase when the temperature is above 50 °C; CG and CO demonstrate better high-temperature performance than CA. Furthermore, the low-temperature deterioration depends greatly on the biochar variety. In the scene of $S(t)$, CO and CG show properties close to virgin asphalt, indicating the controllability of biochar-modified asphalt at the content of 6% by weight of virgin asphalt. The biochar could also influence the VOC emission of modified asphalt. Comprehensively considering high- and low-temperature properties of modified asphalt with biochar properties, it seems like biochar made from biomass with more cellulose is suitable for asphalt modification.

Supplementary Materials: The following supporting information can be downloaded at: https://www.mdpi.com/article/10.3390/ma18071504/s1, Figure S1: Fluorescence microscopic image of biochar modified asphalt; Table S1: Biochar yield in biomass pyrolysis.

Author Contributions: Conceptualization, Y.X.; methodology, Y.X.; validation, Y.Z.; formal analysis, X.Z. and Y.Z.; investigation, X.Z. and Y.Z.; resources, Y.Z.; data curation, X.Z.; writing—original draft preparation, X.Z.; writing—review and editing, Y.X. and Y.Z.; visualization, X.Z.; project administration, Y.X.; funding acquisition, Y.X. All authors have read and agreed to the published version of the manuscript.

Funding: Financial and technical support was supported by Hubei Provincial Natural Science Foundation of China (JCZRYB202400060).

Institutional Review Board Statement: Not applicable.

Informed Consent Statement: Not applicable.

Data Availability Statement: The original contributions presented in the study are included in the article/Supplementary Materials, further inquiries can be directed to the corresponding author.

Conflicts of Interest: The authors declare no conflicts of interest.

Abbreviations

The following abbreviations are used in this manuscript:

CO	Cotton seed
CA	Camelia seed shell
CG	Coffee ground

DSR	Dynamic shear rheometer
BBR	Bending beam rheometer
VOCs	Volatile organic compounds
GC-MS	Gas chromatography–mass spectrometry
SBS	Styrene–butadiene–styrene block copolymer
SBR	Styrene butadiene rubber
PE	Polyethylene
DSC	Differential scanning calorimetry
TD	Thermal desorption
TTS	Time–temperature superposition
WLF	Williams–Landel–Ferry

References

1. Dong, F.; Yang, P.; Yu, X.; Jiang, M.; Wang, S.; Zu, Y.; Chen, B.; Wang, J. Morphology, chemical reaction mechanism, and cross-linking degree of asphalt binder modified by SBS block co-polymer. *Constr. Build. Mater.* **2023**, *378*, 131204.
2. Li, H.; Cui, C.; Temitope, A.A.; Feng, Z.; Zhao, G.; Guo, P. Effect of SBS and crumb rubber on asphalt modification: A review of the properties and practical application. *J. Traffic Transp. Eng. (Engl. Ed.)* **2022**, *9*, 836–863.
3. Yuan, D.; Xing, C.; Jiang, W.; Xiao, J.; Wu, W.; Li, P.; Li, Y. Viscoelastic Behavior and Phase Structure of High-Content SBS-Modified Asphalt. *Polymers* **2022**, *14*, 2476. [CrossRef] [PubMed]
4. Zhang, F.; Yu, J. The research for high-performance SBR compound modified asphalt. *Constr. Build. Mater.* **2010**, *24*, 410–418.
5. Vargas, M.A.; Vargas, M.A.; Sánchez-Sólis, A.; Manero, O. Asphalt/polyethylene blends: Rheological properties, microstructure and viscosity modeling. *Constr. Build. Mater.* **2013**, *45*, 243–250.
6. Davoodi, A.; Aboutalebi Esfahani, M.; Bayat, M.; Mohammadyan-Yasouj, S.E.; Rahman, A. Influence of nano-silica modified rubber mortar and EVA modified porous asphalt on the performance improvement of modified semi-flexible pavement. *Constr. Build. Mater.* **2022**, *337*, 127573.
7. Yan, K.; Chen, J.; You, L.; Tian, S. Characteristics of compound asphalt modified by waste tire rubber (WTR) and ethylene vinyl acetate (EVA): Conventional, rheological, and microstructural properties. *J. Clean. Prod.* **2020**, *258*, 120732.
8. Liu, G.; Zhang, W.; Yang, X.; Ning, Z. Study on the Conventional Performance and Microscopic Properties of PPA/SBS-Modified Bio-Mixed Asphalt. *Materials* **2022**, *15*, 4101. [CrossRef]
9. Babagoli, R.; Rezaei, M. Development of prediction models for moisture susceptibility of asphalt mixture containing combined SBR, waste CR and ASA using support vector regression and artificial neural network methods. *Constr. Build. Mater.* **2022**, *322*, 126430.
10. Hemmati, N.; Yun, J.; Mazumder, M.; Lee, M.-S.; Lee, S.-J. Laboratory Characterization of Asphalt Binders Modified with Styrene Butadiene Rubber (SBR). *Materials* **2021**, *14*, 7666. [CrossRef]
11. Santero, N.J.; Masanet, E.; Horvath, A. Life-cycle assessment of pavements. Part I: Critical review. *Resour. Conserv. Recycl.* **2011**, *55*, 801–809.
12. Wang, T.; Xiao, F.; Zhu, X.; Huang, B.; Wang, J.; Amirkhanian, S. Energy consumption and environmental impact of rubberized asphalt pavement. *J. Clean. Prod.* **2018**, *180*, 139–158.
13. Rondón-Quintana, H.A.; Reyes-Lizcano, F.A.; Chaves-Pabón, S.B.; Bastidas-Martínez, J.G.; Zafra-Mejía, C.A. Use of Biochar in Asphalts: Review. *Sustainability* **2022**, *14*, 4745. [CrossRef]
14. Li, D.; Li, C.; Fan, M.; Shao, Y.; Sun, Y.; Zhang, L.; Zhang, S.; Huang, Y.; Li, B.; Wang, S.; et al. Investigation of property of biochar in staged pyrolysis of cellulose. *J. Anal. Appl. Pyrolysis* **2023**, *172*, 105999.
15. Calixto, G.Q.; Melo, D.M.A.; Melo, M.A.F.; Braga, R.M. Analytical pyrolysis (Py–GC/MS) of corn stover, bean pod, sugarcane bagasse, and pineapple crown leaves for biorefining. *Braz. J. Chem. Eng.* **2022**, *39*, 137–146.
16. Raček, J.; Chorazy, T.; Miino, M.C.; Vršanská, M.; Brtnický, M.; Mravcová, L.; Kučerík, J.; Hlavínek, P. Biochar production from the pyrolysis of food waste: Characterization and implications for its use. *Sustain. Chem. Pharm.* **2023**, *37*, 101387.
17. Kumar, A.; Choudhary, R.; Narzari, R.; Kataki, R.; Shukla, S.K.; Chailleux, E. Evaluation of bio-asphalt binders modified with biochar: A pyrolysis by-product of Mesua ferrea seed cover waste. *Cogent Eng.* **2018**, *5*, 1548534. [CrossRef]
18. Zhang, R.; Dai, Q.; You, Z.; Wang, H.; Peng, C. Rheological Performance of Bio-Char Modified Asphalt with Different Particle Sizes. *Appl. Sci.* **2018**, *8*, 1665. [CrossRef]
19. Ma, F.; Dai, J.; Fu, Z.; Li, C.; Wen, Y.; Jia, M.; Wang, Y.; Shi, K. Biochar for asphalt modification: A case of high-temperature properties improvement. *Sci. Total Environ.* **2022**, *804*, 150194.
20. Hu, C.; Feng, J.; Zhou, N.; Zhu, J.; Zhang, S. Hydrochar from corn stalk used as bio-asphalt modifier: High-temperature performance improvement. *Environ. Res.* **2021**, *193*, 110157.

21. Zhao, S.; Huang, B.; Shu, X.; Ye, P. Laboratory Investigation of Biochar-Modified Asphalt Mixture. *Transp. Res. Rec. J. Transp. Res. Board* **2014**, *2445*, 56–63.
22. Zhou, X.; Moghaddam, T.B.; Chen, M.; Wu, S.; Adhikari, S. Biochar removes volatile organic compounds generated from asphalt. *Sci. Total Environ.* **2020**, *745*, 141096.
23. Li, N.; Jiang, Q.; Wang, F.; Xie, J.; Li, Y.; Li, J.; Wu, S. Emission behavior, environmental impact and priority-controlled pollutants assessment of volatile organic compounds (VOCs) during asphalt pavement construction based on laboratory experiment. *J. Hazard. Mater.* **2020**, *398*, 122904. [PubMed]
24. Zhou, Y.; Shen, C.; Wang, T.; Xue, Y. Inhibition effect of three types of biochar on volatile organic compounds from asphalt: Revealing chemical adsorption as the primary mechanism. *Constr. Build. Mater.* **2023**, *411*, 134322. [CrossRef]
25. Xia, Y.; Zhu, C.; Ouyang, S.; Yang, Y.; Xie, Y.; Deng, T.; Li, L.; Yang, K.; Xiao, Y.; Tsang, C.-W. Thermogravimetric characteristics and evaluation of products during pyrolysis of Camellia oleifera seed residues. *Biomass Convers. Biorefin.* **2024**, *15*, 4677–4693. [CrossRef]
26. Ostadi Moghadam, M.; Moeenfard, M. How does particle size of spent coffee ground affect the physicochemical properties of isolated cellulose nanosphere? *Results Chem.* **2024**, *10*, 101683.
27. Deb Dutta, S.; Patel, D.K.; Ganguly, K.; Lim, K.-T. Isolation and characterization of cellulose nanocrystals from coffee grounds for tissue engineering. *Mater. Lett.* **2021**, *287*, 129311.
28. *ASTM D5*; Standard Test Method for Penetration of Bituminous Materials. ASTM International: West Conshohocken, PA, USA, 2020.
29. *ASTM D36*; Standard Test Method for Softening Point of Bitumen (Ring-and-Ball Apparatus). ASTM International: West Conshohocken, PA, USA, 2014.
30. *ASTM D113-17*; Standard Test Method for Ductility of Asphalt Materials. ASTM International: West Conshohocken, PA, USA, 2017.
31. Zhou, X.; Adhikari, S. Flow-induced crystallization of biochar in bio-asphalt under various aging conditions. *Sci. Total Environ.* **2019**, *695*, 133943. [CrossRef]
32. Yang, S.; Zhu, H.; Tan, Q.; Yang, X.; Chen, Y.; Lei, L. Application of TOR as a secondary modifier for the preparation of crumb rubber modified asphalt with excellent storage stability. *Constr. Build. Mater.* **2024**, *428*, 135863. [CrossRef]
33. Li, Y.; Ma, R.; Wang, X.; Cheng, P.; Chen, Y. Improvement effect of different modifiers on storage stability of high content SBS modified asphalt. *Case Stud. Constr. Mater.* **2024**, *20*, e02820.
34. Al-Abdul Wahhab, H.I.; Dalhat, M.A.; Habib, M.A. Storage stability and high-temperature performance of asphalt binder modified with recycled plastic. *Road Mater. Pavement Des.* **2016**, *18*, 1117–1134.
35. *ASTM D7173-14*; Standard Practice for Determining the Separation Tendency of Polymer from Polymer Modified Asphalt. ASTM International: West Conshohocken, PA, USA, 2014.
36. Han, C.D.; Kim, J.; Kim, J.K. Determination of the Order-Disorder Transition Temperature of Block Copolymers. *Macromolecules* **1989**, *22*, 383–394. [CrossRef]
37. Han, C.D.; Baek, D.M.; Kim, J.K.; Ogawa, T.; Sakamoto, N.; Hashimoto, T. Effect of Volume Fraction on the Order Disorder Transition in Low Molecular Weight Polystyrene-block-Polyisoprene Copolymers. *Marcomolecules* **1995**, *28*, 5043–5062.
38. Yang, Z.; Han, C.D. Rheology of Miscible Polymer Blends with Hydrogen Bonding. *Macromolecules* **2008**, *3*, 15. [CrossRef]
39. *ASTM D6648-16*; Standard Test Method for Determining the Flexural Creep Stiffness of Asphalt Binder Using the Bending Beam Rheometer (BBR). ASTM International: West Conshohocken, PA, USA, 2016.
40. Ren, H.; Qian, Z.; Huang, W.; Li, H.; Liu, Y. Low-temperature thermal cracking performance of waterborne epoxy asphalt emulsion mastic based on bending beam rheometer (BBR). *Constr. Build. Mater.* **2022**, *334*, 127461.
41. Ministry of Commerce, People's Republic of China. *Stationary Source Emission-Determination of Volatile Organic Compounds-Sorbent Adsorption and Thermal Desorption Gas Chromatography Mass Spectrometry Method*; China Environment Publish Group: Beijing, China, 2014; Volume HJ 734-2014.
42. Ariyanti, D.; Purbasari, A.; Sugianto, D.N.; Lesdantina, D.; Widiyanti, M. Rice husk-based magnetic biochar produced via hydrothermal route for petroleum spills adsorption: Characterization, adsorption kinetics, and isotherms. *Adsorption* **2024**, *30*, 2175–2186.
43. Gao, W.; Lin, Z.; Yan, S.; Gao, Y.; Zhang, H.; Hu, X.; Sun, H.; Zhang, S. Preparation of N-, O-, and S-Tri-Doped Biochar through One-Pot Pyrolysis of Poplar and Urea Formaldehyde and Its Enhanced Removal of Tetracycline from Wastewater. *Energies* **2022**, *15*, 8081. [CrossRef]
44. Mastalerz, M.; Drobniak, A.; Briggs, D.; Bradburn, J. Variations in microscopic properties of biomass char: Implications for biochar characterization. *Int. J. Coal Geol.* **2023**, *271*, 104235.
45. Zhang, S.; Cui, Y.; Wei, W. Low-temperature characteristics and microstructure of asphalt under complex aging conditions. *Constr. Build. Mater.* **2021**, *303*, 124408. [CrossRef]

46. Yu, S.; Zhou, J.; Ren, Y.; Yang, Z.; Zhong, M.; Feng, X.; Su, B.; Lei, Z. Excellent adsorptive-photocatalytic performance of zinc oxide and biomass derived N, O-contained biochar nanocomposites for dyes and antibiotic removal. *Chem. Eng. J.* **2022**, *451*, 138959.
47. El-Nemr, M.A.; Ismail, I.M.A.; Abdelmonem, N.M.; El Nemr, A.; Ragab, S. Amination of biochar surface from watermelon peel for toxic chromium removal enhancement. *Chin. J. Chem. Eng.* **2021**, *36*, 199–222. [CrossRef]
48. Gan, X.; Zhang, W. Application of biochar from crop straw in asphalt modification. *PLoS ONE* **2021**, *16*, e0247390. [CrossRef]
49. Mohan, V.B.; Jayaraman, K.; Bhattacharyya, D. Brunauer–Emmett–Teller (BET) specific surface area analysis of different graphene materials: A comparison to their structural regularity and electrical properties. *Solid State Commun.* **2020**, *320*, 114004. [CrossRef]
50. Maziarka, P.; Wurzer, C.; Arauzo, P.J.; Dieguez-Alonso, A.; Mašek, O.; Ronsse, F. Do you BET on routine? The reliability of N2 physisorption for the quantitative assessment of biochar's surface area. *Chem. Eng. J.* **2021**, *418*, 129234. [CrossRef]
51. Al-Ghouti, M.A.; Da'ana, D.A. Guidelines for the use and interpretation of adsorption isotherm models: A review. *J. Hazard. Mater.* **2020**, *393*, 122383. [CrossRef] [PubMed]
52. Thommes, M.; Kaneko, K.; Neimark, A.V.; Olivier, J.P.; Rodriguez-Reinoso, F.; Rouquerol, J.; Sing, K.S.W. Physisorption of gases, with special reference to the evaluation of surface area and pore size distribution (IUPAC Technical Report). *Pure Appl. Chem.* **2015**, *87*, 1051–1069. [CrossRef]
53. Svidrytski, A.; Hlushkou, D.; Thommes, M.; Monson, P.A.; Tallarek, U. Modeling the Impact of Mesoporous Silica Microstructures on the Adsorption Hysteresis Loop. *J. Phys. Chem. C* **2020**, *124*, 21646–21655. [CrossRef]
54. Zhang, H.; Li, J.; Gong, M. Studying the anti-ageing performance of road asphalt based on penetration and ductility. *Proc. Inst. Civ. Eng.—Eng. Sustain.* **2021**, *174*, 199–210. [CrossRef]
55. Yang, S.; Zhu, H.; Li, R.; Yang, X.; Tan, Q.; Chen, Y.; Lei, L. Application of functionalized graphene oxide in the preparation of crumb rubber modified asphalt with excellent storage stability. *Constr. Build. Mater.* **2024**, *450*, 138488. [CrossRef]
56. Iftikhar, S.; Shah, P.M.; Mir, M.S. Potential Application of Various Nanomaterials on the Performance of Asphalt Binders and Mixtures: A Comprehensive Review. *Int. J. Pavement Res. Technol.* **2022**, *16*, 1439–1467. [CrossRef]
57. Ding, Z.; Zhang, J.; Li, P.; Yue, X.; Bing, H. Analysis of viscous flow properties of styrene–butadiene–styrene-modified asphalt. *Constr. Build. Mater.* **2019**, *229*, 116881. [CrossRef]
58. Li, P.; Ding, Z.; Ma, L.X.; Feng, Z.G. Analysis of viscous flow properties of asphalt in aging process. *Constr. Build. Mater.* **2016**, *124*, 631–638.
59. Lu, Q.; Sha, A.; Jiao, W.; Shi, K.; Li, Z.; Chen, Y.; Du, P.; Peng, Z.; Song, R. Waste coffee biochar and bi-oil composite modified rejuvenated asphalt: Preparation, characterization, and performance evaluation. *Constr. Build. Mater.* **2024**, *450*, 138588. [CrossRef]
60. Rosińska, A. Impact of biochar on sorption of polycyclic aromatic hydrocarbons in water. *Desalination Water Treat.* **2023**, *305*, 103–113.
61. Jia, H.; Ye, J.; Wu, Y.; Zhang, M.; Peng, W.; Wang, H.; Tang, D. Evaluation and characterization of biochar on the biogeochemical behavior of polycyclic aromatic hydrocarbons in mangrove wetlands. *Sci. Total Environ.* **2022**, *864*, 161039.
62. Azman, N.A.; Fuzi, S.F.Z.M.; Manas, N.H.A.; Azelee, N.I.W.; Masngut, N. Improved heterocyclic aromatic hydrocarbon compound adsorption using functionalised rice husk biochar. *Bioresour. Technol. Rep.* **2023**, *25*, 101704.
63. Ge, L.; Yao, Y.; Xu, L.; Zhou, Z.; Li, J.; Zhang, X.; Liu, C.; Lv, H. Mitigation of asphalt volatile organic compounds emissions and health hazards using a TiO2-doped biochar composite: Microscopic and physiological insights. *Environ. Technol. Innov.* **2024**, *36*, 103763.

MDPI AG
Grosspeteranlage 5
4052 Basel
Switzerland
Tel.: +41 61 683 77 34

Materials Editorial Office
E-mail: materials@mdpi.com
www.mdpi.com/journal/materials

www.ingramcontent.com/pod-product-compliance
Lightning Source LLC
Chambersburg PA
CBHW041405210926
51850CB00013B/3184